T0330537

Internet of Things and Big Data Analytics for a Green Environment

This book studies the evolution of sustainable green smart cities and demonstrates solutions for green environmental issues using modern industrial IoT solutions. It is a ready reference with guidelines and a conceptual framework for context-aware product development and research in the IoT paradigm and Big Data Analytics for a Green Environment. It brings together the most recent advances in IoT and Big Data in Green Environments, emerging aspects of the IoT and Big Data for Green Cities, explores key technologies, and develops new applications in this research field.

Key Features:

- Discusses the framework for development and research in the IoT Paradigm and Big Data Analytics.
- Highlights threats to the IoT architecture and Big Data Analytics for a Green Environment.
- Present the I-IoT architecture, I-IoT applications, and their characteristics for a Green Environment.
- Provides a systematic overview of the state-of-the-art research efforts.
- Introduces necessary components and knowledge to become a vital part of the IoT revolution for a Green Environment.

This book is for professionals and researchers interested in the emerging technology of sustainable development, green cities, and Green Environment.

Internet of Things and Big Data Analytics for a Green Environment

Edited by
Yousef Farhaoui
Bharat Bhushan
Nidhi Sindhwani
Rohit Anand
Agbotiname Lucky Imoize
Anshul Verma

CRC Press
Taylor & Francis Group
Boca Raton London New York

CRC Press is an imprint of the
Taylor & Francis Group, an **informa** business

A CHAPMAN & HALL BOOK

Designed cover image: © Shutterstock_1983321920

First edition published 2025
by CRC Press
2385 NW Executive Center Drive, Suite 320, Boca Raton FL 33431

and by CRC Press
4 Park Square, Milton Park, Abingdon, Oxon, OX14 4RN

CRC Press is an imprint of Taylor & Francis Group, LLC

© 2025 selection and editorial matter, Yousef Farhaoui, Bharat Bhushan, Nidhi Sindhwani, Rohit Anand, Agbotiname Lucky Imoize and Anshul Verma; individual chapters, the contributors

Library of Congress Cataloging-in-Publication Data
Names: Farhaoui, Yousef, editor. | Bhushan, Bharat, 1989- editor. | Sindhwani, Nidhi, editor. | Anand, Rohit, editor. | Imoize, Agbotiname Lucky, editor. | Verma, Anshul (Computer scientist), editor.
Title: Internet of things and big data analytics for a green environment / edited by Yousef Farhaoui, Bharat Bhushan, Nidhi Sindhwani, Rohit Anand, Agbotiname Lucky Imoize and Anshul Verma.
Description: First edition. | Boca Raton, FL : CRC Press, 2025. | Includes bibliographical references and index. | Summary: "This book studies the evolution of sustainable green smart cities and demonstrates solutions for green environmental issues using modern industrial IoT solutions. It is a ready reference with guidelines and a conceptual framework for context-aware product development and research in the IoT paradigm and Big Data Analytics for a Green Environment. It brings together the most recent advances in IoT and Big Data in Green Environments, emerging aspects of the IoT and Big Data for Green Cities, explores key technologies, and develops new applications in this research field"-- Provided by publisher.
Identifiers: LCCN 2024020759 | ISBN 9781032649825 (hbk) | ISBN 9781032656793 (pbk) | ISBN 9781032656830 (ebk)
Subjects: LCSH: Smart cities--Technological innovations. | Sustainable engineering--Technological innovations. | Internet of things--Industrial applications. | Big data--Industrial applications.
Classification: LCC TD159.4 .I55 2025 | DDC 307.1/416--dc23/eng/20240907
LC record available at https://lccn.loc.gov/2024020759

ISBN: 978-1-032-64982-5 (hbk)
ISBN: 978-1-032-65679-3 (pbk)
ISBN: 978-1-032-65683-0 (ebk)

DOI: 10.1201/9781032656830

Typeset in Times
by SPi Technologies India Pvt Ltd (Straive)

Contents

Preface

The rapid growth in the population density in green cities demands that services and infrastructure be provided to meet the needs of city inhabitants. Thus, there has been an increase in the request for embedded devices, such as sensors, actuators, and smartphones, leading to the considerable business potential for the new era of the Internet of Things (IoT), in which all devices are capable of interconnecting and communicating with each other over the Internet.

This book discusses the various aspects of the IoT and big data for green cities. Specifically, the book explores key technologies and develops new applications in this emerging domain. In addition, the book discusses smart city design and implementation and offers quantitative and qualitative research on the most recent advances in IoT and big data in green environments. The book investigates green engineering technologies for smart cities. Some of the key aspects covered in this book include smart city applications of big data and green IoT-enabling technologies, deep learning techniques in green cities, energy management, waste and materials management in green cities, security and privacy issues in big data and IoT, sustainable and smart healthcare based on IoT and big data analytics (BDA), sustainable growth management of the urban environment using IoT, applications of green technologies in industries, buildings, transportation, application of IoT and big data toward addressing urban design issues, artificial intelligence (AI) techniques to improve power quality for grid-connected renewable energy sources, machine learning (ML) and deep learning in green computing, prospects of AI/ML techniques for green communications, applications of IoT and big data in urban management programs, management, and operations of big data and IoT, and challenges and opportunities for cleaner energy and green environment. In summary, the book is structured into 18 chapters outlined as follows.

Chapter 1 describes IoT-enabled sustainable urban growth management. The necessity of managing urban growth sustainably has gained prominence in a time of increasing urbanization and growing environmental concerns. Cities have thrived as centers of invention, trade, and culture, but they are now under unprecedented resource pressure from population growth and advancing infrastructure. The marriage of the IoT and urban management has opened up novel routes to address these issues and pave the way for a more sustainable and resilient urban future. The interconnected network of sensors and gadgets within the IoT provides real-time data, contributing to the enhancement of urban living conditions. The chapter examines this dynamic intersection thoroughly, exploring the many facets of urban growth and shedding light on how IoT technologies might be cleverly incorporated into urban planning to produce hospitable, environmentally sustainable, and inclusive cities. Last, the chapter focuses on a different facet of managing urban development and offers tactics that show how IoT might revolutionize urban environments.

Chapter 2 elaborates on enhancing IoT data quality validation in medical fields through BDA. The IoT is a term used to describe how smart computing devices are used to connect places and physical objects through the Internet. It is a brand new paradigm that covers practically every aspect of modern life. The IoT, big data, and real-time analytics can be used to create actionable intelligence. IoT and big data are different aspects of the same idea. It is simple to make use of the advantages of IoT to the connection between big data and Internet-connected devices. The industry of software development is under perennial pressure to provide fresh software that operates fairly and competently while maintaining reliability. The question that emerges now is how to construct a high-performance platform to efficiently test the big data and assess an appropriate method to locate the relevant items from big data? The chapter begins with an overview of IoT and big data and discusses overwhelming data volume and testing procedures and some important issues and further research directions.

Chapter 3 discusses ML-based driver assistance system ensuring road safety for smart cities. Technologies around smart city and green computing are gaining more and more interest from diversified workforce areas, and the transportation system is one of them. Transportation vehicles are operating day and night to provide proper support of the need. However, it could be so tiring for the transportation workers, especially the drivers who are driving the vehicle. A slight negligence of a driver may cause huge loss, resulting critical injuries, loss of valuables, or even untimely death. The increasing numbers of road accidents are therefore a big concern. Research works are going on to assist the drivers and increase the security features of vehicles to avoid accidents. The chapter proposes a model which can efficiently detect drivers' drowsiness. The discussion mainly focused on building the learning model. A modified convolution neural network is built to solve the purpose. The model is trained with a dataset of 7000 images of drivers considering open and closed eyes scenarios. For testing purposes, some real-time experiments are done by some volunteer drivers in different conditions, like gender, day, and night. The model is really good for daytime and if the driver is not wearing any glasses. However, with a glass in the eyes and in night condition, the system requires some improvements.

Chapter 4 investigates privacy and security issues in big data and IoT. Big data and the IoT are an essential part of the world of today and tomorrow. Growing technological advancements have led to large-scale integration of communication devices and consequently larger volume of generated data not seen in previous generations. The revolution in IoT and big data has brought with it challenges and issues, particularly concerning privacy and security. The chapter, therefore, attempts to address some of these challenges such as data volume management, network vulnerabilities, and malware. Specifically, the chapter discusses tools and techniques that could be employed to mitigate the effects of these issues on user experience by providing reliable, secure, and protected infrastructure through design, training, continuous research, and others. By understanding the associated security and privacy risks and implementing appropriate measures, large- and small-scale users and organizations can harness the power of big data and IoT to their fullest.

Chapter 5 examines Big Mart Sales Forecasting considering the case of BDA and ML. AI, big data, and business analytics are the most commonly used cognitive tools in the ecospheres today, and they have garnered a lot of attention for their ability to influence organizational decision-making. A customer relationship management (CRM) and enterprise resource planning (ERP) business system, for example, can be integrated with AI solutions through business platform paradigm. In addition to pattern analysis, BDA enables automatic future event forecasting. BDA may revolutionize organizations and create new commercial prospects using AI. The goal of the chapter is to highlight the preventive aspects of using AI in conjunction with BDA in pursuit of digital platforms for business model innovation and dynamics. One particular case study, namely Big Mart Sales Forecasting, was discussed, compared, and analyzed, and the possible obstacles are outlined in terms of business transformation. The chapter provides a roadmap for utilizing AI and BDA to generate commercial value for business and other stakeholders.

Chapter 6 examines the challenges and opportunities for cleaner energy and green environment. The quest for cleaner energy and a better environment has become a critical global concern due to the increasing challenges posed by climate change and environmental degradation. The study examines the many benefits and challenges associated with switching to greener energy sources and establishing a sustainable green environment. Additionally, technology has had a wide range of effects on society and its surroundings and has aided in the growth of more developed economies, such as the global economy of today. Numerous technologies that have altered the lifestyle of people in society and brought comfort have been made possible by science. Examples of these technologies include those found in aircraft, cars, biotechnology, computers, telecommunication, Internet, renewable energy, atomic and nuclear, nanotechnology, and space exploration. Concerns over the sustainability of the environment are necessary to maintain the well-being of individuals in society. The goal of the study is to develop sustainable technologies that can be transformed into green technologies to preserve the environment for coming generations and prevent environmental degradation. Also covered in the study are the opportunities and difficulties associated with green technology in the 21st century for the following areas: education, food and processing, renewable energy, buildings, agriculture, renewable energy, buildings, agriculture, and space exploration.

Chapter 7 sheds light on a text-based approach for product clustering and recommendation in e-commerce using ML. The abundance of options available to customers in today's e-commerce world can make it challenging for them to find the products they want. Recommender systems have become a popular solution to this problem, as they suggest similar products to customers based on their preferences. In the chapter, a three-step approach to developing a recommender system for an online retail store was suggested. The proposed approach involves text preprocessing, clustering, and topic extraction using the DBSCAN algorithm, cosine similarity, and TruncatedSVD, respectively. The performance of the presented approach was evaluated using the Silhouette score, Calinski-Harabasz score, precision at k, and explained variance ratio metrics. Results show that the proposed approach is effective in recommending relevant products to users and provides valuable insights into the underlying structure of the data.

Chapter 8 finds out how multimodal AI and IoT are shaping the future of intelligence. The chapter explores the transformative potential of multimodal AI and the IoT in shaping the future of intelligence. In particular, it highlights the significant learning outcomes such as understanding the role of large language models (LLMs) in multimodal AI, IoT systems, and the concept of embedding in AI. The chapter emphasizes the importance of promoting engineering principles in the development of these technologies and discusses the process of creating live applications using LLMs and generative AI. Furthermore, the chapter provides insights into the usage of multimodal systems in various domains and also highlights the challenges associated with the implementation of these technologies, such as issues of bias, privacy, security, and misinformation. Finally, the chapter concludes by outlining the future technology trends in multimodal AI and IoT systems. The chapter provides a comprehensive understanding of the current landscape and future potential of these technologies, equipping readers with the knowledge to contribute to this rapidly evolving field.

Chapter 9 explores digitally enabled labor market, focusing on the dark side of digital transformation. The unintended negative consequences of technology in workspaces are referred to as its dark side and have caused concerns for academics, managers, and society in general. Although it is widely reported in the literature that digital technologies can improve the performance of organizations, little is known about how digital transformation affects the job market in emerging economies. The chapter examines this relationship. Primary data was captured from Brazilian multinational companies. The study brings expected and unexpected results: (i) the workforce is not substantially affected by digital transformation; and (ii) unexpectedly, the capabilities/skills of the workforce are prominent and the level of stress caused by digital transformation is low. The study is original, fills a gap in the literature and makes significant contributions as follows. First, it helps to better understand how digital transformation affects the workforce in digitally enabled labor markets. Second, it helps to understand which employee capabilities/skills gain or lose importance in a digital scenario. Third, it sheds light on managers regarding the adoption of initiatives to mitigate the perverse effects of digital transformation on the workforce, balancing decent and inclusive work with economic growth (SDG 8). Fourth, it draws attention to the human side of change and helps shape the future of work. Finally, it expands the literature's arguments about the dark side of digital transformation for the workforce.

Chapter 10 investigates AI capability for auditing. Although the potential benefits of AI have been widely published, little is known about the relationship between AI and auditing. The study examines the current state of AI capabilities to achieve audit objectives. Using survey questionnaires, valuable data was collected from company auditors in Brazil. Preliminary results indicate moderate prominence of AI capabilities in achieving audit objectives. The findings suggest that the main challenges to be overcome are related to tangible resources such as data, technologies, and technical human resources. The study fills a gap in the literature and makes significant contributions. Specifically, it expands the arguments of the existing literature on AI capabilities and audit objectives, provides critical insights into where to invest resources to develop AI capabilities toward audit objectives sheds light to managers on which capabilities should be encouraged in decisions to ensure reliability in the information produced for decision-making by stakeholders, and shows managers which paths can still be taken to increase audit performance based on AI.

Chapter 11 beams a laser focus on smart green cities using IoT-based deep reinforcement learning energy management. Modern energy systems, particularly electricity systems, are becoming more intelligent because of the recent development of sophisticated and intelligent measurement tools. Recent communication and hardware technology developments have made global interconnectedness possible through the IoT. The IoT is the foundation for smart grids and smart city green energy management applications. Research on energy management in well-regulated IoT networks still needs to be done, which makes intelligent load forecasting crucial for optimal energy management (EM) in smart cities. Acquiring the best possible energy performance at the building cluster level and reaching energy neutrality by optimizing locally generated renewable energy require accurate short-term (less than a day) energy demand forecasts. However, predictions and analyses of building energy efficiency continue to be centered on the level of the individual structure rather than on building clusters or small neighborhood scales. The chapter presents an IoT-based deep-learning energy management framework and demonstrates that the suggested framework can identify energy use, forecast energy demand in smart cities, and save costs.

Chapter 12 provides an analysis of a technology-driven learning system in higher education, considering a bibliometric analysis. The last two decades have witnessed the use of learning management system (LMS) in higher education and have attracted researchers globally. With the integration of AI, there is transformation in LMS in higher education. After the coronavirus disease (COVID-19) outbreak, the use of AI has brought a revolution in the capabilities of the software of LMS, offering automated process. The emphasis of the chapter is to explore the literature and identify the growth track and geographical distribution of technology-driven LMSs in higher education system. A bibliometric analysis was conducted from 2000 to 2022. After inserting keywords, 6,026 documents were identified, after filtration 744 data were extracted from the Scopus database. A comprehensive analysis was conducted to examine the volume of research publication and citations. The findings unveil the scope for future research on LMSs in higher education.

Chapter 13 presents a time-varying JSON data model and its query language for the management of evolving green things in IoGT-based systems. Internet of Green Things (IoGT) is a technology that allows exchanging information between people and healthy farm things. It provides information like soil moisture, temperature, humidity, and nutrient level via the use of appropriate sensors. IoGT sensor data represents entities which are evolving over time, and several (smart) farming applications require bookkeeping of the entire history of such data. Furthermore, IoGT data could be considered as big data since they satisfy the three Vs of big data (volume, velocity, and variety), and the standard JSON format is considered one of the best data formats to efficiently manage (i.e., model, store, and exchange) big data. Nevertheless, to the authors' best knowledge, there is no standard time-varying data model for IoGT data management, on the one hand, and the JSON format does not provide any explicit and native support for handling temporal data, on the other hand. Hence, in order to manage the evolution over time of big data and time-varying IoGT data, both designers and developers of smart farming applications should proceed in an ad hoc manner. In order to fill this gap, the chapter proposes temporal JSON IoGT data model (TIoGT), a temporal extension of the JSON data model, for an efficient management of temporal and evolution aspects of data and green things in

IoGT-based systems. Moreover, a user-friendly query language is defined for such a model, named query language for TIoGT (QL4TIoGT), to be used for querying time-varying green things and exploiting TIoGT data.

Chapter 14 focuses on enhancing financial transaction security: a deep learning approach for e-payment fraud detection. Mobile payment systems have gained immense popularity, offering convenience and efficiency in conducting financial transactions. However, with the increasing adoption of mobile payments, the risk of fraudulent activities has also risen. Detecting fraudulent transactions in real-time is crucial to protect users and financial institutions from substantial losses. In the chapter, mobile payment fraud detection is addressed through a proposed deep learning-based approach. Complex patterns and relationships between transaction data and customer information are captured by the presented model. The class imbalance problem inherent in credit card fraud detection is tackled using a random undersampling technique and class weights assigned during training. The model is trained using the Adam optimizer with a binary cross-entropy loss function and monitored using early stopping to prevent overfitting. Experimental results demonstrate the effectiveness of the approach, with a test accuracy of 99.72% and an AUC of 97.43%. A promising solution for detecting credit card fraud is provided by the proposed deep learning model, contributing to enhanced security in financial transactions.

Chapter 15 presents an efficient solar radiation prediction through adaptive neighborhood rough set-based feature selection in meteorological streaming data in Errachidia, Morocco. Online streaming feature selection methods have recently gained significant attention for processing high-dimensional data. However, existing approaches often focus on single-label data or treat multi-label data as a combination of single-label datasets, overlooking the integrity of the label set in multi-label scenarios. The chapter proposes a novel online streaming feature selection approach for multi-label learning using the Neighborhood Rough Set model. The proposed method considers feature significance, redundancy, and label space integrity simultaneously. To achieve this, an adaptive neighborhood relation was introduced, eliminating the need for a fixed neighborhood parameter and restructuring the model to process multi-label data directly. In addition, an evaluation criterion was defined to select features important and relative to the label set, along with an optimization objective function for updating the feature subset and filtering out redundant features. Comparative experiments on diverse datasets underscore the advantages of the proposed method. The proposed approach was applied to meteorological streaming data in Errachidia, Morocco, focusing on predicting solar radiation. The results demonstrate the efficacy of the suggested process in handling dynamic environmental conditions for improved solar radiation predictions.

Chapter 16 investigates AI-empowered date palm disease and pest management with emphasis on the current status, challenges, and future perspectives. Oasis farming plays a crucial role in the sustainability of agricultural systems, particularly in arid regions where it represents a primary source of food, income, and ecological balance. Amidst escalating environmental and climatic challenges, there is an urgent need to enhance the resilience and productivity of these systems. The chapter explores the rising trend of AI in date palm farming management, focusing on disease and pest control, a crucial topic in oasis farming development. The chapter provides a detailed review of AI's current state and future potential in this domain, recognizing the importance of date palms and the threats they face. The study appeals to a broad audience to invest in sustainable agriculture and AI development. Additionally, the study directly advances sustainable practices by addressing the proliferating challenges, outlining future directions, and promoting effective AI solutions for date palm protection. The findings suggest that AI can significantly improve the efficiency and effectiveness of agricultural practices, offering a pathway to address current and future challenges in date palm farming. The findings not only aid in safeguarding the sustainability of the oasis but also provide insights that could be applied across the broader context of agricultural technology and food security.

Chapter 17 reviews the trends in green technology for clean energy. The 21st century has seen major technological developments in various industrial sectors. Due to these technological advancement, there is an increase in global energy demand. The rise in energy demand has led to the increased use of fossil fuels for energy consumption. These fossil fuels are limited and they pose major harm to the environment, leading to environmental degradation. The adoption of alternative sources of energy as well as integration of green technology can assist with the conservation of energy resources. These alternative sources are known as renewable energy, and they are obtained from natural sources such as solar, wind, hydropower, and geothermal. Several studies have focused majorly on the environmental, social, and economic impacts of renewable energy technology. The chapter discusses the need for evaluating the impacts of renewable energy technologies on biodiversity and the ecosystem. In addition, the chapter examines the trends and implications of green technology for renewable energy systems.

Chapter 18 explores minimizing fiber transmission degradation through high-quality implementation of fiber optic in highway environments. The fiber optic channel has undoubtedly become the backbone of modern communication networks due to its exceptional bandwidth, low attenuation, distance capability, data security, lightweight, and immunity to electromagnetic interference. However, fiber degradation induced by poor quality implementation remains a colossal challenge to its network performance, reliability, revenue generation, maximum uptime, and high-speed data. The chapter presents a high-quality implementation of a fiber optic channel that guides against fiber degradation using 90.557 Km Galaxy Backbone (NICTIB) Onitsha-Owerri route in Nigeria as a case study using the stepwise method. Results showed a 2.18 dB fiber channel performance improvement thereby engendering optimal fiber optic channel received level. Key findings revealed that performance degradation is avoidable through quality fiber deployment.

Yousef Farhaoui
Department of Computer Science, Faculty of sciences and Technic, Moulay Ismail University, B.P. 509, Boutalamine, Errachidia, Morocco

Bharat Bhushan
School of Engineering and Technology, Sharda University, Greater Noida, Uttar Pradesh, India

Nidhi Sindhwani
Amity School of Engineering and Technology Delhi, Amity University, Noida, India

Rohit Anand
Department of Electronics and Communication Engineering at G.B. Pant Engineering College (Government of NCT of Delhi), New Delhi, India

Agbotiname Lucky Imoize
Department of Electrical and Electronics Engineering, Faculty of Engineering, University of Lagos, Akoka, Lagos, Nigeria

Anshul Verma
Department of Computer Science, Institute of Science, Banaras Hindu University, Varanasi, India

Acknowledgments

We extend our sincere gratitude to the many individuals and organizations whose support and contributions have made this book possible.

First and foremost, we would like to thank our esteemed contributors and authors who have shared their expertise and insights in the fields of Internet of Things (IoT) and Big Data Analytics. Your dedication to advancing knowledge in this critical area of technology and sustainability has been invaluable.

We are grateful to the editorial and review teams for their meticulous work and constructive feedback, which has greatly enhanced the quality of this book. Your commitment to excellence and attention to detail have been essential in shaping the final manuscript.

Our appreciation also goes to the organizations and institutions that provided access to research materials and data.

We would also like to acknowledge the contributions of our families and friends for their unwavering support and encouragement throughout this endeavor. Your patience and understanding have been a source of strength and motivation.

Lastly, we extend our thanks to Taylor & Francis Group, for their professional guidance and support in bringing this book to fruition. Your expertise in publishing has ensured that our work reaches a wider audience.

Thank you all for your invaluable contributions.

Contributors

Ayodele A. Afolabi
University of Lagos
Lagos, Nigeria

Badraddine Aghoutane
Moualy Ismail University
Meknès, Morocco

Lateef Adesola Akinyemi
University of South Africa
Johannesburg, South Africa

Lateef Adesola Akinyemi
Lagos State University
Lagos, Nigeria

Abdullah Saeed Alahmari
Saudi Aramco
Dhahran, Saudi Arabia

Nada K. Alomari
Saudi Aramco
Dhahran, Saudi Arabia

Mourade Azrour
Moualy Ismail University
Meknès, Morocco

Abdullah Mohammad Baaballah
Saudi Aramco
Dhahran, Saudi Arabia

A. Jyothi Babu
Mohan Babu University
Tirupathi, India

Mohamed Khalifa Boutahir
Moualy Ismail University
Meknès, Morocco

Zouhaier Brahmia
University of Sfax
Sfax, Tunisia

Safa Brahmia
University of Sfax
Sfax, Tunisia

Ana Clara Nunes Gomes Cardoso
Fluminense Federal University
Rio de Janeiro, Brazil

Kethellen Santana da Silva
Fluminense Federal University
Rio de Janeiro, Brazil

José Cláudio Garcia Damaso
Fluminense Federal University
Rio de Janeiro, Brazil

Rafael Pires de Almeida
Fluminense Federal University
Rio de Janeiro, Brazil

Ahmad El Allaoui
Moualy Ismail University
Meknès, Morocco

Raouya El Youbi
Sidi Mohamed Ben Abdellah University
Fez, Morocco

Ahmed El Youssefi
Moualy Ismail University
Meknès, Morocco

Promise Elechi
Rivers State University
Port Harcourt, Nigeria

Yousef Farhaoui
Moualy Ismail University
Meknès, Morocco

Debayani Ghosh
Thapar University
Patiala, Punjab, India

Fabio Grandi
University of Bologna
Bologna, Italy

Abdelaaziz Hessane
Moualy Ismail University
Meknès, Morocco

Augustus Ehiremen Ibhaze
University of Lagos
Lagos, Nigeria

Agbotiname Lucky Imoize
University of Lagos
Lagos, Nigeria

Manal Loukili
Sidi Mohamed Ben Abdellah University
Fez, Morocco

M. A. Krishnapriya
SIMATS, Saveetha University
Chennai, India

Parijata Majumdar
Techno College of Engineering Agartala
Agartala, India

Fayçal Messaoudi
Sidi Mohamed Ben Abdellah University
Fez, Morocco

Sanjoy Mitra
Tripura Institute of Technology
Narsingarh
Agartala, India

Ernest Mnkandla
University of South Africa
Johannesburg, South Africa

Ela Umarani Okowa
Rivers State University
Port Harcourt, Nigeria

Selma Regina Martins Oliveira
Fluminense Federal University
Redonda, Rio de Janeiro, Brazil

Ali Omari Alaoui
Moualy Ismail University
Meknès, Morocco

Kingsley Eyiogwu Onu
Rivers State University
Port Harcourt, Nigeria

Rajesh Kumar Pallamala
Chadalawada Ramanamma Engineering
  College
Tirupathi, India

Bishwajeet Prakash
Sharda University
Greater Noida, India

R. Ravindraiah
Madanapalle Institute of Technology &
  Science
Madanapalle, India

Paul Rodrigues
King Khalid University
Riyadh, Saudi Arabia

Apash Roy
Christ University
Bangalore, India

Aishat Titilola Rufai
University of Lagos
Lagos, Nigeria

Adewole Usman Rufai
University of Lagos
Lagos, Nigeria

Anita Singh
Sharda University
Greater Noida, India

Hamed Taherdoost
University Canada West
Vancouver, Canada

Risha Thakur
Sharda University
Greater Noida, India

Sarafudheen M. Tharayil
Saudi Aramco
Dhahran, Saudi Arabia

Niva Tripathy
DRIEMS University College
Cuttack, India

Subhranshu Sekhar Tripathy
KIIT Deemed to be University
Bhubaneswar, India

Editors Bio

Yousef Farhaoui is a professor at Moulay Ismail University, Faculty of Sciences and Techniques, Morocco. He is the Chair of the IDMS Team and the Director of STI Laboratory. He obtained his Ph.D. degree in Computer Security from Ibn Zohr University of Science. He has authored 15 books and several book chapters. He is a senior member of IEEE, IET, ACM, and EAI.

Bharat Bhushan is an assistant professor of the Department of Computer Science and Engineering (CSE) at the School of Engineering and Technology, Sharda University, Greater Noida, India. For the three consecutive years (2021–2023), Stanford University (USA) listed Dr. Bharat Bhushan in the top 2% of scientists list.

Nidhi Sindhwani is currently working as an assistant professor at Amity School of Engineering and Technology Delhi, Amity University, Noida, India. She has done her Ph.D.(ECE) from Punjabi University, Patiala, Punjab, India. She has more than 15 years of teaching experience. She has published 3 book chapters in reputed books,10 papers in Scopus/SCIE Indexed Journals, and 4 patents.

Rohit Anand is currently working as an assistant professor (Group-A Gazetted Officer) in the Department of Electronics and Communication Engineering at G. B. Pant DSEU Okhla-1 Campus (formerly G.B. Pant Engineering College), Government of NCT of Delhi, New Delhi, India.

Agbotiname Lucky Imoize is a lecturer in the Department of Electrical and Electronics Engineering at the University of Lagos, Nigeria. His research interests cover the fields of 6G wireless communication, wireless security systems, and artificial intelligence. He has published 10 books in these fields. He is a Fulbright research fellow, PTDF and DAAD scholar, and Marie Curie researcher. He is the vice chair of the IEEE Communication Society, Nigeria chapter, a registered engineer, and a senior member of the IEEE.

Anshul Verma is currently working as an assistant professor in the Department of Computer Science, Institute of Science, Banaras Hindu University Varanasi, India. His research interests include distributed systems, cloud/edge computing, mobile ad-hoc networks, and formal verification. He is serving as Editor of the Journal of Scientific Research of the Banaras Hindu University.

1 IoT-Enabled Sustainable Urban Growth Management

Hamed Taherdoost
University Canada West, Vancouver, Canada

1.1 INTRODUCTION

In this section, we set the stage by highlighting the challenges posed by urban expansion and the emergence of "smart cities" as a transformative approach. We explore the concept of smart cities, emphasizing their reliance on Internet of Things (IoT) technologies to create intelligent, responsive, and efficient urban environments. Owing to the exponential rise in urban area size and the simultaneous growth in human population, municipalities across the world face an overwhelming array of challenges regarding environmental sustainability, resource management, and the welfare and satisfaction of their residents [1]. In reaction to these difficulties, the idea of "smart cities" has attracted a lot of attention as a paradigm change in urban design heavily emphasizes the use of modern technologies.

Smart cities leverage the data-driven functionalities and interconnectivity of IoT to create urban environments that are intelligent, responsive, and efficient, thereby enhancing sustainability and quality of life [2]. The exponential growth of urban populations in recent decades has placed traditional municipal infrastructures and services under significant strain. Urban policymakers and planners are currently faced with a pressing need to devise innovative approaches that optimize resource utilization while minimizing detrimental environmental impacts, including waste management and energy consumption [3, 4]. Smart cities present a positive perspective on the future, in which IoT technologies form the basis for an interconnected network of sensors, devices, and systems that collect large amounts of data in real-time to enable informed decisions that benefit the city and its residents [5, 6].

The term IoT refers to a hypothetical situation where everything has an identity of its own and can communicate via the Internet or a comparable wide-area network [7]. The seamless integration of physical items with the digital world of information technology is the subject of the IoT concept. Access to the physical world is made possible by computers and other networked devices, and this is useful in both personal and professional contexts. Management will be able to evaluate, plan, and execute actions with the availability of more accurate information [8], and hence they will be able to move from the macro to the micro level with ease. However, IoT will also

DOI: 10.1201/9781032656830-1

significantly improve daily convenience, in addition to streamlining and optimizing corporate process management.

The incorporation of IoT into urban settings has become a crucial facilitator in tackling the complexities associated with metropolitan expansion and administration. Sharma and Jangirala [9] emphasize the pivotal role that IoT solutions fulfill in the development of sustainable smart cities through their ability to enable smooth interconnection, interaction, control, and visibility into diverse urban systems. In addition, the notion of smart city management, which is supported by cutting-edge technologies including intelligent vehicles and wireless sensors, has emerged as a pivotal factor in promoting sustainable urban progress. This chapter explores the profound impact that IoT can have on sustainable urban growth management. It also illuminates how the IoT can be instrumental in tackling the complex issues associated with urban development and administration. This establishes the groundwork for the subsequent sections of this chapter to delve into the diverse facets of sustainable urban growth management facilitated by IoT. This chapter further emphasizes the innovative potential of integrating IoT into sustainable urban growth management, providing insights into transformative approaches for addressing contemporary challenges in urbanization and environmental concerns. This chapter unveils novel perspectives and strategies that extend beyond the current discourse on smart cities, contributing to the ongoing dialogue on the future of resilient and inclusive urban landscapes.

1.1.1 CONTRIBUTIONS OF THE CHAPTER

This chapter makes significant contributions to the discourse on sustainable urban development and growth management. The key contributions can be summarized as follows:

- Comprehensive exploration of urban growth facets: The chapter offers a thorough exploration of various facets of urban growth, addressing the challenges posed by increasing urbanization and environmental concerns.
- Strategic examination of IoT in urban planning: The chapter strategically examines the role of IoT technologies in urban planning. It sheds light on how the interconnected network of sensors and gadgets can enhance urban living conditions, laying the groundwork for an effective integration of IoT into sustainable urban development strategies.
- Practical tactics for urban development management: The chapter presents practical tactics for managing urban development, emphasizing the revolutionary potential of IoT technologies in urban environments. These tactics aim to foster hospitable, environmentally sustainable, and inclusive cities, addressing pressing issues associated with urbanization.
- Emphasis on data-driven decision-making: The chapter underscores the importance of data-driven decision-making, highlighting the role of real-time data provided by IoT in making informed decisions for sustainable urban development. This approach enhances the overall resilience and adaptability of cities in the face of evolving challenges.

- Interdisciplinary approach: The chapter adopts an interdisciplinary approach by bringing together perspectives from urban planning, technology, and sustainability. This holistic viewpoint contributes to a more comprehensive exploration of the challenges and potential solutions in the realm of sustainable urban development, enriching the understanding of how urban growth can be effectively managed with the integration of IoT technologies.

By presenting these contributions in a structured manner, the introduction now clearly highlights what has been added to the chapter that is missing in the existing literature.

1.1.2 CHAPTER ORGANIZATION

To facilitate a clear and logical progression of ideas, this chapter is organized into distinct sections. The introduction sets the stage by highlighting the challenges posed by urbanization and emphasizing the pivotal role of technology, specifically IoT, in addressing these challenges. Following this, Section 1.2, "Sustainable Urban Development and Growth Management," explores the broader context and challenges associated with sustainable urban development, providing a solid foundation for subsequent discussions.

Section 1.3, "Frameworks and Key Technologies," introduces frameworks and key technologies relevant to urban development, offering a theoretical backdrop for the integration of IoT. Section 1.4, "IoT as a Key Enabling Technology," is dedicated to the central theme of the chapter, elucidating how IoT functions as a key enabling technology for sustainable urban development. This section delves into specific applications and benefits.

The chapter concludes with Section 1.5, "Future Trends," forecasting future trends in the intersection of IoT and urban development. This final section provides insights into potential advancements and challenges that may shape the field, offering a forward-looking perspective on the integration of IoT into sustainable urban growth management.

1.2 SUSTAINABLE URBAN DEVELOPMENT AND GROWTH MANAGEMENT

In exploring sustainable urban development and growth management, this section emphasizes the critical need to prioritize effective strategies amid rapid urbanization, recognizing the complex challenges in balancing environmental sustainability, social equity, and economic development within urban areas. If urban areas are to continue to be healthy and sustainable, then growth management and urban development have to take precedence over the issues posed by urbanization. Scholars refer to the process of integrating the expansion of urban areas' economic, environmental, and social subsystems as "sustainable urban growth" [10]. Several different ideas are included in sustainable urban development and growth management, such as inclusive growth, sustainable growth, and wise growth. Urban planning approaches have integrated these ideas to improve the quality of life and lessen the negative impacts of urban sprawl. Particularly in emerging countries, urbanization is essential to laying the groundwork for sustainable development [11]. The main drivers are migration from rural to urban areas, population growth, and then urban expansion.

The effectiveness of growth management initiatives is a crucial consideration in discussions of sustainable urban development. This claim is especially true when taking suburban and urban encroachment mitigation into account [12]. Because of their focus on environmental preservation, land use regulation, and infrastructure development, these programs are essential in fostering sustainable urban growth. Ensuring that social welfare, environmental preservation, and economic development can coexist peacefully in urban areas is crucial; therefore, controlling urban encroachment and growth is a crucial part of this management. Thus, preserving the long-term sustainability of urban regions and addressing the issues raised by rapid urbanization require an understanding of the link between growth management and sustainable urban development.

Given the rapid pace of urbanization and the resultant socioeconomic and environmental consequences, the task of effectively managing sustainable urban expansion poses a significant obstacle. The complexity of overseeing urban expansion is underscored by the functional diversity of cities, presenting administrators and planners of urban areas with a significant challenge. It is critical to prioritize expansion in a way that is both environmentally sustainable and socially equitable, with no additional environmental damage being caused. Cities' provision, administration, and planning of services for their residents comprise a significant proportion of the solution.

Sustainable urban development rethought within the context of an integrated planning and development process is one of the approaches to effectively manage the sustainable urban growth. This entails the implementation of sustainable design methodologies in urban peripheries, which present a unique array of obstacles. Several emerging concepts of urban development management, including the inclusive growth principle, sustainable growth principle, and smart growth principle, have been methodically incorporated into urban planning techniques to improve the quality of life and mitigate the negative effects of urban sprawl. In urban contexts, sustainability is defined as the ability of a municipality to provide accommodation, employment, food, transportation, and a sustainable urban environment while safeguarding the well-being of future generations.

1.2.1 IoT in Smart Cities

Through improved administration, smart cities—made feasible by the IoT—can optimize a variety of services [13]. These amenities include parking, healthcare, transportation, and CCTV cameras. One of the obvious benefits of IoT adoption that may be expected is the rapid transmission and processing of massive amounts of data. Furthermore, there will be a decrease in the expense of managing urban information, and there will be a push for the creation of innovative production and management techniques [14]. In addition, IoT allows machine-to-machine (M2M) connections to be facilitated [15, 16], business process optimization [17], and the integration and interoperability of various architectures. While IoT integration greatly contributes to the development of smart cities, it also presents several difficulties. For example, because smart cities collect and process large amounts of privacy-sensitive data,

TABLE 1.1
Comparison of IoT Applications in Smart Cities

Application	Traffic Management	Energy Efficiency	Security and Safety	Environmental Monitoring
Smart transportation	✔		✔	
Healthcare enhancement			✔	
Energy management		✔		
Public safety and security			✔	
Waste management				
Water management				✔
Smart infrastructure		✔		✔
Urban planning	✔			
Emergency response systems	✔		✔	
Wearable technology			✔	

security and privacy pose significant issues. These issues come up in connection to confidentiality, integrity, non-repudiation, availability, and access control [18, 19]. The dispersed and heterogeneous nature of IoT technology presents a variety of issues, including those related to connectivity, regulatory ambiguity, ethical and sociological concerns, a lack of public knowledge and acceptability, and inadequate government foresight [20].

The vital integration of IoT into transportation networks, intelligent energy management, and municipal governance greatly reduces urban carbon emissions. IoT plays a critical role in smart cities by streamlining transportation networks, controlling energy use, and enhancing local government operations. The aim to reduce carbon emissions and promote sustainability motivates the pursuit of these goals. A concise comparison of IoT implementations in smart cities is presented in Table 1.1, with an emphasis on environmental monitoring, energy efficiency, traffic management, and security.

1.2.2 Smart Energy Management

The intelligent administration of energy, facilitated by IoT, is an essential element in the endeavor to enhance the sustainability and resource efficiency of urban areas. This transition is propelled by the integration of intelligence into traditional power networks via hardware and software enabled by IoT. In addition to increasing the overall resilience of the grid and the efficiency of electricity distribution, these interconnected grids enable real-time monitoring. The integration of intelligent sensors not only facilitates the deployment of demand response systems but also empowers consumers to actively engage in energy conservation efforts during periods of high demand, thereby contributing to the development of a more stable and balanced energy ecosystem.

IoT integration into structures represents an additional vital element of intelligent energy management. Through the deployment of intelligent sensors and automation systems, it is possible to monitor and manage energy-consuming elements such as lighting, HVAC, and other components in real-time. By leveraging data analytics and user preferences, this granular management significantly improves the energy efficiency of buildings, leading to a decrease in both energy consumption and environmental impact. The integration of distributed energy resources (DERs), including energy storage systems and solar panels, into the energy infrastructure increases energy grid diversification, promotes the utilization of renewable energy sources, and bolsters the overall sustainability of urban power systems.

The identification of a comprehensive and integrated approach to urban expansion can be achieved through the harmonious integration of smart energy management into overarching smart city initiatives. Municipalities can create a sustainable and all-encompassing urban ecosystem through the integration of IoT into their transportation, infrastructure, and public service operations. When predictive maintenance, consumer empowerment, and an emphasis on energy efficiency are combined, they collectively contribute to a future where cities not only make intelligent energy usage decisions but also actively strive to achieve a more sustainable and resilient urban environment. The realization of this future is contingent upon our continued emphasis on energy efficiency. With the ongoing advancement of technology, smart energy management will probably play an even more crucial role in shaping the future of cities and the worldwide effort to accomplish urban sustainability. Figure 1.1 illustrates the interconnected elements of IoT-enabled energy administration, emphasizing their role in enhancing the sustainability and resource efficiency of urban areas. The figure depicts the integration of intelligence into traditional power networks, leading to increased efficiency and resilience through real-time monitoring. Intelligent energy management in structures, coupled with the integration of DERs, promotes grid diversification and the utilization of renewable energy sources. The comprehensive approach to urban expansion involves harmoniously integrating smart energy management into overarching smart city initiatives, striving for a more sustainable and resilient urban environment.

1.2.3 TRANSPORTATION SYSTEMS

Intelligent energy management is made possible by IoT, and it is a crucial component of any effort to improve the resource efficiency and sustainability of urban areas. The hardware and software components of IoT enable the integration of intelligence into conventional electricity networks, driving a significant shift. These interconnected networks facilitate real-time monitoring and enhance the overall resilience and distribution efficiency of the power grid. During times of high demand, consumers can actively participate in energy conservation measures thanks to the inclusion of intelligent sensors. This helps create a more stable and balanced energy ecology in addition to making the deployment of demand response systems easier.

Even more important is the integration of IoT technologies into buildings for intelligent energy management. Energy-consuming components, such as HVAC and lights, can be monitored and managed in real-time by implementing automation systems and intelligent sensors. This makes it possible to reduce energy costs. This granular management dramatically reduces energy consumption and the environmental

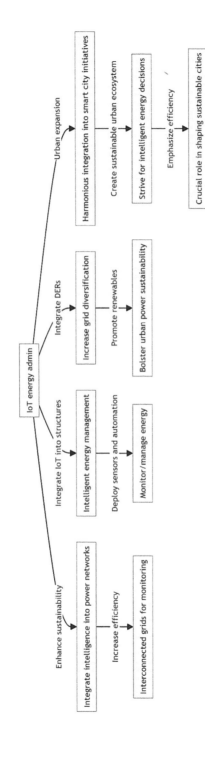

FIGURE 1.1 IoT-driven intelligent energy management.

impact of physical structures by utilizing user preferences and data analytics to improve building energy efficiency. By incorporating DERs into the energy infrastructure, urban power systems become more sustainable overall, the grid becomes more diverse, and the use of renewable energy sources is encouraged. DERs include things like energy storage systems and solar panels.

The identification of a comprehensive and integrated urban expansion strategy can be achieved through the seamless integration of intelligent energy management into broader smart city efforts. Municipal transportation, infrastructure, and public service operations can all benefit from IoT integration, which can help create a robust and long-lasting urban ecosystem. Predictive maintenance combined with consumer empowerment and an energy efficiency emphasis creates a future in which municipalities actively work to create a more sustainable and resilient urban environment in addition to making educated decisions about energy use. Or, to put it another way, this future will come to pass through the combination of energy efficiency, customer empowerment, and predictive maintenance. Maintaining energy conservation as a paramount concern is imperative for achieving this objective. As technology develops, intelligent energy management will most likely play a bigger part in determining how cities will look in the future and in the global effort to attain urban sustainability.

1.2.4 City Administration

Administrators can monitor trends, anticipate issues, and allocate resources more efficiently when they employ sophisticated analytics and data visualization technologies. By using this method, which puts data before people or processes, more informed policies and services are created, and the groundwork for more proactive and efficient governance is laid. Restructuring the organizational structure of municipal governance depends on the incorporation of modern technologies.

Smart governance systems are being used by municipalities at an increasing rate. These platforms allow administrative functions to be seamlessly integrated across numerous departments. This integration improves communication, cooperation, and information sharing, all of which lead to a more cohesive and effective city government. Concurrently, the growth of citizen-focused services is changing how people engage with their local governments. Thanks to the simplicity of use provided by Internet portals and mobile applications, citizens may now directly access information, seek services more easily, and take an active role in decision-making. In addition to improving accountability and transparency, digital transformation creates a governance structure that is adaptable and sensitive to changing conditions.

IoT is becoming increasingly important in terms of how it affects the management and monitoring of urban infrastructure in real-time. Critical infrastructure has a longer lifespan, less downtime, and lower costs when managers are empowered to proactively monitor vital assets using linked sensors. Additionally, technical advancements enable public participation, virtual modeling, and simulations by supplying resources. These initiatives facilitate collaborative urban planning. By ensuring that the expansion of the city aligns with the goals and requirements of its citizens, this collaborative strategy helps foster an inclusive and environmentally sustainable urban environment. Nowadays, smart governance platforms, data-driven decision-making, citizen-centric services, IoT-powered infrastructure monitoring, and redesigned municipal administration are bringing about process simplification, process optimization, and increased public prosperity.

1.2.5 CHALLENGES FOR ADOPTING AND IMPLEMENTING IoT IN SMART CITIES

There has been a lot of research done on the difficulties associated with adopting and deploying IoT in smart cities. Janssen et al. [20] published a report in which they highlighted critical difficulties by employing an integrated MICMAC-ISM strategy. According to the findings of the study, the most significant difficulties are ones involving data security and privacy, business models, data quality, scalability, complexity, and governance. Complexity and a lack of IoT governance were recognized as the primary factors for driving problems, which suggests that there is a requirement to concentrate on dividing complexity into components that can be managed and supported by a governance structure.

For IoT to be successfully used in smart cities, overcoming the problems that were identified in the study is necessary. These problems include security and privacy, business models, data quality, scalability, complexity, and governance. It is necessary to find solutions to these problems to ensure the long-term viability of the smart city concept [21]. In addition, the study stressed how important it is to comprehend the connections that exist between these obstacles to facilitate the growth of smart cities. It suggested an integrated methodology to categorize variables according to their driving strength and dependence, thus giving a systematic framework to examine the dynamics of interactions between various issues. Figure 1.2 visually represents the challenges and solutions associated with adopting and implementing IoT in smart cities. The identified challenges, including data security, business models, data quality, scalability, complexity, and governance, are succinctly presented. The corresponding solutions, such as secure data practices, innovative business models, enhanced data quality measures, scalable infrastructure, simplified system architecture, and an effective governance model, are outlined. The connections between challenges and solutions are highlighted to emphasize the comprehensive approach needed for overcoming obstacles and ensuring the long-term viability of the smart city concept.

1.3 FRAMEWORKS AND KEY TECHNOLOGIES

In navigating the complexities of sustainable urban development, this section delves into the crucial frameworks and key technologies, particularly focusing on the pivotal role of IoT technology, as shown in Table 1.2, to foster intelligent and resourceful urban environments.

Successful development of intelligent and resourceful urban environments necessitates the incorporation of multiple frameworks and critical technologies. IoT technology facilitates this form of sustainable urban development management. Table 1.2 provides an overview of the fundamental IoT technologies that are crucial for

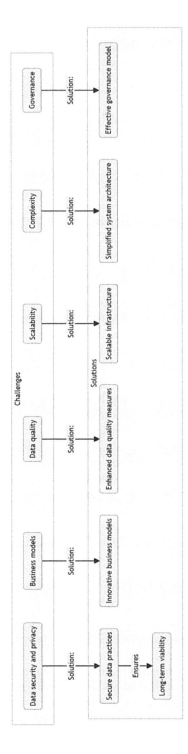

FIGURE 1.2 Challenges and solutions in adopting IoT for smart cities.

TABLE 1.2
IoT Technologies for Sustainable Urban Growth

Category	Framework/ Technology	Use Cases	Challenges
IoT platforms	Microsoft Azure IoT Hub, AWS IoT Core, Google Cloud IoT Core	Device management, data analytics, real-time monitoring	Data privacy, compatibility issues
Smart city frameworks	FIWARE, OMA Lightweight M2M	Urban planning, traffic management, public services	Learning curve, initial setup complexity
Sensor networks	Wireless sensor networks (WSN), LPWAN (LoRa, NB-IoT)	Environmental monitoring, traffic flow optimization, energy management	Deployment cost, network coverage
Data analytics	Big data analytics, machine learning, AI	Predictive maintenance, resource optimization, pattern recognition	Data security, ethical concerns
Connectivity protocols	5G technology, MQTT, CoAP	High-bandwidth applications, low latency communication, massive device connectivity	Infrastructure cost, 5G coverage
Security frameworks	Blockchain, identity and access management (IAM)	Secure transactions, identity verification, data integrity	Scalability, energy consumption (blockchain)
Smart grid technology	Smart meters	Demand response, grid stability, renewable energy integration	Initial implementation cost, compatibility issues
Urban mobility solutions	Intelligent transportation systems (ITS), smart parking	Traffic management, parking optimization, public transportation enhancement	Implementation cost, public acceptance
Environmental monitoring	Air quality monitoring, water quality sensors	Pollution control, early warning systems, ecosystem preservation	Sensor calibration, maintenance costs
Citizen engagement platforms	Mobile apps and portals	Community feedback, service requests, emergency alerts	Digital divide, privacy concerns

achieving sustainable urban expansion. The analysis particularly emphasizes the use cases and challenges associated with these technologies.

1.3.1 FRAMEWORK FOR SUSTAINABLE SMART CITIES

Both the function of information technology (IT) in sustainable smart cities and the framework for resource management require the integration of diverse technologies to establish urban environments that are interconnected, efficient, and sustainable. By facilitating the collection, analysis, and administration of data in real-time, IoT

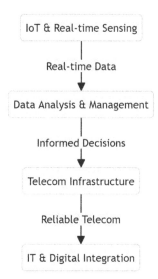

FIGURE 1.3 Components of a smart sustainable city framework.

significantly contributes to this framework's objective of enhancing resource utilization and improving the overall quality of life for citizens [22]. In addition, secure and dependable telecommunications infrastructure is required to support the vast quantity of ICT-based services in smart, sustainable cities. Typical framework components consist of the following (Figure 1.3):

1. IoT and real-time sensing: The IoT facilitates the acquisition of real-time data from a multitude of sensors and devices that are strategically placed throughout the urban environment, thereby enhancing the efficiency of resource administration and decision-making processes.
2. Data analysis and management: Real-time data reception, analysis, and management by cloud-based IoT applications assist enterprises, municipalities, and citizens in making more informed decisions that enhance the overall quality of life.
3. Telecommunication infrastructure: The provision of interoperable, dependable, secure, and stable telecommunication infrastructure is critical to facilitate the vast array of information and communication technology-based services offered in smart sustainable cities.
4. Integration of IT and digital technology: The integration and adoption of sustainable technology and practices at the city scale can be significantly accelerated through the utilization of IT and digital technology.

By utilizing IT to manage resource supply and demand, reduce energy consumption, and mitigate the effects of negative externalities such as climate change, the framework seeks to attain sustainability. The establishment of sustainable smart cities necessitates the formulation of technical criteria, international standards, and processes that facilitate the coordinated advancement of IoT technologies.

1.3.2 Smart City Management

Intelligent vehicle and wireless sensor management are examples of cutting-edge technologies utilized by smart city management to enhance the quality of life in metropolitan areas. The concept of smart cities involves the integration of various technologies to enhance the efficiency, environmental friendliness, and overall standard of living for the inhabitants of a municipality. Smart buildings, smart energy grids, IoT, and smart mobility and transit are among the most important technologies utilized in the administration of smart cities. The aforementioned technologies facilitate the administration and remote control of urban facilities and devices, in addition to the monitoring of energy, water, air, refuse, and transportation in real-time. Furthermore, these technologies facilitate the monitoring of transportation, energy, water, and air quality, as well as refuse management, in real-time. Through the application of cutting-edge sensors, data analytics, and integrated technology, the implementation of smart city technologies additionally aims to enhance public safety, minimize pollution, and optimize the utilization of available resources.

In smart city administration, the integration of IoT devices with traffic control systems powered by artificial intelligence is one of the most crucial technologies. Real-time optimization and analysis of traffic are made possible by this integration, resulting in a reduction in congestion and an increase in traffic flow. Smart energy systems, which facilitate the personalization of energy distribution, the rationalization of power distribution, and the prompt detection of anomalies to ensure the sustainability of the energy supply, are an additional crucial technology. Furthermore, smart cities promote environmental sustainability and social inclusivity by enhancing waste management, public safety, and urban accessibility through the utilization of intelligent vehicles and wireless sensors [23].

Metropolises, including Helsinki, London, and Singapore, exemplify the effective implementation of state-of-the-art technologies in the administration of smart cities. Implementing initiatives such as Smart Nation, these cities have endeavored to integrate digital elements and information and communication technologies into the daily routines of their inhabitants to enhance the standard of living in major metropolitan regions [23]. Furthermore, with the increasing demand for technologically advanced urban environments, it is expected that the advancement of smart cities will have a substantial impact on the trajectory of construction and infrastructure in the coming years.

1.3.3 IoT-Enabled Smart Sustainable Cities

IoT-enabled intelligent and sustainable communities face a multitude of challenges that demand innovative solutions across domains such as transportation, healthcare, industry, and mobility. The integration of IoT technology into urban settings has enabled the development and automation of one-of-a-kind services and complex applications that benefit a wide range of municipal stakeholders. Two of the challenges are integrating the disparate IoT technologies that serve as the foundation and achieving the required level of scalability [24]. The development of smart cities enabled by IoT [24] has been propelled by the increasing urbanization of the global population and the need to resolve a vast array of urban challenges.

The integration of IoT technologies significantly improves the functionality of smart city mobility services, rendering them indispensable in the transportation and mobility sector. An example of how technologies associated with big data platforms and analytics have been implemented to enhance smart city mobility services is provided later [25]. Furthermore, establishing infrastructure that enables the detection of critical urban parameters and the regulation and activation of components to assist with a wide array of application domains, including but not limited to traffic, public transportation, and parking, is of the utmost importance [24].

Smart cities facilitated by IoT are intended to promote the growth of intelligent industry and manufacturing. This entails the implementation of IoT technology to enhance resource efficacy, minimize environmental harm, and optimize industrial processes. One application of IoT technologies [4] is the monitoring and optimization of the surrounding environment and energy consumption in industrial contexts.

Furthermore, in the domain of healthcare, IoT-enabled intelligent and sustainable communities place paramount importance on the creation of cutting-edge applications and services that improve the provision of healthcare. As an illustration, smart city services may encompass unified platforms that streamline the management of personal health records, promote the utilization of public transportation, and aggregate information about immunization and birth registration [24].

1.3.4 PRINCIPLES OF SUSTAINABLE URBAN ECOSYSTEMS DEVELOPMENT

The development and functioning of sustainable urban ecosystems are influenced by a variety of concepts and theories. These principles are crucial in promoting the sustained welfare of both humanity and the planet, while also reducing ecological imprints in urban regions and optimizing the utilization of natural resources.

Urban sustainability is a complex challenge aimed at enhancing the livability of an area while advancing the long-term well-being of both the planet and its residents. This is achieved through the effective use of natural resources and waste management in urban areas [26]. Among the most fundamental ecological principles is the idea that urban environments are ecosystems, distinguished by their spatial diversity, dynamism, and the interplay of natural and human processes. Following these tenets is necessary to guarantee the creation and ongoing upkeep of sustainable urban ecosystems.

The 1992 Hannover Principles not only offer a framework for sustainable urban development, but also emphasize how important it is to design sustainably [27]. This page includes a link to the Hannover Principles. In addition, several suggestions that align with the principles of sustainable urban ecosystem development have been created by the Sustainable Sites Initiative and several progressive landscape architects. These guidelines define environmental functions that correlate to services rendered in urban environments, hence emphasizing the significance and validity of the useful recommendations made by these projects.

Moreover, biophysical constraints and limited resources provide global obstacles to the attainment of urban sustainability. This emphasizes the necessity of an all-encompassing approach that prioritizes the aspects of equality, the environment, and the economy [28]. To meet the opportunities and difficulties that develop in the process of advancing urban sustainability theory and practice, a holistic approach is needed [28].

The incorporation of many conceptual frameworks and guiding principles into theories about the establishment and functioning of sustainable urban ecosystems is intended to advance the long-term welfare and satisfaction of urban residents as well as the planet as a whole. Urban-relevant ecological concepts are given priority in these theories, such as the realization that cities are ecosystems with spatial diversity, dynamism, and the interplay of natural and human processes. The goal of urban sustainability is to minimize waste generation and maximize the efficient use of natural resources within a metropolitan region to promote the long-term well-being of both humans and the planet [26]. Several innovative landscape architects and the sustainable sites initiative have written a set of doable recommendations that follow the principles controlling the growth of sustainable urban ecosystems. The criteria described above emphasize how crucial ecosystem services and functions are in urban settings.

To clarify the ecological and systemic elements inherent in urban sustainability, several writers have investigated the accepted presumptions and principles guiding the construction and operation of sustainable urban ecosystems. Furthermore, an examination of the theories and approaches related to sustainable urban ecosystems has been carried out. As a result, it is now clear that to confront obstacles and seize opportunities in the growth of urban sustainability theory and action, a systemic approach that integrates economic, environmental, and equitable views is necessary [27, 28]. These theories emphasize how critical it is to recognize global biophysical constraints and limited resources. They also emphasize the significance of regional collaborations, citizen participation, and sustainable urban governance [26].

1.3.5 URBAN SUSTAINABILITY INDICATOR FRAMEWORKS

Indicator frameworks for urban sustainability often use several core indicators and subject areas. A wide range of topics and areas of concentration are covered by these theme categories. Themes, such as education, health, housing, safety and security, equity (social and economic), the environment, and the economy, are found in the frameworks that were studied [29]. Thematic categories are employed to evaluate many aspects of urban sustainability, including social, economic, and environmental aspects [29–31].

The social dimension of sustainability encompasses various topic categories, such as fairness (social and economic), safety and security, education, health, and housing. Furthermore, a wide range of topics are covered by the thematic categories found in the reviewed frameworks, which reflects the wide range of topics and themes that are included in urban sustainability [29].

Within the four dimensions of sustainability—environmental, economic, social, and governance—a list of widely accepted theme categories has been compiled as a result of the content analysis of the studied frameworks [29]. This shows that the theme categories are arranged by the sustainability dimensions, offering an all-inclusive framework for evaluating urban sustainability. Figure 1.4 outlines a streamlined process for evaluating urban sustainability, covering dimension definition, theme identification, social exploration, theme analysis, and comprehensive framework compilation.

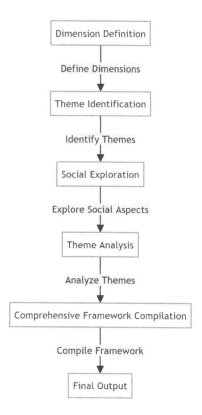

FIGURE 1.4   Conceptual flow of urban sustainability evaluation.

1.4   IoT AS A KEY ENABLING TECHNOLOGY FOR SUSTAINABLE URBAN DEVELOPMENT

In the era of transformative urban development, this section underscores the pivotal role of IoT as a key enabling technology, reshaping conventional cities into intelligent and interconnected ecosystems. With a focus on sustainability, IoT technology empowers municipalities to address environmental challenges, optimize resource management, and foster citizen engagement, heralding a paradigm shift toward efficient, habitable, and environmentally conscious urban environments. Conventional cities are undergoing a paradigm shift in municipal development as IoT transforms them into intelligent, interconnected ecosystems. Cities can reduce carbon emissions, monitor and manage energy consumption, and implement sustainable practices by utilizing IoT and data analytics. By monitoring water usage, waste management, and air quality with IoT-enabled sensors, municipalities can implement targeted environmental protection measures. The significance of IoT in propelling urban development cannot be overstated, given that it empowers municipalities to promote citizen engagement, optimize resource utilization, and enhance sustainability.

With over 50% of the global populace residing in urban regions, municipalities are utilizing IoT technology to establish connections between individuals, infrastructure, and devices to tackle a variety of issues, including traffic congestion, air pollution, waste management, and power grids. Through the utilization of IoT technologies, municipalities can effectively oversee their expanding populations by enhancing the standard of living, optimizing urban services and operations, bolstering competitiveness, and attending to economic, social, environmental, and cultural requirements. By incorporating IoT into smart city initiatives, how we reside, labor, and engage in urban surroundings may be fundamentally transformed, resulting in decreased expenditures, enhanced productivity, and a more environmentally sustainable trajectory.

Collaboration between governments, technology providers, businesses, and citizens is vital for the advancement of IoT-driven urban development in the coming years. Ecosystems that encourage collaboration will promote co-creation, innovation, and sustainable urban development. In the course of our investigation into the notion of smart city IoT, we shall examine the potential for enhanced urban environments that are sustainable, efficient, and habitable as a result of the convergence of IoT and smart city initiatives [4]. By incorporating IoT into the advancement of smart cities, the capacity of interconnected devices and data-driven insights can be utilized to create urban environments that are more sustainable, livable, and efficient.

The significance of converging AI, IoT, and big data technologies in tackling environmental sustainability challenges is underscored in the literature review concerning environmentally sustainable smart cities. Environmental sustainability challenges can be addressed in novel ways through the convergence of AI, IoT, and big data technologies; however, this convergence also entails environmental costs, ethical risks, and regulatory complexities. The exponential expansion of smart cities that prioritize environmental sustainability can be ascribed to the accelerated decarbonization and digitalization initiatives, which are propelled by the development of data-driven technologies. Incorporating technology solutions, compact urban design strategies, and eco-city design strategies, the proposed framework exemplifies the multifaceted approach necessary to realize environmentally sustainable smart cities [32].

1.4.1 SMART INFRASTRUCTURE AND RESOURCE MANAGEMENT IN URBAN DEVELOPMENT

The effective administration of intelligent resources and infrastructure, facilitated by IoT, is a critical element of sustainable urban development. Within this revolutionary approach, interconnected devices and sensors operate in perfect harmony to optimize the efficiency of urban systems and maximize the utilization of existing resources. Real-time monitoring in smart energy infrastructure is made possible by IoT sensors, which facilitates load balancing, the detection of inefficiencies, and the integration of renewable energy sources. This domain holds considerable importance in terms of influence. This not only enhances energy resilience but also substantially diminishes carbon footprints, thereby contributing to the sustainability of power infrastructures as a whole.

Intelligent transportation systems (ITSs) are an additional critical element of urban development enabled by IoT. Urban areas possess the capability to implement dynamic traffic management systems, which enhance routing efficacy and mitigate congestion by utilizing real-time data supplied by sensors mounted on roadways and vehicles. In the transportation sector, IoT finds extensive utility in various domains, encompassing intelligent parking mechanisms, traffic congestion mitigation, fuel conservation, and pollutant reduction. Furthermore, IoT assumes a critical role in the administration of refuse and water. IoT enables sensors to monitor water quality, detect breaches, and optimize waste collection routes; consequently, these functions foster environmental consciousness and sustainable resource management.

Furthermore, IoT facilitates the monitoring and maintenance of critical infrastructure components. Predictive maintenance, facilitated by ongoing monitoring, enables the timely identification of issues affecting infrastructure such as buildings, roads, and bridges. This effectively reduces repair expenses and ensures minimal downtime. The remote monitoring and administration functionalities of IoT enable centralized infrastructure management by city managers, thereby enhancing the effectiveness of public services, energy usage, and lighting. IoT-driven solutions facilitate the progressive implementation of technology by municipalities to evolving demands, owing to their adaptability and capacity for expansion. Consequently, this results in the development of urban infrastructures that are more resilient, effective, and ecologically conscious.

1.4.2 ENVIRONMENTAL MONITORING AND CONSERVATION

The integration of IoT-connected technologies holds significant transformative potential, especially when applied to the domains of environmental monitoring and conservation. Urban areas possess the capacity to acquire data in real-time via IoT devices, thereby facilitating the resolution of environmental issues and the promotion of sustainable practices. IoT facilitates air quality monitoring, which empowers municipalities to detect areas with high concentrations of pollutants, implement focused interventions, and establish timely warning systems. These advancements collectively contribute to the enhancement of public health and the mitigation of adverse environmental impacts caused by industrial and vehicular activities. IoT-driven management provides green spaces, which are vital for the health of urban residents, with sensors that track soil conditions, sunlight exposure, and moisture levels; this enables effective irrigation and maintenance. Intelligent waste management systems, which utilize IoT-enabled receptacles, revolutionize waste management by monitoring bin levels and optimizing collection routes. Furthermore, these systems promote recycling practices and aid in the efficient management of refuse at a reasonable cost.

Furthermore, there has been a noticeable emergence of the influence of IoT technologies in the domain of water quality monitoring. This feature aids in the maintenance of a dependable and secure water supply through the prompt identification of anomalies or contaminants. These technologies contribute to more extensive conservation endeavors beyond urban regions through the utilization of satellite and remote sensor technology for monitoring ecosystem, tracking deforestation, and assessing the effects of urbanization on adjacent natural habitats. Municipalities are empowered to make informed decisions regarding sustainable development with the data collected by IoT devices. Additionally, these data contribute to the preservation of vital ecosystems and the diversity of life on our planet.

1.4.3 Data-Driven Decision-Making and Citizen Engagement

Urban sensors, devices, and IoT generate enormous quantities of data that enable city administrators and planners to make informed decisions that benefit not only the environment but also the quality of life for residents. The capacity of the IoT to accumulate and analyze vast quantities of data stands out as one of its most crucial contributions to the urban development process. By employing this data, urban planners can gain valuable insights into a multitude of facets of urban existence, including but not limited to waste management, energy consumption, and traffic patterns. By employing sophisticated analytics techniques, decision-makers can enhance the effectiveness of resource allocation, optimize urban planning methodologies, and expedite infrastructure development. This methodology, informed by empirical evidence, not only enhances the overall sustainability of the city but also strengthens its long-term capacity to withstand and adapt to evolving urban challenges.

Furthermore, IoT technologies transcend mere data collection capabilities by enabling authentic citizen engagement. Residents are enabled to actively participate in the urban development process through the utilization of interconnected devices and applications. For instance, members of the public possess the capacity to furnish real-time air quality data, detect infrastructure malfunctions, and participate in community-oriented initiatives. Engaging in bidirectional communication not only furnishes decision-makers with crucial, primary perspectives on the desires and apprehensions of the general public, but also fosters a sense of community pride and ownership among the general public. The utilization of IoT-enabled devices is critical for public health and safety. Implementing intelligent sensors capable of real-time monitoring and response to emergencies can facilitate an efficient and prompt evacuation across an entire metropolitan area. Furthermore, the acquisition of health-related data through sensors and monitoring devices contributes to the implementation of a proactive strategy for addressing urban health issues.

1.5 FUTURE TRENDS

As a means of enhancing sustainability and quality of life, the integration of IoT into urban infrastructure is among the forthcoming developments in IoT-enabled sustainable urban growth management. Intelligent, sustainable cities are utilizing IoT to enhance public safety, resource management, environmental conditions, and mobility, among other facets of urban administration. Smart cities have the potential to substantially contribute to sustainability by optimizing public resources, improving service quality, and reducing costs through the effective utilization of IoT applications [25, 33].

IoT is an indispensable component of data-driven environmental management, which is the foundation of smart city initiatives. IoT-enabled sensors amass a diverse range of data about factors such as energy consumption, atmospheric conditions, air quality, and water levels [5]. This empowers decision-makers to make informed choices that promote sustainable development and efficient use of resources.

Sustainable urban growth management facilitated by IoT also requires the integration of intelligent mobility solutions. This encompasses the implementation of data analytics and IoT devices to optimize traffic flow, alleviate congestion, and enhance public transportation systems. Moreover, using intelligent energy management systems, IoT technologies can facilitate the construction of energy-efficient infrastructure and structures, thereby contributing to sustainability as a whole. Moreover, forthcoming developments in sustainable urban growth management will involve the proliferation of IoT functionalities aimed at bolstering public safety and security. To achieve this, emergency response solutions, predictive analytics, and IoT-enabled surveillance systems are implemented to create safer urban environments.

1.6 CONCLUSION

In summary, this chapter elucidates the intricate landscape of sustainable urban development and growth management, with a specialized focus on the catalytic role of the IoT. Through a meticulous examination of urban growth complexities, this chapter provides pragmatic strategies and underscores the indispensable nature of data-driven decision-making in urban planning. The integration of an interdisciplinary approach, amalgamating insights from urban planning, technology, and sustainability, augments the depth of our comprehension regarding the multifaceted challenges confronting contemporary cities. Within this holistic framework, IoT emerges as a pivotal enabler, offering prospects for the establishment of cities characterized by both hospitality and environmental sustainability. Looking forward, this chapter anticipates forthcoming trends in the convergence of IoT and urban development, accentuating the imperative for continual innovation and adaptive methodologies. The potential of IoT in urban development remains expansive, with emerging trends indicating a continued integration of advanced technologies. The convergence of IoT with artificial intelligence, sustainable energy solutions, and community engagement holds promise for creating even more resilient, intelligent, and inclusive urban spaces. This foresight underscores the importance of ongoing innovation and adaptive methodologies in navigating the evolving landscape of sustainable urban growth.

REFERENCES

1. Samih, H., Smart cities and internet of things. *Journal of Information Technology Case and Application Research*, 2019. **21**(1): p. 3–12.
2. Barrionuevo, J.M., P. Berrone, and J.E. Ricart, Smart cities, sustainable progress. *IESE Insight*, 2012. **14**(14): p. 50–57.
3. Hiremath, R.B., et al., Indicator-based urban sustainability—A review. *Energy for Sustainable Development*, 2013. **17**(6): p. 555–563.
4. Yigitcanlar, T. and D. Dizdaroglu, Ecological approaches in planning for sustainable cities: A review of the literature. *Global Journal of Environmental Science and Management*, 2015. **1**(2): p. 159–188.
5. Rathore, M.M., A. Ahmad, and A. Paul. IoT-based smart city development using big data analytical approach. In *2016 IEEE International Conference on Automatica (ICA-ACCA)*. 2016. IEEE.
6. Tanwar, S., S. Tyagi, and S. Kumar. The role of internet of things and smart grid for the development of a smart city. In *Intelligent Communication and Computational Technologies: Proceedings of Internet of Things for Technological Development, IoT4TD 2017*. 2018. Springer.

7. Uckelmann, D., M. Harrison, and F. Michahelles, *Architecting the internet of things.* 2011. Springer Science & Business Media.

8. Al-Fuqaha, A., et al., Internet of things: A survey on enabling technologies, protocols, and applications. *IEEE Communications Surveys & Tutorials*, 2015. **17**(4): p. 2347–2376.

9. Sharma, P. and S. Jangirala, Internet of Things for sustainable urbanism. In *Journal of Physics: Conference Series*. 2022. IOP Publishing.

10. Perveen, S., M. Kamruzzaman, and T. Yigitcanlar, Developing policy scenarios for sustainable urban growth management: A Delphi approach. *Sustainability*, 2017. **9**(10): p. 1787.

11. Keiner, M., Sustainable development and urban management in developing countries: the case of Africa. In *The real and virtual worlds of spatial planning*. 2004, Springer. p. 43–59.

12. Ewing, R., et al., Growth management effectiveness: A literature review. *Journal of Planning Literature*, 2022. **37**(3): p. 433–451.

13. Tiwari, P., P.V. Ilavarasan, and S. Punia, Content analysis of literature on big data in smart cities. *Benchmarking: An International Journal*, 2019. **28**(5): p. 1837–1857.

14. Zhang, C., Design and application of fog computing and Internet of Things service platform for smart city. *Future Generation Computer Systems*, 2020. **112**: p. 630–640.

15. Marques, G., et al., Indoor air quality monitoring systems for enhanced living environments: A review toward sustainable smart cities. *Sustainability*, 2020. **12**(10): p. 4024.

16. Gazis, V., A survey of standards for Machine-to-Machine and the Internet of Things. *IEEE Communications Surveys & Tutorials*, 2016. **19**(1): p. 482–511.

17. Leminen, S., et al., Industrial internet of things business models in the machine-to-machine context. *Industrial Marketing Management*, 2020. **84**: p. 298–311.

18. Javadzadeh, G. and A.M. Rahmani, Fog computing applications in smart cities: A systematic survey. *Wireless Networks*, 2020. **26**(2): p. 1433–1457.

19. Taherdoost, H., Security and Internet of Things: Benefits, challenges, and future perspectives. *Electronics*, 2023. **12**(8): p. 1901.

20. Janssen, M., et al., Challenges for adopting and implementing IoT in smart cities: An integrated MICMAC-ISM approach. *Internet Research*, 2019. **29**(6): p. 1589–1616.

21. Moosavi, N. and H. Taherdoost. Blockchain and Internet of Things (IoT): A disruptive integration. In *Proceedings of the 2nd International Conference on Emerging Technologies and Intelligent Systems*. 2023. Springer International Publishing, Cham.

22. Mishra, P. and G. Singh, 6G-IoT framework for sustainable smart city: Vision and challenges. In *Sustainable Smart Cities: Enabling Technologies, Energy Trends and Potential Applications*. 2023, Springer. p. 97–117.

23. Vodák, J., D. Šulyová and M. Kubina, Advanced technologies and their use in smart city management. *Sustainability*, 2021. **13**(10): p. 5746.

24. Bauer, M., L. Sanchez and J. Song, IoT-enabled smart cities: Evolution and outlook. *Sensors*, 2021. **21**(13): p. 4511.

25. Belli, L., et al., IoT-enabled smart sustainable cities: Challenges and approaches. *Smart Cities*, 2020. **3**(3): p. 1039–1071.

26. National Academies of Sciences, E. and Medicine, *Pathways to urban sustainability: challenges and opportunities for the United States*. 2016.

27. Cepeliauskaite, G. and Z. Stasiskiene, The framework of the principles of sustainable urban ecosystems development and functioning. *Sustainability*, 2020. **12**(2): p. 720.

28. Childers, D.L., et al., Advancing urban sustainability theory and action: Challenges and opportunities. *Landscape and Urban Planning*, 2014. **125**: p. 320–328.

29. Michalina, D., et al., Sustainable urban development: A review of urban sustainability indicator frameworks. *Sustainability*, 2021. **13**(16): p. 9348.

30. Halla, P. and A. Merino-Saum, Conceptual frameworks in indicator-based assessments of urban sustainability—An analysis based on 67 initiatives. *Sustainable Development*, 2022. **30**(5): p. 1056–1071.
31. Merino-Saum, A., et al., Indicators for urban sustainability: Key lessons from a systematic analysis of 67 measurement initiatives. *Ecological Indicators*, 2020. **119**: p. 106879.
32. Bibri, S.E., et al., Environmentally sustainable smart cities and their converging AI, IoT, and big data technologies and solutions: an integrated approach to an extensive literature review. *Energy Informatics*, 2023. **6**(1): p. 9.
33. Bibri, S.E., The IoT for smart sustainable cities of the future: An analytical framework for sensor-based big data applications for environmental sustainability. *Sustainable Cities and Society*, 2018. **38**: p. 230–253.

2 Enhancing IoT Data Quality Validation in Medical Field through Big Data Analytics

Rajesh Kumar Pallamala
Chadalawada Ramanamma Engineering College, Tirupathi,
India

Paul Rodrigues
King Khalid University, Abha, Saudi Arabia

A. Jyothi Babu
Mohan Babu University, Tirupathi, India

R. Ravindraiah
Madanapalle Institute of Technology & Science,
Madanapalle, India

2.1 INTRODUCTION

Internet of Things (IoT) has emerged as one of the 21st century's most significant technologies in the past few years. Now that commonplace items like vehicles, baby monitors, thermostats, and kitchen appliances can be connected to the Internet through embedded devices, communication between people, processes, and things may happen seamlessly. IoT is a worldwide network of connected computing devices with the capacity to sense, gather, and share data. Large datasets from multiple sources that are too big to handle with conventional methods are referred to as "big data." The network of physical objects, or "things," that are implanted with sensors, software, and other technologies in order to communicate and exchange data with other devices and systems over the Internet is known as Internet of Things [1].

Connecting intelligent objects, or things, to the Internet in a transparent manner is the fundamental tenet of IoT. This results in a data exchange across everything and provides users with information in a more secure manner. In 2020, according to

Cisco Systems, there were 50 billion Internet-connected devices worldwide. It was also expected that many physical objects, such as computers and sensor actuators, would have unique addresses and the capacity to securely transfer data from routine daily tasks to private medical records. Physical objects may communicate and gather data with little assistance from humans thanks to cloud computing, mobile technologies, big data, analytics, and low-cost computation. Digital systems are able to record, monitor, and modify every interaction between connected objects in this hyper-connected world. The digital and physical realms collide, yet they work together [2].

The use of IoT technology in industrial settings, particularly in relation to the instrumentation and management of sensors and devices that use cloud technologies, is referred to as industrial IoT. A great illustration of IoT may be seen in this Titan use case PDF. Machine-to-machine (M2M) communication has been employed recently by industry to accomplish wireless automation and control [3]. But, as cloud computing and related technologies (such analytics and machine learning) gain traction, industries can accomplish a new automation layer and, as a result, develop new sources of income and business models.

2.1.1 KEY CONTRIBUTIONS OF THE CHAPTER

The following are the significant contributions of this chapter:

i. It examines how the big data ecosystem helps the medical IoT data can be collected, managed, maintained, and tested massive amount of data.
ii. It investigates multiple types of big data and identifies the difference between outmoded and restrained analysis methods [4].

For the early prediction of certain diseases, a framework based on explainable deep learning has been suggested, and a real-world medical dataset has been utilized to assess the effectiveness of the proposed framework.

2.1.2 CHAPTER ORGANIZATION

A dynamic, worldwide network infrastructure that can self-configure is built upon standard, interoperable communication protocols. In this world, both virtual and physical objects have identities, physical characteristics, and virtual personalities. They also use intelligent interfaces and are seamlessly integrated into information networks and wireless networks, frequently exchanging data related to users and their surroundings. IoT technology has many uses in the medical and healthcare industry, including smart sensor integration, remote monitoring, and medical device integration. It enhances the way a doctor provides care for patients while keeping them safe and well. A brief literature review of existing methods is presented in Section 2.2 followed by the proposed methodology (Section 2.3), results (Section 2.6), and conclusions. The remainder of the chapter is structured with an overview of IoT and healthcare in Sections 2.1 and 2.2. Recent developments in healthcare technology are covered in Section 2.3. IoT and healthcare present a number of issues, which are covered in Section 2.4. Numerous services and applications are brought about by this innovation, which are discussed in Section 2.5.

2.2 RELATED WORKS

The life cycle of the big data analytical process is for the development and verification processes. Teams need to thoroughly comprehend how big data will affect the design [5], configuration, and operation of IoT systems as well as different data formats. Since data is continually growing, there have always been problems with their analysis, storage, transfer, and visualization. To address these problems, a number of studies from different fields of study were examined and contrasted [6].

S.B. Baker et al. [8] described the IoT research in healthcare and provided a framework for IoT-based healthcare systems, which can be deployed for general systems as well as narrowly focused monitoring applications. They also reviewed non-invasive, wearable sensors with a focus on those that measure vital signs, blood pressure, and blood oxygen level. Two types of communication protocols, long range and short range, were compared to evaluate their suitability for medical purposes. This chapter represents to failure the size of the data is very less in MB's and the columns and attributes collection was showing very less communication. Aceto Y. et al. [9] provided a comprehensive overview of IoT-based healthcare systems and offered a summary of the underlying technologies and smart healthcare devices in IoT. Specific implementation techniques and approaches to ontology-based resource management, big data management, and knowledge management are also addressed but the ecosystem of collecting the IoT data from device system is not presented in this chapter.

Ashlesha et al. [10] implemented the HDFS method to test the datasets having a size ranging from few terabytes to many petabytes. They tested complex data using work nodes that can easily fit into big data and can be analyzed with comfort. They used low-quality data and the performance of this system ceases with the rise of data volume. The frequent crash of their application has happened even with a stretch of the data to a minimum limit. This system does not even meet the least strands since it fails to come across Service-Level Agreements (SLAs) and did not explore their system performance using statistical investigations. Hoger K et al. [11] conducted a performance check among Scala and Java in Apache Spark MLlib. They intended to incorporate standard data into big data to analyze business intelligence reports. The data transfer is done from a local disk to the ETL process. Every dataset is treated as a node and allowed for testing. The authors created only one node for testing which cannot be considered a basic strand of big data. Even though they considered 5 Vs, the proper implementation is not described well and further, it consumes more time for testing. S. Nachiyappan et al. [12] developed a software verification of database testing which depends on the quality of automation testing. This has risen the complexity of high-level database processing and testing has become complicated which in turn has led to intensive workload. Multi-model structure validation can handle the big data validation and testing for the future forecast. Table 2.1 presents an overview of related publications [8–12] that use machine learning-based models, summarizing their scope and focus, important constraints, and a comparison with this chapter.

TABLE 2.1

Limitations of Some Related Works

Year	Reference	Focus and Coverage	Limitations	Comparison with This Chapter
2017	S.B. Baker et al. [8]	Focused on accelerated Hadoop MapReduce considering two optimization techniques	There is a need to assess the performance of the models for the large number of datasets and their sizes	This chapter presents a detailed statistical analysis of the acquired throughput data through performance status quality reporting at the different user equipment terminal locations
2016	Aceto Y. et al. [9]	Provided a comprehensive overview of IoT-based healthcare systems and offered a summary of the underlying technologies and smart healthcare devices in IoT	Big data management and knowledge management are also addressed but the ecosystem of collecting the IoT data from device system is not presented in this chapter	The current chapter examined the performance of the projected learning-based models and smart healthcare devices in IoT with big data implementations
2018	Ashlesha et al. [10]	Implemented the HDFS method to test the datasets having a size ranging from few terabytes to many petabytes	They used low-quality data and the performance of this system ceases with the rise of data volume. The frequent crash of their application has happened even with a stretch of the data to a minimum limit	We can test complex data using Hadoop work nodes that can easily fit into big data and can be analyzed with comfort
2019	Hoger K et al. [11]	Conducted a performance check among Scala and Java in Apache Spark MLib. Intended to send standard data into big data to analyze business intelligence reports	They created only one node for testing which cannot be considered a basic strand of big data. Even though they considered 5 Vs, the proper implementation is not described well and further	We used 5 Vs, the proper implementation to describe well and further, it consumes more time for analyzing the dataset by using big data
2019	S. Nachiyappan et al. [12]	Developed a software verification of database testing which depends on the quality of validations	This has risen the complexity of high-level database processing and testing has become complicated which in turn has led to intensive workload	Multi-model structure validation can handle the big data validation and testing for the future forecast

2.2.1 PROBLEM STATEMENT

It has been noted that most current methods only use IoT for big data analysis, even though it is powerful enough to handle small datasets. Most efforts focused mostly on handling semi-structured and unstructured data as well as processing and evaluating structured data. This makes it impossible to manage and maintain information in the medical field using traditional methods in the modern world. The authors gave a vague account of their works and neglected to define the IoT approaches. Neither they disclosed the test versions, tools, or procedures nor did they publish the test results. Since many of them are incapable of using nodes and data clusters, they are unable to manage all kinds of data formats. The authors of [8], [7–10] used Hadoop as a key dataset for their analyses. The technique is hampered by Hadoop's intrinsic shortcomings, such as lengthy simulation, testing, and loading times. The authors even experienced poor analysis of test report. The mathematical model of the proposed methods [7, 8, 11] was poorly explained. The method used by S. Nachiyappan et al. [12] cannot manage testing both functionally and non-functionally, and Hoger et al. [11] only handled conventional structural data. Checkpoint forecasts are not without issues as illustrated by Hoger et al. [11]. However, the test technique for the suggested DRE approach was not covered by Ashlesha et al. [10]. They failed to provide any statistical support for their claims and failed to show how their work was actually implemented.

2.3 PROPOSED METHODOLOGY

IoT and big data analysis have already gained a lot of importance and are vividly used in business platforms. It helps in the evaluation as well as the performance checking of different commercial trends. The analyst needs to verify the effective processing of terabytes of data during testing. The data quality checking is a significant factor and is distinguished in terms of accuracy, validity, consistency, conformity, duplication, etc. [12]. The analyst verifies the successful processing of terabytes of knowledge using commodity clusters and other supportive components. It demands a high level of testing skills because the processing should be done exceptionally firm. Data processing depends upon batch, real-time, and interactive processes. Before testing the appliance, it is necessary to see the standard of the test engineer's knowledge about big data testing [13]. It includes testing various features like conformity, accuracy, duplication, consistency, validity, and data completeness.

The DB2 files are loaded into HDFS [14] upon the completion of the validation process (see Figure 2.1). Utilizing Hadoop MapReduce operations to process input files from several sources is fraught with issues. Among these include improper data aggregation during the reduction phase, inappropriate output data format, improper coding during the MapReduce phase, and mapper spill size, which all hinder the dataset's proper operation on numerous nodes even when only one node is involved. It is possible to confirm that the Hadoop MapReduce technique can modify the number of maps and tasks as needed. To verify the creation of output files and DB2 data processing, the output file format must comply with the specifications, examine, and contrast the formats of the input source files and the resultant file [15].

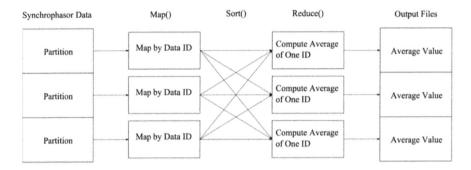

FIGURE 2.1   Hadoop MapReduce ecosystem for medical data.

Each node's business logic configuration is checked and validated against the others. Once the reduction phase is finished, the details are combined and verified. Next, the key-value pair [16] that was produced during the MapReduce stage is verified, and close attention is paid to the shuffling process. The observable performance will be obtained with the use of minimal map output key-map output value, compression of mapper output, and minimalization of mapper output, and this may be achieved by filtering out records on the mapper side rather than the reducer side. The mapper's output size is dependent on memory during the shuffle phase, disk input, and network input. The intermediate output of the mappers is compressed before sending it to the reducers to minimize network traffic and disk I/O from reading and transferring files [17].

Figure 2.1 shows the four stages of the MapReduce process. Let us now discuss these steps in more detail. At this point, a MapReduce operation is performed on the initial data in order to get the intended result. Huge datasets can be compressed using a data processing technique [18] called "MapReduce" to produce useful aggregated results and handle the required business [19, 20].

- Utilize the list of clean text (full text + title + abstract) from each of the prefiltered documents to train the BM25 model.
- Create a list of customized synonyms for the original question, additional keywords, frequent medical terms, and extra keywords to construct the question's keywords.
- Utilize the subtask's keywords in conjunction with the corpus to get a list of the highest ranked documents.
- To find the top-ranked phrases that correspond to the keywords in the question, apply text summarization to each of the top-ranked documents.
- To determine the overall top-ranked sentences for the subtask, rank the top sentences in each document.
- As the general summary, show the top two sentences from each of the top-ranked documents.
- Display the final findings in the form of an HTML table containing the text summary and the rankings of the top-ranked documents.

2.4 RESULTS AND DISCUSSIONS

Hadoop Distributed File System (HDFS [21]) facilitates the input data which is available to multiple nodes. The input data is extracted from multiple sources in a semi-structured or unstructured manner and loaded into HDFS by splitting files into multiple numbers and process. These multiple files using map and reduce operations finally extract the data generated from HDFS and store them in downstream systems. While working with huge data and processing them at multiple nodes, there are chances of losing quality data and inducing unrelated data, which in turn results in quality issues at each stage in the testing process. Data functional testing involves the testing of the information in three phases:

1. Validation of Hadoop pre-processing,
2. Validation of Hadoop MapReduce process, and
3. Validation of extracted data and store in downstream systems shown within the Hadoop MapReduce architecture [22].

Table 2.2 shows the sample data that was gathered from various medical sources in order to analyze the test suites. It explains the various IoT system formats and how data may be fed into data staging and source files such as semi-structured format conversion. To send information from the source to the destination, data can be poured into an FTP task. The data sources are fetched and stored in data warehouses using the SQL Server Integration Service (SSIS) and Power BI tools [23]. The semi-structured IoT [24] data on the local storage must be imported into the TestingWhiz application program. Each of the 72,345 rows of semi-structured (.csv) data utilized in this example is linked to 21 attributes. The test project's Test suite needs to load and process this file as a string operation, using 1.5 GB of memory storage. The testing approaches that are offered assess the attributes of the data source by utilizing variables such as set, compare, length, text, string, and data type. This process occurs within the test command wizard, and Figure 2.1 illustrates the functional flow. For data validation, the data is additionally permitted to enter Test editor. The test commands are generated after the data is allowed into the Test editor. These commands can be used to write messages, perform, receive messages, click, if conditions, open page, and infer the input or output [25].

Utilizing Python with Hadoop semi-structured [26] data is kept in the HDFS for business logic analysis. The Excel formatted source data is kept in the local system directory under the filename "F:\Bigdata\patientMonitoring.csv." For instance, the target file is loaded with the source file for the suggested validation. Important steps in the process are record mismatch, row validation, column validation, and deduplication. The data type of the attributes needed to implement the source to target mapping and the lists of operations carried out on a sample example using Hive queries. In this example, the semi-structured sample data collection, called "\DB1," needs to have its schemas certified. It has a field called "Users" with sample five columns, frame.time_delta, frame.len, tcp.srcport, and mqtt.clientid, are taken for MapReduce process for examination.

TABLE 2.2
Sample Medical Data

frame.time_delta	frame.time_relative	frame.len	ip.src	ip.dst	tcp. srcport	tcp.payload
0	0	105	10.5.126.141	10.5.126.56	35161	30:0f:00:0a:47:6c:75:63:6f:6d:65:74:65:72:31:32:31
0.000249	0.000249	105	10.5.126.143	10.5.126.56	34237	30:0f:00:0a:47:6c:75:63:6f:6d:65:74:65:72:31:32:37
0.000105	0.989377	85	10.5.126.143	10.5.126.56	34237	30:0f:00:0a:47:6c:75:63:6f:6d:65:74:65:72:31:32:38
4.50E-05	0.989422	78	10.5.126.145	10.5.126.56	46623	30:11:00:0d:42:6c:6f:6f:64:50:72:65:73:73:75:72:65:37:34
3.90E-05	0.989461	77	10.5.126.147	10.5.126.56	45663	30:10:00:0c:49:6e:66:75:73:69:6f:6e:50:75:6d:70:31:30
3.60E-05	0.98957	78	10.5.126.145	10.5.126.56	38937	30:0a:00:07:41:69:72:46:6c:6f:77:30
3.50E-05	0.989605	77	10.5.126.147	10.5.126.56	46653	30:0a:00:07:41:69:72:46:6c:6f:77:30
4.50E-05	0.98965	87	10.5.126.142	10.5.126.56	43335	30:10:00:0c:50:75:6c:73:6f:78:69:6d:65:74:65:72:36:38
1.30E-05	0.989663	80	10.5.126.146	10.5.126.56	39593	30:10:00:0c:50:75:6c:73:6f:78:69:6d:65:74:65:72:37:38
0.000219	2.961477	80	10.5.126.146	10.5.126.56	33387	30:08:00:03:45:43:47:31:33:36
0.00016	2.961637	86	10.5.126.148	10.5.126.56	36573	30:07:00:03:45:4d:47:35:36
0.000128	2.961765	86	10.5.126.148	10.5.126.56	39979	30:08:00:03:45:43:47:31:38:31
1.997911	4.959676	78	10.5.126.145	10.5.126.56	46623	30:07:00:03:45:4d:47:35:37
1.998696	40.96055	78	10.5.126.145	10.5.126.56	46623	30:0a:00:07:41:69:72:46:6c:6f:77:30

2.4.1 Validation of Hadoop MapReduce Process through IoT

As the term BIG [27] data implies, handling a lot of data can be difficult. Apache Hadoop comes with a built-in programming model called "MapReduce." On the cluster, it will process the data in parallel. Python grating Hadoop enables smooth interface with other well-known large data processing frameworks, such as Apache Spark. Being fast and versatile, Spark offers a cluster computing technology that is readily linked with Hadoop. The Single Node Cluster in Docker is used in the Hadoop implementation. Using the following commands, the text files (.csv) are moved from the local file system to HDFS [28, 29].

> *$ docker start -i paitentmonetring*
> *hduser@localhost:~/examples$ hdfs dfs -put *.csv input*

The lines from stdin (standard input) will be read by the mapper. Using the standard output (stdout), Hadoop will deliver a stream of data read from HDFS to the mapper. The mapper will read every line that is sent through the stdin, removing any non-alphanumeric characters and splitting the text into a Python list. Finally, it will produce the string "word\t1," which is a pair (work,1). The output is then delivered back to the data stream via stdout (print). Every input (line) from the stdin will be read by the reducer, which will then count each repeated word (raising the counter for this word) and deliver the result to the stdout. Iteratively running the operation will continue until no more inputs are received in the stdin [30].

> *import numpy as np*
> *import pandas as pd*
> *from IPython.display import HTML*
> *from sklearn.manifold import TSNE*
> *from sklearn.feature_extraction.text import CountVectorizer*
> *from sklearn.feature_extraction.text import TfidfTransformer*
> *from sklearn.metrics.pairwise import cosine_similarity*
> *from nltk.stem import WordNetLemmatizer*
> *import ipywidgets as widgets*
> *from ipywidgets import interact, interact_manual*
> *from rank_bm25 import BM25Okapi*
> *%%javascript*
> *// Disable scrolling in jupyter notebook*
> *IPython.OutputArea.prototype._should_scroll = function(lines)turn false;*

After reading each input (line) from the stdin, the reducer counts each repeated word by raising the counter for that word and outputs the result to the stdout. Until there are no more inputs in the stdin, the process will be carried out iteratively. Using the .csv files found in /user/hduser/input (HDFS), mapper.py, and reducer.py, the following command will start the MapReduce process. The distributed file system's /user/hduser/output will include the result and the MapReduce process described in Table 2.3.

TABLE 2.3
MapReduce Process

Input Sampling	Average Map Time (s)	Average Shuffle Time (s)	Average Merge Time (s)	Average Reduce Time (s)	Average Total Time (s)
frame.time_delta	18	16.4	7.3	1.66	43.36
frame.len	9.7	10.09	3.7	4.8	28.29
tcp.srcport	32.3	30.9	10.21	17.07	90.48
mqtt.clientid	16,2	13.2	6.2	1.2	34.33
health.class	23.5	21.3	2.5	3.1	78.31

hduser@localhost:~/examples$ hadoop jar $HADOOP_HOME/share/hadoop/tools/lib/hadoop-streaming-3.3.0.jar -mapper mapper.py -reducer reducer.py -input /user/hduser/input/.txt -output /user/hduser/output*

This is the architecture for distributed processing and storage in big data clusters [20] using commodity technology. HDFS splits files into 1.5 GB blocks, which are subsequently dispersed throughout the cluster's nodes. Hadoop MapReduce transfers code to the nodes that hold the required data, as shown in Figure 2.2. The nodes then carry out parallel processing. Unlike conventional architecture, parallel processing frequently makes use of a parallel file system. The data locality is used in this method. It is composed of many little side projects that are grouped together under the distributed computing infrastructure heading [31]. This is the framework for big data

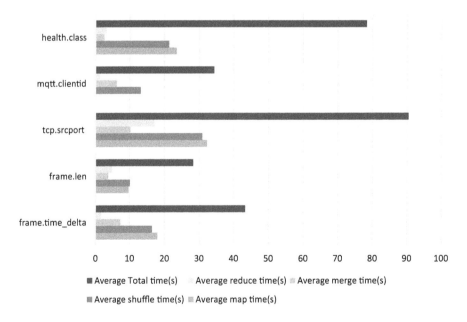

FIGURE 2.2 MapReduce time.

clusters using commodity hardware for distributed processing and storage. Files are divided into 5 GB blocks by HDFS, which then distribute the blocks throughout the cluster's nodes.

Hadoop MapReduce processes data by transferring code to the nodes that contain the necessary data. The nodes then carry out parallel processing. In contrast to traditional architecture, parallel processing often relies on a parallel file system. This method makes use of data locality. It is made up of numerous tiny side projects that fall under the umbrella of distributed computing infrastructure. Hadoop primarily includes. Large amount of data storage poses a number of challenges; despite the drives' enormously enlarged storage capacities, the speed at which data is read from them has not significantly improved. Utilizing a lot of hardware also causes a lot of issues because it might result in the likelihood of hardware failure. Replication involves making redundant copies of the same data on several devices so that in the event of a failure, a copy of the data is still available. Combining the data being read from many devices is a major issue. Although distributed computing offers numerous solutions to this issue, it still remains a difficult one. Hadoop can easily manage these issues. The HDFS addresses the issue of failure, and the MapReduce programming paradigm addresses the issue of merging data. By offering a programming style that deals with computation involving keys and values, MapReduce lessens the issue of disk reads and writes. Hadoop thus offers an analysis system and a trustworthy shared storage. HDFS handles the storage, whereas MapReduce handles the analysis.

The speed at which data is read from big volumes of data storage has not improved much, despite the drives' widely increased storage capacities. Large amount of data storage provides a variety of issues. Using a lot of hardware increases the risk of hardware failure, which leads to a lot of problems. This is avoided through replication, which is a process of creating redundant copies of the same data across multiple devices so that in the case of a failure, a duplicate of the data is still accessible. One big problem is combining the data that is being read from multiple devices. Distributed computing provides several answers to this problem, but it is still a challenging one. These problems are easily handled using Hadoop [32]. The MapReduce programming paradigm tackles the problem of integrating data, whereas HDFS tackles the problem of failure. MapReduce mitigates disk reads and writes by providing a programming style that handles computation involving keys and values. Thus, Hadoop provides a reliable shared storage system along with an analysis system. MapReduce does the analysis, whereas HDFS manages storage [21].

Figure 2.3 shows that TCP/IP vulnerabilities have been shown to impact 75% of healthcare businesses. These vulnerabilities have, on average, the greatest diversity of vulnerable vendors (12), the greatest number of vulnerable devices (almost 500), and the highest diversity of vulnerable device categories. Compared to other verticals, healthcare businesses are around five times more vulnerable to TCP/IP vulnerabilities [33]. There are 259 vulnerable suppliers and 79 different categories of devices that are susceptible. In healthcare businesses, printers, VoIP, infusion pumps, networking equipment, and building automation systems are the most frequently encountered vulnerable device types. Patient monitors, point-of-care diagnostic systems, and infusion pumps are the three most often found categories of medical devices that are vulnerable. These and other susceptible devices frequently connect

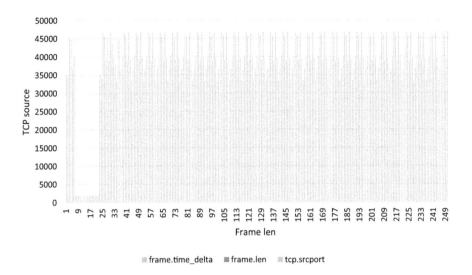

FIGURE 2.3 TCP frame length analysis.

to the same areas of a company's network, which raises the possibility and severity of incidents.

Figure 2.4 illustrates the Hadoop logic applied to the nodes to verify the working status of the proposed work. The data validation process is done to ensure the expected generation of the output. Table 2.3 presents the *scale of the factor* with the MapReduce scheduling process for each stage or node and input data will be accessible for read, write, shuffle, and analyze [34]. The Master can handle 798 tasks per stage with a multiplicative order of "2". On the other hand, when 1200 tasks per stage are presented, the multiplicative factor rises to "3". This represents that it is not

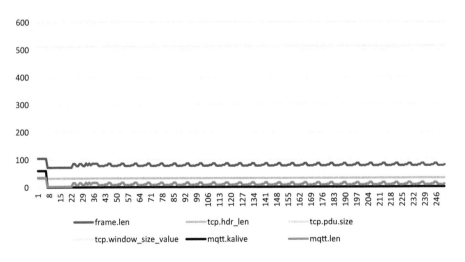

FIGURE 2.4 Medical data monitoring.

ready to schedule them to the workers. It is observed that if the Master is not limited by the number of tasks per stage, the workers will become CPU bound prior to memory bound. So, to scale gracefully, the Master memory needs to be scaled up to ensemble with the number of available workers. The testing process is smoothly done even if the number of nodes increases. Therefore, it ensures optimal system performance irrespective of resource dimensions scaled in even with the rise of the problem.

Figure 2.4 shows that IoT has significantly monitoring the healthcare and will continue to do so for years to come. Individuals with medical conditions may be able to have devices with sensors inserted under their skin. When a patient's health drops too low, the sensors in the gadgets will store previous data and send information to their IoT devices or mobile phones. Patients will be able to determine when they are most likely to experience low health levels both now and in the future. When they need to pull patient monitor data for examinations and checkups, medical professionals will have access to it. Healthcare providers can receive alerts from the wearable devices in the event that a patient has a heart attack, palpitations, arrhythmia, or stroke. The quick dispatch of ambulances can therefore be the difference between life and death [34].

If the IoT and big data encounter any critical failures, the architecture data is distributed across various nodes and stages in the big data environment. Chances of node letdowns like name node failure, data node failure, and network failure are typically observed. The Hadoop enterprise solution must be capable of automatic recovery from node failures to proceed with the processing. So, failover analysis is an important area of focus in big data implementations to validate the recovery process. This even ensures that data processing is resumed properly when switched to some other node after a node failure. Metrics attained during failover testing are the Recovery Time Objective (RTO) and Recovery Point Objective (RPO) metrics. These metrics are to be studied by the testers for any further issues.

2.5   CONCLUSION

This chapter presents an in-depth analysis of IoT and big data using the process along with output validation. It elucidates about the intricacies and hardships involved in challenging semi-structured data using the traditional procedures. The description of the above topic is discussed in Section 2.2. The study encompasses how the IoT can be used to analyze the semi-structured data from Hadoop sample data successfully. The suggested method saves time, lowers expenses, boosts productivity, and sharpens accuracy. It is evident that patients will be able to determine when they are most likely to experience low health levels both now and in the future. When they need to pull patient monitor data for examinations and checkups, medical professionals will have access to it. The graphical report gives a brief overview of the in-depth analysis of the methods that were suggested to elevate the response time, graphical user interface, and customer data needs. This graphical evaluation demonstrates the suggested approach's usability, effectiveness, robustness, firmness, and support for sophisticated analytics.

2.6   FUTURE SCOPE

Global data has shown no signs of slowing down since they began to increase significantly a decade ago. The findings for the chosen site are, indeed, encouraging. There is a ratio of work ahead of us in terms of specific language advancements that may be developed: (1) expand the medical dataset to reduce the risk of over fitting and increase the quantifiable importance of the results; (2) examine the many aspects of unstructured format dataset, such as legislative concerns, race, religion, and socio-economic status; (3) make use of the data provided by IoT to enhance the order by expanding the component space with profile sensor data, a list of followers, and geo-location, among other things. Large amounts of data are challenging to analyze since they are in unstructured, semi-structured, and organized formats. As a result, no one understands how data is entered into a database.

REFERENCES

[1] Stewart, S.; Burns, D. WebDriver. W3CWorking Draft. 2020. Available online: https://www.w3.org/TR/webdriver/ (accessed on 29 June 2020).

[2] Pietraszak, M. Browser Extensions. W3C Community Group Draft Report. 2020. Available online: https://browserext.github.io/browserext/ (accessed on 29 June 2020).

[3] Kredpattanakul, K.; Limpiyakorn, Y. Transforming JavaScript-Based Web Application to Cross-Platform Desktop with Electron. In *International Conference on Information Science and Applications*, Springer, 2018; pp. 571–579.

[4] Leotta, M.; Stocco, A.; Ricca, F.; Tonella, P. Pesto: Automated Migration of DOM-Based Web Tests Towards the Visual Approach. *Software Testing, Verification & Reliability.* 2018, 28, e1665. [CrossRef]

[5] García, B.; Gortázar, F.; Gallego, M.; Hines, A. Assessment of QoE for Video and Audio in WebRTC Applications Using Full-Reference Models. *Electronics.* 2020, 9, 462.

[6] Meyappan, R.; Nachiyappan, S., Nair, D. Towards an Automated Testing Framework for Big Data. *International Journal of Engineering and Advanced Technology (IJEAT).* December 2019, 9(1S3) 450.

[7] Omar, H. K.; Jumaa, A. K. Big Data Analysis Using Apache Spark MLlib and Hadoop HDFS with Scala and Java. *Kurdistan Journal of Applied Research (KJAR).* June 2019, 4(1). https://doi.org/10.24017/science.2019.1.2

[8] Baker, S. B.; Xiang, W.; Atkinson, I. Internet of Things for Smart Healthcare: Technologies, Challenges, and Opportunities. *IEEE Access.* 2017, 5, 26521–26544. https://doi.org/10.1109/ACCESS.2017.2775180

[9] Yin, Y.; Yan Zeng; X. Chen; Y. Fan, The Internet of Things in Healthcare: An Overview. Journal of Industrial Information *Integration.* 2016, 1, 3–13. https://doi.org/10.1016/j.jii.2016.03.004

[10] Ashlesha, S.; Nagdive, R.; Tugnayat, M. Overview on Performance Testing Approach in Big Data. *International Journal of Advanced Research in Computer Science.* Nov–Dec, 2014, 5(8), 165–169.

[11] Omar, Hoger Khayrolla; Jumaa, Alaa Khalil. Big Data Analysis Using Apache Spark MLlib and Hadoop HDFS with Scala and Java. *Kurdistan Journal of Applied Research (KJAR).* June 2019, 4(1). https://doi.org/10.24017/science.2019

[12] Nachiyappan, S.; Justus, S. Automated Testing for Big Data Environment using Multi-Model Structure Validation. *International Journal of Engineering and Advanced Technology (IJEAT).* June 2019, 8(5), 340.

[13] Malik, D.; Goel, P. K. *A Brief about Big Data, It's Technology and Challenges.* 2020, 22, 01–05.

[14] Reimer, A. P.; Madigan, E. A. Veracity in Big Data: How Good Is Good Enough. *Health Informatics Journal.* 2019, 25(4), 1290–1298.

[15] Hariri, R. H.; Fredericks, E. M.; Bowers, K. M. Uncertainty in Big Data Analytics: Survey, Opportunities, and Challenges. *Journal of Big Data.* 2019, 6(1), 1–16.

[16] Jin Wang et al. Big Data Service Architecture: A Survey. *Journal of Internet Technology.* 2020, 21(2), 393–405.

[17] Zhihui Lu et al. IoTDeM: An IoT Big Data-Oriented MapReduce Performance Prediction Extended Model in Multiple Edge Clouds. *Journal of Parallel and Distributed Computing.* 2018, 118, 316–327.

[18] Jifu Guo; Chunlin Huang; Jinliang Hou A Scalable Computing Resources System for Remote Sensing Big Data Processing Using GeoPySpark Based on Spark on K8s. *Remote Sensing.* 2022. 14(3), 521.

[19] Ahmad Latifian. How Does Cloud Computing Help Businesses to Manage Big Data Issues. *Kybernetes.* 2022. 51(6), 1917–1948

[20] Jansen, M. On-Board-Unit Big Data Analytics: from Data Architecture to Traffic Forecasting. Doctoral Dissertation, Katholieke Universiteit Leuven, 2022.

[21] Saravanan, V.; Hussain, F.; Kshirasagar, N. Role of Big Data in Internet of Things Networks. *Research Anthology on Big Data Analytics, Architectures, and Applications,* IGI Global, pp. 336–363, 2022.

[22] Cavicchioli, R.; Martoglia, R.; Verucchi, M. A Novel Real-Time Edge-Cloud Big Data Management and Analytics Framework for Smart Cities. *Journal of Universal Computer Science.* 2022, 4, 3–26.

[23] Belcastro, L. et al. Programming Big Data Analysis: Principles and Solutions. *Journal of Big Data.* 2022, 6-9.

[24] Murugan, S. et al. Impact of Internet of Health Things (IoHT) on COVID-19 Disease Detection and Its Treatment Using Single Hidden Layer Feed Forward Neural Networks (SIFN). *How COVID-19 is Accelerating the Digital Revolution: Challenges and Opportunities.* Springer International Publishing, pp. 31-50, 2022.

[25] Bansal, S.; Kumar, D. IoT Ecosystem: A Survey on Devices, Gateways, Operating Systems, Middleware and Communication. *International Journal of Wireless Information Networks.* 2020, 27, 340–364. https://doi.org/10.1007/s10776-020-00483-7

[26] Mohd Aman, A. H.; Hassan, W. H.; Sameen, S.; Attarbashi, Z. S.; Alizadeh, M.; Latiff, L. A. IoMT amid COVID-19 Pandemic: Application, Architecture, Technology, and Security. *Journal of Network and Computer Applications.* 2021, 174. https://doi.org/10.1016/j.jnca.2020.102886

[27] Awad, A. I.; Fouda, M. M.; Khashaba, M. M.; Mohamed, E. R.; Hosny, K. M. Utilization of Mobile Edge Computing on the Internet of Medical Things: A survey. *ICTExpress.* 2022. https://doi.org/10.1016/j.icte.2022.05.006.

[28] Stavropoulos, T. G.; Papastergiou, A.; Mpaltadoros, L.; Nikolopoulos, S.; Kompatsiaris, I. Iot Wearable Sensors and Devices in Elderly Care: A Literature Review. *Sensors.* 2020, 20. https://doi.org/10.3390/s20102826

[29] Pratap Singh, R.; Javaid, M.; Haleem, A.; Vaishya, R.; Ali, S. Internet of Medical Things (IoMT) for Orthopaedic in COVID-19 Pandemic: Roles, Challenges, and Applications. *Journal of Clinical Orthopaedics and Trauma.* 2020, 11, 713–717, https://doi.org/10.1016/j.jcot.2020.05.011

[30] Dimitrov, D. V. Medical Internet of Things and Big Data in Healthcare. *Healthcare Informatics Research.* 2016, 22, 156–163. https://doi.org/10.4258/hir.2016.22.3.156

[31] Aceto, G.; Persico, V.; Pescapé, A. Industry 4.0 and Health: Internet of Things, Big Data, and Cloud Computing for Healthcare 4.0. *Journal of Industrial Information Integration*. 2020, 18. https://doi.org/10.1016/j.jii.2020.100129

[32] Qadri, Y. A.; Nauman, A.; Zikria, Y. B.; Vasilakos, A. V.; Kim, S. W. The Future of Healthcare Internet of Things: A Survey of Emerging Technologies. *IEEE Communications Surveys and Tutorials*. 2020, 22, 1121–1167, https://doi.org/10.1109/COMST.2020.2973314

[33] Qi, J.; Yang, P.; Min, G.; Amft, O.; Dong, F.; Xu, L. Advanced Internet of Things for Personalised Healthcare Systems: A Survey. *Pervasive and Mobile Computing*. 2017, 41, 132–149. https://doi.org/10.1016/j.pmcj.2017.06.018

[34] Sun, W.; Cai, Z.; Li, Y.; Liu, F.; Fang, S.; Wang, G. Security and Privacy in the Medical Internet of Things: A Review. *Security and Communication Networks*. 2018, 2018. https://doi.org/10.1155/2018/5978636

3 Machine Learning-Based Driver Assistance System Ensuring Road Safety for Smart Cities

Apash Roy
Christ (Deemed to be University), Bangalore, India

Debayani Ghosh
Thapar University, Patiala, India

3.1 INTRODUCTION

Technologies around smart city and green computing are gaining more and more interest from diversified workforce areas. With the advancement of various fields, the demand for transportation has become high. To meet the demand, the vehicles are running day and night. This is really tiring for the transportation workers, especially the drivers who are driving the vehicle. A slight negligence of a driver may cause huge loss. The increasing number of road accidents is therefore a big concern. According to the road accident report published on the website of Ministry of Road Transport and Highways of India, there were 366138 road accidents in India in the year 2020, as updates last till date [1]. One of the major causes is less alertness of the driver, especially during nighttime. With the advancement of automated vehicle technology, the research is going on different aspects of safety too. The approach to detect driver's drowsiness is one among them. Here, we are in search of an automated system that can detect drowsiness and alert the driver. Machine learning is used in diversified areas like handwritten character recognition [2–6], medical image processing [7–13], computer vision [14, 15], and data science [16–18].

3.1.1 CONTRIBUTIONS OF THE CHAPTER

In this chapter, a neural network-based model is proposed, which can efficiently detect driver's drowsiness and alert the driver. The work mainly focuses on building the learning model. A modified convolutional neural network (CNN) is built to solve

DOI: 10.1201/9781032656830-3

the purpose. The model is trained with a dataset of 7000 images of open and closed eyes. For testing purpose, some real-time experiments are done by some volunteer drivers in different conditions, like gender, day, and night, and the model is really good for daytime and if the driver is not wearing any glass. But with a glass in the eyes and in night condition, the system needs improvements.

3.1.2 CHAPTER ORGANIZATION

The chapter starts with the introduction and literature review of the field. Then the methodology is proposed by discussing the overall system and the actual used model. The experiment result with the system is discussed thereafter with the discussion of dataset, trials, tables, and charts. Finally, the content is concluded and all the references are listed.

3.2 STUDY OF THE LITERATURE

The smart city needs safety all around. Road safety being one of the major concerns, a huge work is going on in the area. Over the recent years, detection of driver fatigue has generated quite a lot of research interest and can be broadly classified into four categories [19]. The first category deals with methods based on physiological signals obtained from the driver, for example, electroencephalograph (EEG), electrocardiograph (ECG), and electrooculogram (EOG) signals [20, 21]. It has been shown that these methods have good predictive ability. However, obtaining clean datasets in this case offers considerable challenge in designing such methods [22, 23]. The second category relies on the behavior of the driver such as decrease in the grip strength on the steering wheel or the lack of ability to control the steering wheel; both provide a measure for the driver's fatigue [24]. The departure of the vehicle from the intended trajectory, that is, deviation of the vehicle state can also be a good measure for the driver's fatigue and forms the foundation for the third category [25]. Finally, the driver's drowsiness can also be detected through physiological reactions from the driver, such as closed eyes over duration, which is the focus of the current work. The frame-wise facial expressions and their ratios are used for detecting the drowsiness [26]. Another work based on facial expressions [27] claims 95.58% sensitivity and 100% accuracy for online detection with SVM classifier. Eye closure and yawning ratios are also used for facial expressions, and they are classified through machine learning algorithm to detect drowsiness [28]. Considering eyes as the only facial expression to detect sleepiness is sufficient and is established in more studies [29, 30].

3.3 PROPOSED METHODOLOGY

In this section, first the whole system will be discussed, followed by a discussion of each of the details. As a whole, a camera will record video of the driver and send each frame to a pretrained machine learning model. If the model finds the driver's eyes are closed for some predefined time, it will raise an alarm. There are three main

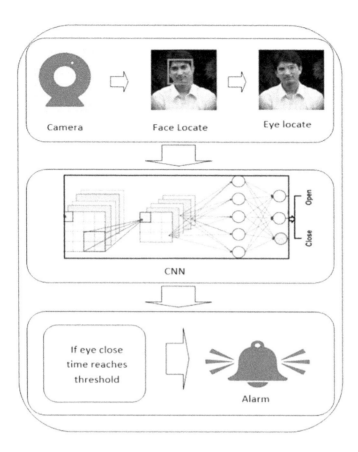

FIGURE 3.1   The system at a glance.

parts: video capturing part, classification part, and alarming part. Also, there is some pre-processing step before classifying the video frame.

3.3.1   THE OVERALL SYSTEM

The methodology is shown in Figure 3.1. The first stage consists of a webcam that captures the real-time video of the driver. It locates the face as the first region of interest (ROI) and then the eyes as the second region of interest. The second stage consists of a previously trained convolutional learning model that is used to the classifier. The job of the classifier is to classify the state of the eyes as "open" or "closed." The final stage of the proposed system is to ring an alarm if the eyes are found closed for some threshold time (3 seconds in this case).

3.3.2   PRE-PROCESSING

Before producing the input, the image is converted to gray scale and basic transformations like translation, rotation, and scaling are applied. To translate the original coordinate (X, Y) to a new translated coordinate (X_t, Y_t) by the shift coordinate (T_x, T_y), the basic rule of translation is performed as follows:

$$\begin{bmatrix} X_t \end{bmatrix} = \begin{bmatrix} X \end{bmatrix} + \begin{bmatrix} T_x \end{bmatrix}$$
$$\begin{bmatrix} Y_t \end{bmatrix} = \begin{bmatrix} Y \end{bmatrix} + \begin{bmatrix} T_y \end{bmatrix} \tag{3.1}$$

To rotate the images from coordinate (X, Y) to the desired angle θ with new coordinate (X_r, Y_r), the basic scaling technique is used as follows:

$$\begin{bmatrix} Xr, Yr \end{bmatrix} = \begin{bmatrix} X, Y \end{bmatrix} \begin{bmatrix} \cos\theta & \sin\theta \\ -\sin\theta & \cos\theta \end{bmatrix} \text{ for positive rotation}$$

$$= \begin{bmatrix} X, Y \end{bmatrix} \begin{bmatrix} \cos\theta & \sin\theta \\ -\sin\theta & \cos\theta \end{bmatrix} \text{ for negative rotation} \tag{3.2}$$

Scaling is achieved by multiplying the original coordinates (X, Y) with a scaling factor (S_x, S_y), which is calculated by comparing the desired input size and the given input image size. The desired coordinate points (X_s, Y_s) are calculated as follows:

$$\begin{pmatrix} X_s \\ Y_s \end{pmatrix} = \begin{pmatrix} X \\ Y \end{pmatrix} \begin{bmatrix} S_x & 0 \\ 0 & S_y \end{bmatrix} \tag{3.3}$$

Finally, the image matrix of 24×24 size is produced as input to the input layer of the model.

3.3.3 THE MODEL

The CNN model is used here for classifying the images of eyes as open or closed. A CNN model is a feedforward neural network that consists of the following layers: (i) input layer, (ii) hidden layer, and (iii) output layer. The hidden layer consists of convolution layers activated by ReLU function and pooling layers. In particular, the workflow of our CNN model is shown in Figure 3.2.

The hidden layer consists of three convolution layers, each activated by ReLU function to extract significant features from the input images to rectify the feature maps. Note that the ReLU activation function is as follows:

$$f(x) = \max(0, x). \tag{3.4}$$

Each convolution layer is succeeded by a pooling layer that reduces the dimensionality of the feature map. Here, we have implemented a max pool operation in the pooling layer.

In the first convolution layer, the input data is convoluted with 32 filters of 3×3 size with a stride of 1. So, it produces a matrix of 22×22 elements in output. Then a

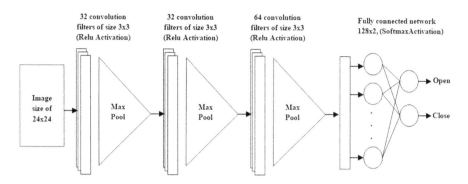

FIGURE 3.2 The neural network model.

max pool layer with 2 × 2 filter and stride 2 is used to reduce the dimensionality of the matrix. Some dropouts are also used. In the same manner, another two convolutional layers with 32 and 64 filters are used. Finally, a flatten layer is used to prepare vector input for a fully connected network, which is activated using SoftMax activation function.

$$f(X)i = \exp(X_i) \bigg/ \sum_{j=1}^{n} \exp(X_j) \qquad (3.5)$$

The fully connected layer classifies the eyes as open or closed based on the input feature map. The CNN model is first pretrained with a set of 70,000 eye images for classification. Then, in real time, the same CNN model is used to classify the state of the eyes from the frames of the video, captured through the webcam.

3.4 RESULTS AND DISCUSSION

In this section, the implementation, used dataset, and the results obtained during real-time trials are discussed. Implementation done by Python programming language and the dataset used for training are taken from the Internet source and included in our own data. Tabulated and visualized result data gives a clear understanding.

3.4.1 IMPLEMENTATION

The implementation is done in a real-time manner. It captures real-time video frames through camera and spots the eyes as the region of interest. Using the pretrained CNN model, it classifies if the eyes are opened or closed and gives an alarm sound if the eyes are found closed for a predefined threshold time (10 seconds in our case). A webcam is used to capture the image, and several Python packages are used: OpenCV to detect the face and eyes, Keras to build the model, TensorFlow as Keras used it as backend, and finally pygame to play alarm sound.

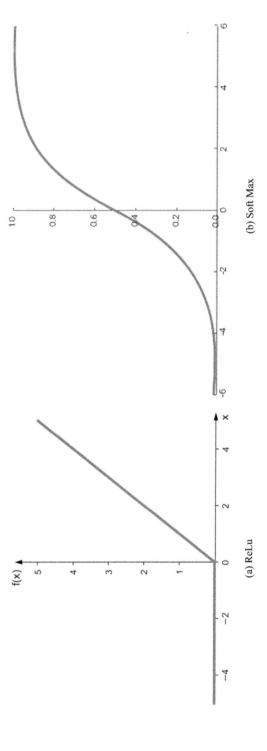

FIGURE 3.3   Activation functions: (a) ReLu and (b) SoftMax.

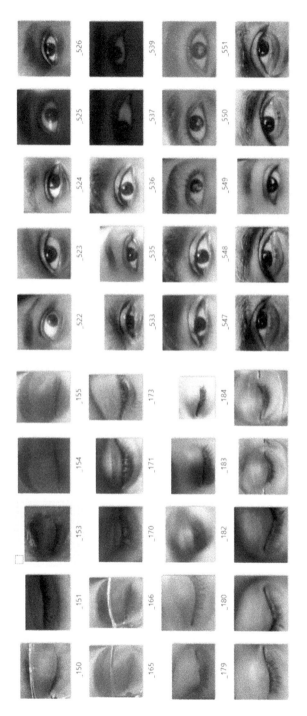

FIGURE 3.4 Sample images from the dataset for training the system.

3.4.2 DATASET

The CNN model is trained with a dataset of 70,000 eye images and is found on the website: https://www.kaggle.com/datasets/serenaraju/yawn-eye-dataset-new. It consists of open and closed eye images taken from different persons of both genders. Various light conditions are also covered.

3.4.3 TRIALS

A total of 1000 trials by 10 volunteer drivers (7 males and 3 females) are taken into consideration. The test is taken in day and night times and with or without glass. Observed output during the trial is tabulated in Table 3.1 and Figure 3.5 shows the summary of

TABLE 3.1
Observed Output during Trials

Driver	Sex	No. of Trial	Success in Day with Glass	Success in Day without Glass	Success in Night with Glass	Success in Night without Glass	Total Trial	Total Success
D1	M	25	23	25	20	19	100	87
D2	M	25	24	25	18	21	100	88
D3	M	25	24	25	19	21	100	89
D4	M	25	21	25	19	21	100	86
D5	M	25	23	25	18	22	100	88
D6	M	25	22	25	16	22	100	85
D7	M	25	25	25	19	20	100	89
D8	F	25	25	25	18	21	100	89
D9	F	25	22	25	19	21	100	87
D10	F	25	24	25	16	21	100	86
Total	—	250	232	250	181	210	1000	874
%	—	—	92.8	100	72.4	84	—	87.4

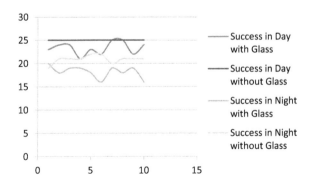

FIGURE 3.5 Summary of trials.

the results. With a good light condition during daytime, and if the driver is without glass, the system shows an outstanding result of 100% accuracy. But, if the driver is wearing glass, or the light condition is not appropriate during nighttime, system seeks refinement. However, considering all scenarios for day and night, driver with or without glass, male or female, the overall accuracy is 87.4%, which is quite encouraging.

3.5 CONCLUSIONS

There are a huge number of road accidents recorded due to drowsiness of drivers. To overcome this, people are looking for an automated system that can detect and alert the driver if there is drowsiness. In this chapter, one such approach is proposed. A simple architecture with a camera, a CNN trained with huge dataset, is used to detect the drowsiness. The camera captures the real-time video; the system spots the face and finally the eyes as the region of interest. Then the eyes region is fed into CNN, which is pretrained to classify if the eyes are closed or open. If the eyes are found closed for 3 seconds at-a-stretch, the system rings an alarm and the driver gets alert. The trials are showing an outstanding result without a glass and in daytime. But when the light condition is not appropriate at night or the driver is with a glass, the system needs refinement, which is the focus of our future work.

REFERENCES

[1] https://morth.nic.in/road-accident-in-india. Accessed on 04/09/2023.
[2] Jagdish Kumar, and Apash Roy, DograNet – A Comprehensive Offline Dogra Handwriting Character Dataset, *Journal of Physics*, vol. 2251, p. 012008, 2022. https://doi.org/10.1088/1742-6596/2251/1/012008
[3] A. Roy and D. Ghosh, Pattern Recognition based Tasks and Achievements on Handwritten Bengali Character Recognition, *2021 6th International Conference on Inventive Computation Technologies (ICICT)*, Coimbatore, India, 2021, pp. 1260–1265, IEEE. https://doi.org/10.1109/ICICT50816.2021.9358783
[4] Apash Roy, Handwritten Bengali Character Recognition – A Study of Works during Current Decade, *Advances and Applications in Mathematical Sciences*, vol. 18, no. 9, July 2019, pp. 867–875.
[5] Jagdish Kumar and Apash Roy, Dogra Handwritten Text Recognition Using Machine and Deep Learning Models. *European Chemical Bulletin*, vol. 12, Special Issue 1(Part-B), 2023, pp. 985–996. https://doi.org/10.31838/ecb/2023.12.s1.104
[6] J. Kumar and A. Roy, Handwritten Text Recognition for Regional Languages of Indian Subcontinent. In: Mathur, G., Bundele, M., Tripathi, A. and Paprzycki, M. (Eds.), *Proceedings of 3rd International Conference on Artificial Intelligence: Advances and Applications. Algorithms for Intelligent Systems*. Springer, Singapore, 2023. https://doi.org/10.1007/978-981-19-7041-2_19
[7] Swapnil Deshmukh, and Apash Roy, Early Detection of Diabetic Retinopathy using Vessel Segmentation based on Deep Neural Network. *Proceedings of International Conference on Recent Advances in Materials, Manufacturing and Machine Learning (RAMMML-2022)*, Nagpur, 26–27 April 2022.
[8] Samia Mushtaq, Apash Roy, and Tawseef Ahmed Teli, A Comparative Study on Various Machine Learning Techniques for Brain Tumor Detection Using MRI, *Global Emerging Innovation Summit (GEIS-2021)*, 2021, pp. 125–137, Bentham Science Publishers. https://doi.org/10.2174/97816810890101210101

[9] S. V. Deshmukh and A Roy, An Empirical Exploration of Artificial Intelligence in Medical Domain for Prediction and Analysis of Diabetic Retinopathy: Review, *Journal of Physics: Conference Series*, vol. 1831, p. 012012. https://doi.org/10.1088/1742-6596/1831/1/012012

[10] Swapnil V. Deshmukh, Apash Roy and Pratik Agrawal, Retinal Image Segmentation for Diabetic Retinopathy Detection using U-Net Architecture. *International Journal of Image, Graphics and Signal Processing (IJIGSP)*, vol. 15, no. 1, 2023, pp. 79–92. https://doi.org/10.5815/ijigsp.2023.01.07

[11] S. V. Deshmukh and A. Roy, Retinal Blood Vessel Segmentation Based on Modified CNN and Analyze the Perceptional Quality of Segmented Images. In: Woungang, I., Dhurandher, S. K., Pattanaik, K. K. and Verma, A. (Eds.), Advanced Network Technologies and Intelligent Computing. ANTIC 2022. Communications in Computer and Information Science, vol 1798. Springer, Cham, 2023. https://doi.org/10.1007/978-3-031-28183-9_43

[12] Nisar Ahmad Kangoo and Apash Roy, *A Review on Deep Learning Techniques for Detecting COVID-19 from X-rays and CT Scans*. Recent Advances in Computing Sciences, 1st Edition 2023, CRC Press, 2023 (eBook ISBN 9781003405573). https://www.taylorfrancis.com/chapters/edit/10.1201/9781003405573-24/review-deep-learning-techniques-detecting-covid-19-rays-ct-scans-nisar-ahmad-kangoo-apash-roy

[13] S. V. Deshmukh and Apash Roy, *Early Detection of Diabetic Retinopathy Using Vessel Segmentation Based on Deep Neural Network*. Recent Advances in Material, Manufacturing, and Machine Learning, 1st Edition 2023, CRC Press, 2023 (eBook ISBN 9781003370628). https://www.taylorfrancis.com/chapters/edit/10.1201/9781003370628-78/early-detection-diabetic-retinopathy-using-vessel-segmentation-based-deep-neural-network-deshmukh-apash-roy

[14] U. I. Musa and A. Roy, Marine Robotics: An Improved Algorithm for Object Detection Underwater. *Indian Journal of Computer Graphics and Multimedia*, vol. 2, no. 2, 2023, pp. 1–8. https://doi.org/10.54105/ijcgm.c7264.082222

[15] A. Roy and D. Ghosh, Real-Time Driver Drowsiness Detection System Using Machine Learning. In: Tripathi, A. K., Anand, D. and Nagar, A. K. (Eds.), *Proceedings of World Conference on Artificial Intelligence: Advances and Applications. WWCA 1997. Algorithms for Intelligent Systems*. Springer, Singapore, 2023. https://doi.org/10.1007/978-981-99-5881-8_37

[16] N. A. Kangoo, M. Sharma and A. Roy, Dataset for Classifying English Words into Difficulty Levels by Undergraduate and Postgraduate Students, Data in Brief. https://doi.org/10.1016/j.dib.2023.109744

[17] N. A. Kangoo and A. Roy, Supervised Machine Learning Text Classification: A Review. In: Yadav, A., Nanda, S. J. and Lim, M. H. (Eds.), *Proceedings of International Conference on Paradigms of Communication, Computing and Data Analytics. PCCDA 2023. Algorithms for Intelligent Systems*. Springer, Singapore, 2023. https://doi.org/10.1007/978-981-99-4626-6_53

[18] A. Roy and D. Ghosh, Learning-Based Data Science Model for Car Price Prediction. In: Mukhopadhyay, S., Sarkar, S., Mandal, J. K. and Roy, S. (Eds.), *AI to Improve e-Governance and Eminence of Life*. Studies in Big Data, vol. 130. Springer, Singapore, 2023. https://doi.org/10.1007/978-981-99-4677-8_10

[19] L. Wang, X. Wu and M. Yu, Review of Driver Fatigue/Drowsiness Detection Methods. *Journal of Biomedical Engineering*, vol. 24, no. 1, p. 245248, 2007.

[20] G. Borghini, L. Astol, G. Vecchiato, D. Mattia and F. Babiloni, Measuring Neuro-physiological Signals in Aircraft Pilots and Car Drivers for the Assessment of Mental Workload, Fatigue and Drowsiness. *Neuroscience and Biobehavioral Reviews*, vol. 44, p. 5875, 2014.

[21] T. Myllylä, V. Korhonen, E. Vihriälä et al., Human Heart Pulse Wave Responses Measured Simultaneously at Several Sensor Placements by Two MR-Compatible BRE Optic Methods. *Journal of Sensors*, vol. 2012, Article ID 769613, 8 pages, 2012.

[22] M. Simon, E. A. Schmidt, W. E. Kincses et al., EEG Alpha Spindle Measures as Indicators of Driver Fatigue Under Real Traffic Conditions. *Clinical Neurophysiology*, vol. 122, no. 6, p. 11681178, 2011.

[23] S. K. L. Lal and A. Craig, Reproducibility of the Spectral Components of the Electroencephalogram During Driver Fatigue. *International Journal of Psychophysiology*, vol. 55, no. 2, p. 137143, 2005.

[24] B. T. Jap, S. Lal, P. Fischer and E. Bekiaris, Using EEG Spectral Components to Assess Algorithms for Detecting Fatigue. *Expert Systems with Applications*, vol. 36, no. 2, p. 23522359, 2009.

[25] C. X. Huang, W. C. Zhang, C. G. Huang and Y. J. Zhong, Identication of Driver State Based on ARX in the Automobile Simulator. *Technology & Economy in Areas of Communications*, vol. 10, no. 2, p. 6063, 2008.

[26] Rohith Chinthalachervu, Immaneni Teja, M. Ajay Kumar, N. Sai Harshith and T. Santosh Kumar, Driver Drowsiness Detection and Monitoring System using Machine Learning. *Journal of Physics: Conference Series*, vol. 2325, p. 012057, 2022. https://doi.org/10.1 088/1742-6596/2325/1/012057

[27] J. Haribabu, T. Navya, P. V. Praveena, K. Pavithra and K. Sravani, Driver Drowsyness Detection Using Machine Learning. *Journal of Engineering Sciences*, vol. 13, no. 06, June 2022.

[28] N. Prasath, J. Sreemathy and P. Vigneshwaran, Driver Drowsiness Detection Using Machine Learning Algorithm. *2022 8th International Conference on Advanced Computing and Communication Systems (ICACCS)*, pp. 01–05, 2022. https://doi.org/10.1109/ICAC CS54159.2022.9785167

[29] S. Cheerla, D. P. Reddy and K. S. Raghavesh, Driver Drowsiness Detection using Machine Learning Algorithms. *2022 2nd International Conference on Artificial Intelligence and Signal Processing (AISP)*, pp. 1–6, 2022. https://doi.org/10.1109/AISP53593.2022.9760618

[30] Al Redhaei, Y. Albadawi, S. Mohamed and A. Alnoman, Realtime Driver Drowsiness Detection Using Machine Learning. *2022 Advances in Science and Engineering Technology International Conferences (ASET)*, pp. 1–6, 2022. https://doi.org/10.1109/ ASET53988.2022.973480

4 Privacy and Security Issues in Big Data and Internet of Things

Promise Elechi, Kingsley Eyiogwu Onu,
and Ela Umarani Okowa
Rivers State University, Port Harcourt, Nigeria

4.1 INTRODUCTION

This section introduces the chapter and shows why the subject of privacy and security of big data and IoTs should be of serious concern to one and all. Key contributions of this chapter are included in this section. Also included is the organization of the rest parts of the chapter.

The rise of big data and the Internet of Things (IoT) has revolutionized our world, generated vast amounts of data, and created a network of interconnected devices. While these advancements offer immense potential for progress in various sectors, they also raise significant concerns about privacy and security. This chapter explores the key privacy and security challenges associated with big data and the IoT, highlighting the vulnerabilities and potential risks and discussing possible solutions to mitigate these threats.

Many things in the modern world are being interconnected through the technology of the IoT. The healthcare sector, the business sector, etc. have all adopted the IoT to improve their services. To underscore how rapidly the use of the IoT is growing globally, it was reported in Hassija et al. [1] that IoT inter-connected devices would rise from 8.4 billion in 2020 to 20.4 billion by the end of 2022. And Kott and Linkov [20] estimated that by 2030, nearly 50 billion IoT-enabled devices will be used. Radio-frequency identification (RFID) tags, smart watches, heartbeat sensors, smartphones, etc. are some of the interconnected things [2]. The IoT is simply a network made up of smart things (objects). Dhuha and Jhanjhi [3] consider the IoT to be the key aspect of the present expansion of Internet services to house different kinds of objects. The objects or devices that make up the IoT are capable of data collection and data sharing anytime without the need for human intervention [4].

As a result of the use of the IoT, the size of the dataset gathered is never seen in the history of Science and Technology. Moura and Serrão [17] therefore observed that the size of gathered data due to the application of the IoT is far beyond what the available software tools are capable of capturing, managing, and analyzing on a timely basis, stressing that these data must be transformed into a form that is valuable for easy analysis.

50   DOI: 10.1201/9781032656830-4

The use of the IoT has a significant impact on our daily endeavors socially, economically, commercially, etc. The monetary impact of the use of the IoT is so huge that the authors of [2] estimated that the revenue due to the use of the IoT would jump to over 4 trillion US dollars by the year 2025 from a mere 892 billion US dollars back in 2018, just within 7 years. Interestingly, the use of the IoT can significantly improve the comfort, speed, and ease of the lives of customers, as noted by Sarker et al. [18]. As promising as this sounds, several issues in several aspects arise due to the current expansion of networks of the IoT. We have issues in the management of IoT devices, management of data, security and privacy issues, computational issues, etc. If the issues are not adequately addressed, they may jeopardize the full application of the IoT in the future, thereby depriving people of the full benefits of this promising technology.

As previously mentioned, the application of the IoT has led to astronomical rise in available data. Such huge data is termed big data. Within seconds, big data comes from different sources and in different formats like video files, images, and figures. Given what is involved in big data, we portray big data with 6V properties. The 6Vs are Value, Volume, Variety, Velocity, Veracity, and Variability, as depicted in Figure 4.1. Frankly speaking, these 6Vs with which big data is characterized bring about serious challenges, yes challenges in terms of data confidentiality, data integrity, data availability, data authenticity, data efficiency, etc. The different challenges are sometimes

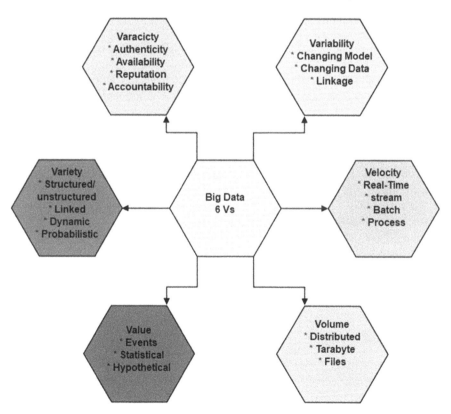

FIGURE 4.1 Big data 6Vs.

grouped into Privacy and Security issues. The privacy aspect deals with how to handle people's data in the right way as outlined in certain policies, regulations, and guidelines. This issue arises because it is possible for people to easily use others' private information wrongly for selfish reasons. The security aspect deals with how to protect people's data from those who intend to steal the data and from cyberattacks [19].

Data value is highly important and could be lost if hackers were to get access to the data, tamper with it, or steal vital information. Therefore, the confidentiality of data must be taken seriously so that the value of data will be preserved. Data integrity, which is another area of interest, relates to the accuracy of the data. When data is not accurate due to security breaches, integrity is lost and whatever analysis is done on the inaccurate data will lead to false and dangerous results. Data availability involves having access to the data when it is needed. It could be very frustrating if data cannot be accessed at the appropriate time. In some cases, this could result in serious harm, especially in the healthcare sector. For data authenticity, it is required that the very source of the data be reliable and trustworthy. When data is requested, authenticity demands that the person requesting such data be verified as to his authority to make such a request or whether he is impersonating someone else [19].

4.1.1 KEY CONTRIBUTIONS OF THE CHAPTER

The major contributions of this chapter are as outlined below:

i. The chapter holistically looked at the challenges of big data and the solutions provided.
ii. Challenges on the IoTs in the areas of privacy and security were discussed in detail, and ways to deal with the challenges were highlighted.

4.1.2 CHAPTER ORGANIZATION

The remaining sections of this chapter are organized as follows:

Section 4.2 covers related works on security and privacy issues of big data and the IoT. Section 4.3 focuses on the challenges of big data to information privacy and security. Section 4.4 presents big data challenges and security. Section 4.5 presents solutions to identified big data challenges. Section 4.6 discusses the IoT layers. Section 4.7 presents privacy and security issues in IoT layers. Section 4.8 focuses on the challenges of the IoT layers. Section 4.9 takes a look at the challenge of lack of sufficient testing and updates. Section 4.10 discusses the home intrusion challenge. In Section 4.11, the IoT security challenge is presented. Section 4.12 looks at IoT latency and capacity challenges, and Section 4.13 concludes the chapter.

4.2 RELATED WORK

Section 4.2 considers what other researchers have done in the subject area. Works that provide background information, context, or insights into the specific area of this current work are explored, with the strengths and weaknesses of the key reviewed works highlighted.

The words big data refer to large amounts of datasets collected from different sources such as healthcare, markets, and banks. Many researchers define big data in different ways depending on their perspective. For example, Refs. [5, 6, 7, 8] characterize big data in terms of 4 Vs: Value, Velocity, Volume, and Variety.

It was recommended in Ref. [9] that to access data within a cloud, techniques for encryption and decryption should be used. To ensure security, the approach adopts encryption based on ciphertext-policy-attribute where bilinear maps, attributes-based encryption, and elliptic curve cryptography are relied upon. Che et al. [10] identified some of the objects that fall under the IoT as laptops and many other important components.

The authors in Ref. [11] proposed an authentication method based on clusters. Since the enhancement of security and privacy requirements for the IoT is very important, Ullah et al. [12] proposed a method that provides security for information transmission on the IoT based on WLAN. Performance evaluation of the recommended method was carried out and found to be effective in securing data transmission.

Kouicem et al. [13] noted that for the security of transmitted data, there must be encryption nodes. This usage, coupled with access permission control, ensures that every transmitted data is protected and secured. Combining access permission control and encryption nodes based on the algorithm of advanced encryption standards enhances the security of transmitted data. To ascertain the effectiveness of the method, the authors carried out an experiment based on data transmission, and the result was better data security. Schäfer and Rossberg [14] studied the security of both wireless and fixed networks, giving attention to various issues and challenges and how to overcome them. Vasilomanolakis et al. [15] strongly recommended that any process of connecting IoT devices must have a security-improved interface for data confidentiality. Network security entails integrity, confidentiality, availability, authenticity, etc. Vasilomanolakis et al. [15] explained that any established connection must be with the entity already authenticated.

Reyna et al. [16] studied the integration of blockchain with the IoT and identified a serious issue such as how the channel between the sensor node and Internet host is secured. This means that how security is provided between the sensor node and the Internet host can lead to security issues. It was therefore mentioned that some attackers can easily bring malicious nodes into a network and cause them to function alongside the network. When this happens, the malicious nodes can cause very serious damage to the network.

Based on the identity management perspective in IoT-enabled devices, Guo et al. [21] reviewed different trust-evaluation schemes. Trust evaluation schemes that were rarely and regularly visited were analyzed and the analysis was presented. Based on the gathered data, the authors identified a trust evaluation system already established. In Ref. [22], a study on the possible understanding of trust was undertaken as it relates to the IoT. The authors provided a perspective that enables practitioners to come up with the right trust management scheme (TMS) within techno-service situations. Pourghebleh et al. [23] reviewed already existing schemes on the IoT and proposed a scheme based on four classes, pointing out the pros and cons of the schemes in each of the four classes. The authors of [24, 25, 26, 27] independently looked at IoT trust and reputation, with attention focused on different trust evaluation schemes standards of availability, trustworthiness, and reliability; different threat vulnerabilities; and trust management.

The authors of [36, 37] looked at the issues of security and privacy on the IoT, whereas Assa-Agyei al. [37] focused attention particularly on the healthcare sector. In Ref. [55], however, the expected details of solution strategies neither provided security and privacy issues in big data nor outlined adequate solutions to the problem. In Refs. [28, 29, 30, 31], TMSs specifically designed for the IoT were studied. The design used in the development of various TMSs was critically examined and compared. The schemes were used to categorize centralized trusted models and distributed models. Particularly, Alabaa et al. [31] proposed a TMS that executes a multi-agent scheme as well as evaluates the efficiency of three different techniques. The authors of [41, 42, 43, 44, 45] all considered various methods to address the challenges of the IoT (Table 4.1).

TABLE 4.1
Related Work Comparison

S/N	Past Work	Pros	Cons	Contribution to the Current Work
1	Alferidah et al. [55]	Discussed detailed IoT security issues and based on layers of IoT gave an analysis of IoT attacks	Effective methods of protecting IoT devices from attacks are not adequately presented	Beyond the detailed discussion of security and privacy issues in big data and IoT, effective methods of preventing these attacks are highlighted
2	Assa-Agyei et al. [37]	Adequately utilized questionnaire method and obtained 100% response level; successfully identified key issues and challenges in IoTs	Did not provide current ways of mitigating the issues and challenges	In addition to the detailed discussion of security and privacy issues in big data and IoT presented, effective methods of preventing these attacks are highlighted
3	Riaz et al. [54]	Successfully identified current issues and challenges of big data	Solutions to the identified issues are lacking	Dealt with both big data and Internet of Things security and privacy issues and challenges and provided detailed solutions to the issues
4	Fu et al. [36]	Outlined key effects of security and privacy issues, and the technical aspect was well-presented	The provided solutions to the issues of security and privacy are not adequate	The chapter holistically looked at the challenges of big data and the solutions provided Challenges on Internet of Things (IoTs) in the areas of privacy and security were discussed in detail, and ways to deal with the challenges were highlighted
5	Our current work	Effectively discussed the security and privacy issues in both big data and IoT in detail	Adequate solutions to the security and privacy issues of big data and IoTs are provided	The chapter holistically looked at the challenges of big data and the solutions provided Challenges on Internet of Things (IoTs) in the areas of privacy and security were discussed in detail, and ways to deal with the challenges were highlighted

4.3   CHALLENGES OF BIG DATA TO INFORMATION PRIVACY AND SECURITY

There are many challenges associated with big data, especially in the areas of information security and privacy. In this section, these challenges with regard to information privacy and security of big data will be presented.

Numerous challenges exist concerning privacy and security in big data. Some of these privacy challenges include:

- Data collection and aggregation: Big data often involves the collection and aggregation of personal information from various sources, such as social media, online transactions, and loyalty programs. This raises concerns about individuals' privacy as their data can be used to form detailed profiles and track their online and offline activities.
- Data sharing and monetization: Companies often collect and share user data with third parties for marketing, advertising, and other purposes. This lack of transparency and control over data sharing can lead to misuse of personal information and potential discrimination.
- Data retention and disposal: Big data systems often store massive amounts of data for extended periods, raising concerns about data retention practices and potential leaks. The lack of clear guidelines and regulations for data disposal can lead to the retention of unnecessary and sensitive data.
- Data anonymization and de-identification: While anonymizing or de-identifying data can mitigate privacy risks, these techniques are not fool-proof and can be re-identified by combining information from different sources. This highlights the limitations of anonymization techniques and the need for more robust privacy-preserving solutions.

Security challenges in big data include:

- Data breaches: Big data environments are often targeted by cybercriminals due to the vast amount of valuable information they contain. Data breaches can expose sensitive personal information and financial data, leading to identity theft, fraud, and other financial losses.
- Unauthorized access and insider threats: The complex nature of big data systems and the need for multiple parties to access data increase the risk of unauthorized access and insider threats. These threats can include privileged users abusing their access or malicious actors gaining access through vulnerabilities in the system.
- Data manipulation and injections: Malicious actors can attempt to manipulate data stored in big data systems to alter analytics results, inject misinformation, or disrupt operations. Such attacks can have severe consequences for organizations and individuals relying on big data insights.
- Lack of visibility and control: The distributed and diverse nature of big data environments can make it challenging to monitor and control data access and activities effectively. This lack of visibility can hinder the detection of suspicious behavior and timely response to security incidents.

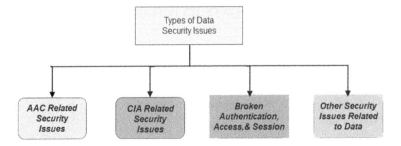

FIGURE 4.2   Types of data security issues.

Since the advent of IoT technology, the number of devices that are connected to the Internet and one another has increased drastically, and this has resulted in an astronomical increase in the size or volume of data, including data storage facilities, just to mention a few. Besides, this has introduced new challenges as far as information security is concerned. This is so because available security measures like firewalls are incapable of providing security for big data [17]. Besides, addressing the various challenges of big data requires knowing different computational difficulties and techniques associated with big data analysis, security threats, etc. Again, computational techniques that are used with little or no problems at all when a small dataset is involved perform poorly if the dataset is big (that is with big data) [35]. Hence, a new security measure must be put in place and implemented (Figure 4.2).

4.4   BIG DATA CHALLENGES AND SOLUTIONS

Big data challenges refer to the difficulties and obstacles associated with managing, processing, and extracting valuable insights from large and complex datasets. The challenges of big data include dealing with the complexity of diverse data sources and ensuring data quality and security.

Specifically, the challenges of big data will be presented in this section.

According to Ref. [17], the challenges of big data can be grouped into four different security categories:

1. Data management focuses on securing the origin or source of data and its storage.
2. Infrastructure security deals with securing all possible distributed computations with the help of MapReduce.
3. Reactive and integrity security monitors attacks and anomalies in big data in real-time.
4. Data privacy involves a pattern of data mining that ensures the preservation of data privacy.

In more specific terms, other challenges include:

1. Data variety
2. Data volume

3. Veracity of data
4. Data velocity
5. Data variability
6. Data availability
7. Data quality, etc.

A brief look at these challenges will be considered next.

4.4.1 DATA MANAGEMENT CHALLENGES

Data management challenges can be best divided into other different groups such as provenance or source of data, securing data storage, and transaction logs. As the interconnection of different devices results in the availability of big data, source-enabled programming environments have generated a huge amount of source graphs. Carrying out an analysis of such huge source graphs as a way of detecting meta-data dependencies for any confidentiality or security applications is rather computationally difficult. In addition, storing transaction and data logs most often help the information technology (IT) manager to be aware of the kind of data that is moved and when the movement of the data occurs because such storage is done manually in a multitiered storage media. This process allows him to have full control over the data. However, in this era of big data, the use of auto-tiering for the storage of big data is practically simple and faster. However, the issue is that in this auto-tiering, the track of the data storage area is not kept. Hence, securing data storage areas is very difficult.

4.4.2 INFRASTRUCTURE SECURITY

Infrastructure security in big data refers to the practices and measures put in place to protect the underlying technology, systems, and resources that handle and process large volumes of data. This includes securing servers, networks, storage systems, and other components of the infrastructure. Key aspects of infrastructure in big data involve data encryption, access controls, authentication mechanisms, network security, and monitoring for potential vulnerabilities or suspicious activities. The goal is to ensure the confidentiality, integrity, and availability of data within the big data environment. In infrastructure security, they are what we call distributed programming frameworks. To process a large amount of data, the frameworks use the idea of parallel storage and computation. An example is the MapReduce framework that separates the input file into different lumps, while the lumps are read by the map reader.

4.4.3 DATA VARIETY

Data variety refers to the diverse types and formats of data that exist within a dataset or a data ecosystem. In the context of big data, it highlights the presence of various data sources and the different structures in which data is stored. Data comes in a variety of formats from web pages, videos, images, email messages, social media, etc. In most cases, the data obtained are simply raw, unstructured, or semi-structured.

With these different datasets and different formats, the complexity of data arises. Therefore, handling data variety is a serious challenge. This is especially so since traditional relational databases may not easily handle a wide range of data formats.

4.4.4 VOLUME OF DATA

In big data, data volume refers to the sheer size or quantity of data generated, processed, and stored within a given dataset or system. Big data is characterized by the immense volume of data that surpasses the capabilities of traditional data processing tools. This data can range from terabytes to petabytes and beyond. Currently, the volume of data in and out of some systems can be described as oceans and mountains. Kumar [35] reported in a study that it was estimated by the year 2020, the volume of data must go beyond zeta bytes. Daily, the volume of data is increasing. Kumar [35] mentions mobile phones and social media as the leading generators of these huge data volumes. Dealing with this huge data is challenging.

4.4.5 DATA VELOCITY

Velocity is the rate of change of displacement. Data velocity in big data, however, refers to the speed at which data is generated, collected, processed, and made available for analysis. It emphasizes the rapid pace at which data is flowing into a system. Real-time or near-real-time processing is often crucial for handling high data velocity. In modern systems, the speed or velocity with which data arrives and leaves a system is alarming. Responding to such a high flow of data, especially without appropriate or even adequate technology, makes doing so challenging. The concept of data velocity is particularly relevant in today's fast-paced, interconnected world where data is continuously generated from various sources, such as social media, sensors, and online transactions.

4.4.6 DATA QUALITY/RELEVANCE

Data quality is the accuracy, reliability, and fitness of the use of data in a given context. High-quality data is free from errors, inconsistencies, and inaccuracies, making it reliable for analysis and decision-making. Data relevance, on the other hand, focuses on the appropriateness of data for a particular purpose. Relevant data aligns with the objectives of the analysis and decision-making process. Ensuring data relevance involves selecting and using data that directly contributes to achieving the desired outcomes. Normally, for easy analysis and the right inference to be made, data must be relevant to the issues; if not, no right results can be inferred. However, getting to determine data quality and the relevance of the data to a given case is a serious challenge.

4.5 SOLUTIONS TO IDENTIFIED BIG DATA CHALLENGES

Navigating the complex landscape of big data challenges requires solutions that address the identified challenges in big data. The exploration into the solutions of big data involves innovative technologies, advanced analytics, or strategic approaches. This section will present solutions to the big data challenges that have been identified.

4.5.1 SOLUTION TO THE CHALLENGE OF DATA VOLUME

To address the challenge of data volume, we recommend the use of Apache Hadoop, Robust Hardware, Spark, and Grid Computing. Apache Hadoop is a tool designed to handle big data volume. Apache Hadoop facilitates the use of a network of various PCs (or computers) to address the problems related to mountains of data. It uses the MapReduce Programming Model to process big data. The modules in Apache Hadoop are designed with a key assumption that failures in system's hardware are possible and therefore the framework should handle the failures automatically. Robust hardware is designed to address the issues of volume because there is PPP (Powerful Parallel Processing) and increased memory to swallow very quickly large volumes of data. Then, too, the use of the Spark platform creates high-performance gains to tackle large data volumes. In grid computing, different servers are interconnected using a network of high speed. In the end, grid computing is seen to have the capacity for high storage as well as excellent processing power. Applying the above-mentioned approaches can help to address large data volume challenges.

4.5.2 SOLUTION TO THE CHALLENGE OF BIG DATA VARIETY

To address the challenges associated with big data variety, the following should be utilized: Apache Hadoop, Online Analytical Processing Tools, SAP HANA, and Redundant Physical Infrastructure. The key function of Apache Hadoop is to handle large data within the shortest possible time. To achieve this feat, Apache Hadoop divides data between different systems to process the data. Therefore, a map is created of the data content in the Hadoop for easy accessibility. The Online Analytical Processing Tool is used for processing data; and for establishing connections between data; and its lagging time when processing large data is always very low [35]. OLAP (Online Analytical Processing) tools gather data into logical patterns so that accessing it will be done easily. The main disadvantage of OLAP tools is that all data sent to it is processed, whether it is relevant or not. SAP HANA is a platform that is used for carrying out real-time analytics and applications. It is used in the cloud or as a piece of on-premises equipment. The use of Redundant Physical Infrastructure is important because such infrastructure promotes the execution of big data. However, the use of Redundant Physical Infrastructure is done after the requirements against availability, scalability, and performance are determined and established.

If the system must operate 24 hours every day, then expensive infrastructure must be available; computing power together with storage capacity must be taken care of. The performance of the system and infrastructural cost have a direct proportionality relationship: an increase in one leads to an increase in the other. It should be noted, however, that a particular situation determines whether one of the tools or a combination of all the tools by way of an algorithm to synchronize all possible varieties of data could provide the much-expected solution. We should not expect a generalized result for every problem.

4.5.3   SOLUTIONS TO THE CHALLENGE OF BIG DATA VELOCITY

To address the challenge of big data velocity, the following ought to be implemented: use of (1) flash memory, (2) sampling data, (3) transactional database, and (4) expansion of private cloud with a hybrid model.

Primarily, flash memory is used to catch data, particularly in dynamic solutions capable of analyzing data as cold if the data are not accessed frequently, and as hot if the data are highly accessed. Sampling data (data sampling) is a process that is used to handle situations where data volume and/or velocity is high. This approach involves statistical methods that not only select data but also manipulate and analyze data. The use of transactional databases is highly encouraged to address the problem of big data velocity. It can undo any database transaction that was not carried out appropriately. Transactional databases respond quickly to decision-making and are very much equipped with analytics in real-time. The fourth approach is the expansion of the private cloud through a combination of models (hybrid models) that permits erupting for further computational or calculative power required for the analysis of data to peak software or hardware to address high-velocity data needs. Using hybrid systems together with storage and cloud computation can address the velocity issues of big data.

4.5.4   SOLUTIONS TO THE BIG DATA QUALITY CHALLENGE

Big data algorithms and data visualization can be used to address the problem of the quality and relevance of big data. Though the concept of quality and relevance in big data is not new, it became prominent when the storage of these big data became serious. Big data algorithms are very good at managing, cleaning, and maintaining data. In addition, data cleaning using big data algorithms is not too difficult to handle. A person can choose to use already existing big data algorithms or make his/her algorithms clean data for use. Data visualization helps us to ascertain very clearly wherein lie non-required data and outliers. It is recommended that data surveillance, control, or the process of information management should be activated to guarantee clean data.

4.5.5   SOLUTIONS TO THE CHALLENGE OF BIG DATA PRIVACY AND SECURITY

To address the challenge of big data privacy and security, the following should be ensured/considered:

1. Data protection: Every aspect of data life or stage should be protected through proper encryption. Perhaps an attribute-based encryption can be considered.
2. Communication protection: During communication, information or data should be properly protected so that integrity and confidentiality can be guaranteed.
3. Security monitoring in real-time: Access to available data should be monitored in real-time to make sure that no unauthorized persons gain access. Implementation of threat intelligence will be an added protection against intruders.

4. Data anonymization: Data should be anonymous. Sensitive data should be kept away from the easy reach of unauthorized persons. And before data analysis and sharing is initiated, every sensitive data ought to be taken out of the collected records [35].

5. Putting adequate access control policy in place: There should be policies created such that only authorized persons can have access [38].

6. Examination of cloud providers: Having big data stored in the cloud is good but it is also important that the cloud provider carry out regular security audits, and there should be an assurance that if the cloud provider defaults in protecting the data stored in the cloud, adequate compensation should be paid.

7. Use of authentication methods: A method that verifies the person trying to have access to the data is called an authentication method. It helps to protect the system and data against unauthorized individuals.

8. Logging: This method is used to easily detect failures, attacks, or any unusual behaviors. Big data collects and manages incident (event) data. Activities are logged, and from the logged files when something goes wrong, either due to hacking or failures, it is easily detected. Therefore, there should be periodic audits of the system [39].

4.6   INTERNET OF THINGS (IoTs) LAYERS

The IoT operates through a layered architecture, each layer plays a distinct role in the functioning of connected devices and systems. This section of the chapter will focus on five layers.

There are different layers of the IoT, but five key layers will be considered here: the application layer, network layer, middleware or transport layer, business layer, and the perception layer. Each of the layers has its peculiar function. The very first layer is the perception layer where we have various sensors, physical objects, and a host of other things; the second layer is often the network layer where we have various network generations, transmission, etc.; the third layer is the transport or middleware layer where information is not only stored but also processed, in addition to other functions performed by the middleware layer; the fourth layer is the application layer; and lastly the business layer, as depicted in Figure 4.3–4.5.

- **The Perception Layer**
  The perception layer is a very important layer of the IoT layers due to the information it contains. It is often divided into two parts: perception network and perception node [31]. Data collection is performed by the perception node while any instruction that has to do with managing data, sending data, etc. is performed by the perception network [32]. It was noted in Ref. [33] that whatever data is collected in this very layer, the perception layer could undergo processing before passing the data to the next layer, the network layer. Since the IoT nodes are quite often vulnerable to attacks, Fan et al. [34] developed a node for security in the perception layer.

FIGURE 4.3   Different forms of data variety.

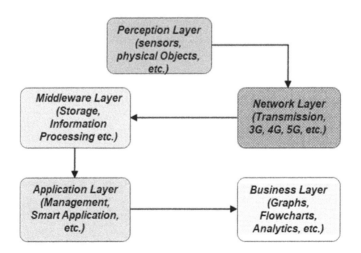

FIGURE 4.4   Different layers of the IoT.

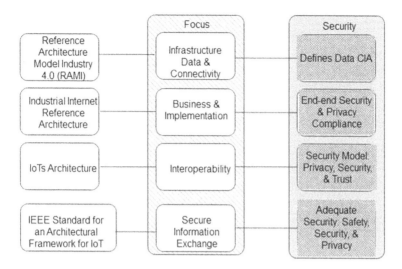

FIGURE 4.5 IoT architecture.

The sensors in the perception layer include sensors such as temperature sensors, vibration sensors, and sound sensors. Data is collected using various sensors in the perception layer.

- **Network Layer**

 The network layer is the second layer of the IoT layers. This layer takes data gathered by the perception layer and passes it on to the middleware layer for further action on the gathered data. To pass such data to the middleware layer, the network layer makes use of available technologies like 3G, 4G, and Wi-Fi. The network layer also ensures that all data transferred to it are kept confidential, and it passes it confidentially to the middleware layer.

- **Middleware Layer**

 The middleware layer sits between the network layer and the application layer, managing communication and data flow. In the middleware layer, you have features that enable data storage, data processing, data computation, etc. Therefore, data is stored in the middleware layer and transferred to devices based on the devices' address and name. It can make decisions. However, the decisions must be based on the permutations carried out on the data from the sensors.

- **Application Layer**

 The management of applications is done by the application layer, as the name suggests. However, the management is done based on the obtained information from the previous layer, the middleware layer. Examples of these applications include putting OFF or ON a device, sending email, activating security systems, and alarms. It can be said that this layer often involves applications, interfaces, and services that provide value to end users or other systems.

- **Business Layer**
  As portrayed in the pictorial representation of IoT layers, the business layer makes flowcharts, draws graphs, and makes an analysis of results before sending them to various customers. This is carefully done to promote the good reputation of the business organization.

4.7 PRIVACY AND SECURITY ISSUES IN IoT LAYERS

Security and privacy issues in each layer of the IoT must be identified to be in a clear position to address them. Therefore, this section of the chapter will be devoted to the privacy and security issues of IoT layers.

4.8 CHALLENGES ON THE INTERNET OF THINGS (IoTs)

Like big data, the IoT has its challenges that must be addressed for efficient performance and for the benefit of all those who use the IoT. This section will present the challenges.

Privacy challenges on the IoT include the following:

- Data collection and sensor surveillance: IoT devices collect a vast array of data about their users' environment, including location, physical activity, and even personal conversations. This constant monitoring raises concerns about privacy and the potential for data to be used for unwanted surveillance or profiling.
- Insecure devices and network vulnerabilities: Many IoT devices lack robust security features and are vulnerable to hacking attacks. These vulnerabilities can allow attackers to gain control of devices, steal data, or even use them as part of coordinated botnet attacks.
- Data sharing and third-party access: IoT devices often share data with various third-party platforms and services without clear transparency or user consent. This raises concerns about the potential misuse of data and the lack of control users have over their information.
- Privacy regulations and compliance: The rapidly evolving nature of the IoT landscape poses challenges for regulatory bodies to keep pace with technological advancements and establish comprehensive data protection frameworks. This lack of clear regulations can leave users vulnerable to privacy violations.

Some security challenges on the IoT are:

- Insecure device security: Many IoT devices have weak passwords, unpatched vulnerabilities, and lack basic security features, making them easy targets for cyberattacks.
- Unprotected communication channels: Data communication between IoT devices and cloud platforms often occurs over unencrypted channels, making it vulnerable to interception and manipulation.

- Malware and botnets: IoT devices are increasingly targeted by malware designed to steal data, disrupt operations, or launch large-scale botnet attacks.
- Physical attacks and tampering: IoT devices can be physically attacked or tampered with to gain access to sensitive data or manipulate their functionality.

Some of these challenges will be highlighted and the possible solution(s) to the challenges will be outlined.

4.9 CHALLENGE OF LACK OF SUFFICIENT TESTING AND UPDATES

Some manufacturers of IoT devices are much more interested in the money they are making from their products than in the interest of their customers. Hence, they fail to carry out regular application upgrades. As a result, computers connected to the IoT devices but running obsolete applications can be vulnerable to different kinds of ransomware attacks among other attack types. Besides, after an upgrade is carried out and a system transfers information or data to the cloud system, there might be downtime often. If there is no encryption of the link at the right time, hackers can easily access the upgrade files because they will be very vulnerable.

4.9.1 SOLUTION TO LACK OF SUFFICIENT TESTING AND UPDATES

A key to solving the security issues in the IoT is carrying out automatic updates from time to time, that is, regularly. To some extent, the vendors have a responsibility to monitor every software to ensure quick discovery of possible attacks. A detailed measure may include implementing the following: enabling over-the-air updates so as to allow firmware updates for IoT devices; conduct regular security audits to identify vulnerabilities in IoT devices; implement continuous monitoring of IoT devices to detect any anomalies or suspicious activities; use secure boot mechanisms and strong authentication methods; use end-to-end encryption communication between IoT devices and backend systems to protect data from interception; engage the services of cybersecurity experts to conduct thorough security assessments and receive guidance on the best practices for securing IoT devices.

4.10 HOME INTRUSION CHALLENGE

There are challenges associated with smart homes or houses due to the rapid increase in the applications of the IoT in different households. In many homes, there are different IoT devices whose protection mechanisms are too weak. Despite that, these IoT devices with poor protection mechanisms are used in many households to broadcast Internet protocol (IP) addresses. This makes it quite easier for hackers to use Shodan search methods to get the computer address.

4.10.1 SOLUTION TO HOME INTRUSION CHALLENGE

Different methods can be used to protect against the challenge of home intrusion. To address vulnerabilities in the home intrusion challenge for IoTs, the following solutions can be implemented: *password policies*: the use of strong password polices for all IoT devices must be enforced; *encryption standards*: implement end-to-end encryption for communication between IoT devices and the central hub; *firmware updates*: on a regular basis release and apply firmware updates to patch vulnerabilities; *network segmentation*: segment the home network to isolate IoT devices from critical systems so that even if one device is compromised, all sensitive information in all devices will not be leaked; *user education*: conduct regular user awareness programs to educate relevant persons about potential threats and best practices; *multi-factor authentication*: implement a multi-factor authentication for accessing IoT devices and associated apps. If possible, no two IoT devices should have the same password, and passwords should be changed as regularly as possible. It is also good to avoid the use of generic passwords.

4.11   INTERNET OF THINGS DATA SECURITY CHALLENGE

IoT data security faces challenges such as insecure devices, lack of standardization, insufficient encryption, and potential vulnerabilities in communication protocols. Additionally, the massive volume of data generated by IoT devices poses privacy concerns, requiring robust authentication and access control measures to safeguard sensitive information. Ongoing efforts focus on addressing these issues to enhance the overall security of IoT ecosystems. Due to the application of the IoT, large amounts of data are gathered quickly, processed, and stored. Some of the data is personal and therefore needs to be secured and protected from intruders or hackers. However, providing security for the data is a serious challenge.

4.11.1   THE SOLUTION TO THE CHALLENGE OF INTERNET OF THINGS DATA SECURITY

Addressing IoT data security challenges involves implementing a multi-faceted approach that includes the following: *device security*: ensuring that devices have strong authentication, secure boot processes, and regular software updates to patch vulnerabilities; *encryption*: implementing end-to-end encryption to protect data during transmission and storage; *network security*: use firewalls and secure communication protocols to protect against unauthorized persons and use network segmentation to isolate IoT devices from critical systems; *standardization*: carefully promote industry-wide standards for IoT security to ensure consistency and interoperability across devices and their data; *data integrity*: implement measures to vary the integrity of data, such as checksums or digital signatures, to detect and prevent tampering; *security monitoring*: employ continuous monitoring top detect abnormal activities or security breaches promptly. Using intrusion detection systems and anomaly detection can be very helpful in this regard too; *user education*: users should be regularly educated about IoT security risks and best practices, fostering a security-conscious mindset; and use of secure sockets layer (SSL) protocol. SSL certification can be used for encrypting and therefore securing user data online. Again, the use of encryption-based wireless protocol should be used as this will protect wireless protocol [40].

4.12 INTERNET OF THINGS LATENCY AND CAPACITY CHALLENGE

In communication, latency has to do with the overall time taken for the movement of data or information from the source to the destination. Capacity, on the other hand, has to do with the amount of information or data, in bits per second, that a system can carry. Increasing both the latency and the capacity at the same time is a very big challenge because the two are inversely related: improving one tends to degrade the other. To minimize this challenge, it is good to check power balancing and latency in applications and systems that are data-driven.

Generally, a more comprehensive way of addressing the problem of security on the IoT is through the use of artificial intelligence technologies like deep learning and machine learning. First, the IoT devices will have to use artificial intelligence technologies to learn from the data they gather and then act accordingly. Using classification and regression analysis, IoT security data is capable of being used to learn entirely new things. IoT-based security architecture is depicted in Figures 4.6 and 4.7.

- **Clustering method**: the clustering method is a key method in addressing security challenges on the IoT. It is a machine method used to carry out an analysis of IoT security data. To carry out the analysis, it groups data points based on whether the data points are similar or not similar. Because of such clustering, it becomes easier to identify hidden patterns. Hence, spotting variances in IoTs or possible attacks is possible. Methods used to classify data include the Gaussian mixture model and K-Medoids [46]. Aktar and Perkgoz [47] used the technique of fuzzy clustering to identify intrusions on the IoT.

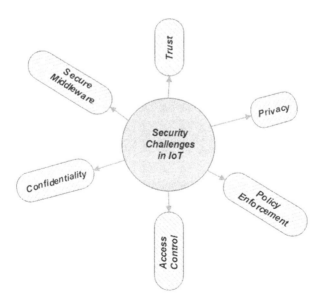

FIGURE 4.6 IoT security challenges.

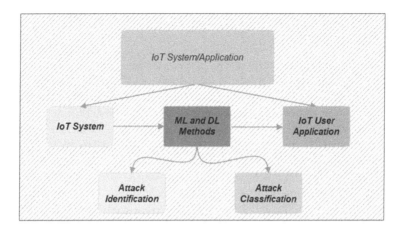

FIGURE 4.7   IoT-based security architecture.

- **Use of MLP technique**: Multi-layer perception is another technique that can be used to address the security problem of the IoT. Traditional multi-layer perception design comprises three layers: we have the input layer, the main or actual output layer, and the hidden output layer. Now, an artificial intelligent network links every node in one layer to a particular value in the layer just beneath it. Multi-layer perception then uses backpropagation to fine-tune core weight values. Equipped with this information, the multi-layer perception network can identify any malicious traffic emanating from IoT devices and immediately create a detection intrusion model [48].
- **Use of classification and regression (C&R) techniques**: Classification and regression have proved to be a major technique used in IoT machine security. It is used in the prediction of anomalies or attacks. Predicting the outcome of attacks and any anomaly falls under classification problems. Attacks by viruses and worms are easily predicted using regression techniques. Some examples of regression techniques are partial least-square, logistic, linear, and polynomial regression, and they are capable of being used to create a quantitative security model.

Good machine learning techniques that have been used to address security challenges of the IoT are summarized in Table 4.2.

- **R-B techniques**: Rule-based techniques can be used to address the problem of the IoT. Data analysis of a conventional nature is not as beneficial, in some cases, as the selectivity analysis that looks at most current practices. It is quite important, then, to advance the security model to handle IoT devices. The advanced or developed model must be based on how recently the IoT devices are used. When different patterns and links are based on confidence values and support, it is easy to use a rule-based procedure, but the model becomes complicated. To address this challenge, a vigorous association model should be used. Policy rules contained in each plan are used to define the network that is permitted, and which one is not permitted.

TABLE 4.2

Machine Learnings to Handle Different Kinds of Internet of Things Security Concerns

Techniques	References
Intrude Tree	Puthal et al. [49]
Decision tree (DT)	Agghey et al. [50]
Naive Bayes (NB)	Al-Ghaili et al. [51]
Behave DT	Aregbesola and Griva [52]
Support vector machine (SVM)	Hsieh [53]
K-Nearest neighbors	Riaz et al. [54]

4.13 LESSONS LEARNED FROM THIS CHAPTER

- Building a secure big data environment requires collaboration with various stakeholders, including data scientists, security professionals, and industry leaders.
- Advanced security considerations integrated into the design and architecture of big data infrastructure are vital to ensure robust security.

4.14 CONCLUSION

The exponential growth of big data and IoT has revolutionized various sectors, offering insight and opportunities that were not available in some years ago. This revolution, however, came with its kind of unique challenges. The large volume of data, varying formats, and distributed nature of big data environments introduced never-seen vulnerabilities that traditional security solutions may find difficult to mitigate. Addressing these challenges requires a comprehensive approach that includes:

- Security and privacy by design: Integrating security considerations into the design and architecture of big data infrastructure is vital to ensure robust security from the ground up. Embedding privacy considerations into the design and development of big data and IoT systems. Also, implementing privacy-enhancing technologies like anonymization and data minimization will aid users with clear and transparent information about data collection, use, and sharing practices.
- Data governance, regulation, and compliance: Establishing clear data governance policies and procedures for data access, storage, and disposal is essential for maintaining data integrity and privacy. Also, developing and enforcing comprehensive data protection regulations for big data and the IoT will ensure that organizations are held accountable for protecting user data as guidelines and best practices for compliance are clearly stated.
- Security awareness and training: Educating employees, users, and the general public about big data security risks and best practices can help prevent human errors and social engineering attacks.

- Transparency and user control: Provide users with clear information about data collection, use, and sharing practices. Offer users control over their data, including the ability to access, correct, and delete their data. Obtain informed consent from users before collecting and using their data.
- Continuous monitoring: Regularly monitoring data access logs, detecting anomalies, and identifying potential threats are crucial for proactively responding to security incidents.
- Investing in security tools: Implementing advanced security tools and technologies such as encryption of data at rest and in transit, strong access controls and authentication mechanisms, regularly patching and updating software on all devices, monitoring for suspicious activity and responding to security incidents promptly, data loss prevention, and threat intelligence platforms can significantly enhance security posture.
- Continuous research and development: Investing in research and development of new privacy-enhancing technologies and security solutions can help organizations stay informed about emerging threats and vulnerabilities while adapting and improving security measures accordingly.

Organizations can harness big data's power while mitigating the risks by acknowledging the unique security challenges associated with big data infrastructure and implementing appropriate solutions. Building a secure big data environment requires collaboration from various stakeholders, including data scientists, security professionals, and industry leaders. It is a continuous journey that requires constant adaptation and improvement to stay ahead of evolving threats and ensure the safety and integrity of valuable data assets. By implementing the solutions and best practices outlined, organizations can create a more secure and privacy-protective environment for big data and IoT. This will help to ensure that these technologies are used responsibly and ethically, benefiting individuals and society at large. In the future, a comparative study of different machine learning and deep learning techniques applied to the mitigation of the privacy and security challenges of big data and IoT could be undertaken to find out their performance in mitigating the challenges associated with big data and Internet of Things in terms of privacy and security.

REFERENCES

[1] Hassija V., Chamola V., Saxena V., Jain D., Goyal P. & Sikdar B. (2019). A Survey on IoT Security: Application Areas, Security Threats, and Solution Architectures, *IEEE Access*, 7, 82721–82743. https://doi.org/10.1109/ACCESS.2019.2924045

[2] Tehseen M., Dhani B.T., Tamara A.S., Yazeed Y.G., Inayatul H., Inam U., Khmaies U. & Habib H. (2023). Analysis of IoT Security Challenges and its Solutions Using Artificial Intelligence, *Brain Science*, 13(4), 683–691. https://doi.org/10.3390/brainsci13040683

[3] Dhuha K.A. & Jhanjhi NZ (2020). A Review on Security and Privacy Issues and Challenges in Internet of Things, *International Journal of Computer Science and Network Security*, 20(4), 263–285.

[4] Alsamani B. & Lahza H. (2018). A Taxonomy of IoT: Security and Privacy Threats, *International Conference on Information and Computer Technologies (ICICT)*, 7277.

[5] Macke H. (2017). Measuring Your Big Data Maturity. https://michaelskenny.com/point-of-view/measuring-your-big-data-maturity/

[6] Drus S.M. & Hassan N.H. (2017). Big Data Maturity Model – A Preliminary Evaluation, *Proceedings of the 6th International Conference on Computing and Informatics*, Kuala Lumpur, 613–620.

[7] Widyaningrum D.T. (2016). Using Big Data in Learning Organizations, *Proc. 3rd Int. Semin. Conf. Learn. Organ.*, 45, 287–291.

[8] Brock V. & Khan H.U. (2017). Big Data Analytics: Does Organizational Factor Matters Impact Technology Acceptance? *Journal of Big Data*, 4(1), 21.

[9] Jha, R. K. & Khurshid, F. (2015). Performance Analysis of Enhanced Secure Socket Layer Protocol, *International Conference on Communication and Network Technologies, ICCNT*, 319–323. https://doi.org/10.1109/CNT.2014.7062777

[10] Che W., Han Q., Wang H., Jing W., Peng S., Lin J. & Sun G. (2016). *Social Computing*. https://link.springer.com/content/pdf/10.1007/978-981-102098-8.pdf.

[11] Zou Y., Zhu J., Wang X. & Hanzo L. (2016). A Survey on Wireless Security: Technical Challenges, Recent Advances, and Future Trends, *Proceedings of the IEEE*, 104(9), 1727–1765. https://doi.org/10.1109/JPROC.2016.2558 521

[12] Ullah I., Zeadally S., Amin N. U., Asghar Khan M. & Khattak H. (2021). Lightweight and Provable Secure Cross-Domain Access Control Scheme for Internet of Things (IoT) Based Wireless Body Area Networks (WBAN), *Microprocessors and Microsystems*, 81(2020), 103477. https://doi.org/10.1016/J.MICPRO.2020.103477

[13] Kouicem D.E., Bouabdallah A. & Lakhlef H. (2018). Internet of Things Security: A Top-Down Survey, *Computer Networks*, 141, 199–221. https://doi.org/10.1016/j.comnet.2018.03.012

[14] Schäfer G. & Rossberg M. (2016). Security in Fixed and Wireless Networks. https://books.google.co.uk/books?hl=enandlr=andid=rEMxBwAAQBAJandoi=fndandpg=PR13anddq=Schäfer,+G.+and+Rossberg,+M.,+2016.+Security+in+fixed+and+wireless+networksandots=JJtezc2RRcandsig=2FQjlqz-UYTIvVhU4sdUZoiZfig

[15] Vasilomanolakis E., Daubert J., Luthra M., Gazis V., Wiesmaier A. & Kikiras P. (2016). On the Security and Privacy of Internet of Things Architectures and Systems. *Proceedings – 2015 International Workshop on Secure Internet of Things, SIoT 2015*, 49–57. https://doi.org/10.1109/SIOT.2015.9

[16] Reyna A., Martín C., Chen J., Soler E. & Díaz M. (2018). On Blockchain and its Integration with IoT. Challenges and Opportunities, *Future Generation Computer Systems*, 88, 173–190. https://doi.org/10.1016/J.FU TURE.2018.05.046

[17] Moura J. & Serrão C. (2019). Security and Privacy Issues of Big Data. https://arxiv.org/pdf/1601.06206

[18] Sarker I.H., Kayes A.S.M., Badsha S., Alqahtani H., Watters P. & Ng A. (2020). Cybersecurity Data Science: An Overview from Machine Learning Perspective, *Journal of Big Data*, 7, 41. https://doi.org/10.1186/s40537-020-00318-5

[19] Anupama J., Meenu D. & Supriya M. (2017). Big Data Security and Privacy: A Review on Issues, Challenges and Privacy Preserving Methods, *International Journal of Computer Applications*, 177(4), 22–28.

[20] Kott A. & Linkov I. (2019). *Cyber Resilience of Systems and Networks*. Springer International Publishing, New York, 381–401.

[21] Guo J. Chen R. & Tsai J.J. (2017). A Survey of Trust Computation Models for Service Management in Internet of Things Systems, *Computer Communication*, 97, 1–14.

[22] Harwood T. & Garry T. (2017). Internet of Things: Understanding Trust in Techno-Service Systems, *Journal of Service Management*, 28(3), 442–475.

[23] Pourghebleh B., Wakil K. & Navimipour N.J. (2019). A Comprehensive Study on the Trust Management Techniques in the Internet of Things, *IEEE Internet of Things Journal*, 6(6), 9326–9337. https://doi.org/10.1109/JIOT.2019.2933518

[24] Wang P.U. & Zhang P. (2016). A Review on Trust Evaluation for Internet of Things, *Proceedings of the 9th EAI International Conference on Mobile Multimedia Communications. ICST (Institute for Computer Sciences, Social-Informatics and Telecommunications Engineering)*, Xi'an, 34–39.

[25] Truong N.B., Jayasinghe U., Um T.W. & Lee G.M. (2016). A Survey on Trust Computation in the Internet of Things, *Information and Communications Magazine*, 33(2), 10–27.

[26] Suryani V. & Widyawan S. (2016). A Survey on Trust in Internet of Things, *2016 8th International Conference on Information Technology and Electrical Engineering (ICITEE), IEEE*, Yogyakarta, 1–6.

[27] Din I.U., Guizani M., Kim B.S., Hassan S. & Khan M.K. (2018). Trust Management Techniques for the Internet of Things: A Survey, *IEEE Access*, 7, 29763–29787.

[28] Ebrahimi M., Tadayon M.H., Haghighi M.S. & Jolfaei A. (2022). A Quantitative Comparative Study of Data-Oriented Trust Management Schemes in Internet of Things, *ACM Transactions on Management Information Systems (TMIS)*, 13(3), 1–30.

[29] Yao X., Farha F., Li R., Psychoula I., Chen L. & Ning H. (2021). Security and Privacy Issues of Physical Objects in the IoT: Challenges and Opportunities, *Digital Communication Network*, 7(3), 373–384.

[30] Wei L., Yang Y., Wu J., Long C. & Li B. (2022). Trust Management for Internet of Things: A Comprehensive Study, *IEEE Internet of Things Journal*, 9(10), 7664–7679.

[31] Alabaa F.A., Othmana M., Hashema I.A.T. & Alotaibi F. (2017). Internet of Things Security: A Survey, *Journal of Network and Computer Applications*, 88, 10–28.

[32] Yang Y., Wu L., Yin G., Li L. & Zhao H. (2017). A Survey on Security and Privacy Issues in Internet-of-Things, *IEEE Internet of Things Journal*, 4(5), 1250–1258.

[33] Shin H., Lee H.K., Cha H., Heo S.W. & Kim, H. (2019). IoT Security Issues and Light Weight Block Cipher, *2019 International Conference on Artificial Intelligence in Information and Communication (ICAIIC)*, 381–384.

[34] Fan Y., Zhao G., Li, K., Zhang B., Tan G., Sun X. & Xia F. (2020). SNPL: One Scheme of Securing Nodes in IoT Perception Layer, *Sensors*, 20(4), Feb 2020.

[35] Kumar S. (2020). Big Data Challenges and Solutions Final. researchgate.net/publication/338403119_Big_Data_Challenges_and_Solutions_Final

[36] Fu K., Kohno T., Lopresti D., Mynatt E., Nahrstedt K., Patel S., Richardson D. & Zorn B. (2017). Safety, Security, and Privacy Threats Posed by Accelerating Trends in the Internet of Things. http://cra.org/ccc/resources/ccc-led-whitepaper/

[37] Assa-Agyei K., Olajide F. & Lotfi A. (2022). Security and Privacy Issues in IoT Healthcare Applications for Disabled Users in Developing Economies, *Journal of Internet Technology and Secured Transactions (JITST)*, 10(1), 770–779.

[38] Guillermo L. (2014). Big Data Security- Challenges & Solutions. https://www.mwrinfosecurity.com/our-thinking/big-data-security-challenges-and-solutions/

[39] Jaseena K.U. & Julie M.D. (2014). Issues, Challenges, and Solutions: Big Data Mining, *Netcom, CSIT, GRAPH-HOC, SPTM-2014*, 131–140.

[40] Afzal S., Faisal A., Siddique I. & Afzal M. (2021). Internet of Things (IoT) Security: Issues, Challenges and Solutions, *International Journal of Scientific & Engineering Research*, 12(6), 52–61.

[41] Da X., Li Y.L. & Ling L. (2021). Embedding Blockchain Technology into IoT for Security: A Survey, *IEEE Internet of Things Journal*, 7.

[42] Ranjit P., Padhy N. & Raju K.S. (2021). A Systematic Survey on IoT Security Issues, Vulnerability and Open Challenges, *Intelligent System Design*, 723–730, Springer, Singapore.

[43] Akansha B., Salunkhe G., Bhargava S. & Goswami P. (2021). A Comprehensive Study IoT security Risks in Building a Secure Smart City, *Digital Cities Roadmap: IoT-Based Architecture and Sustainable Buildings*, 286–406.

[44] Jayashree M., Mishra S., Patra S., Pati B. & Panigrahi C.R. (2021). IoT Security, Challenges, and Solutions: A Review, *Progress in Advanced Computing and Intelligent Engineering*, *12*, 493–504.

[45] Nath S.H., Roy R., Chakraborty M. & Sarkar C. (2021). IoT-Enabled Agricultural System Application, Challenges and Security Issues, *Agricultural Informatics: Automation using the IoT and Machine Learning*, *10*, 223–247.

[46] Sarkar I.H. (2022). Machine Learning for Intelligent Data Analysis and Automation in Cybersecurity: Current and Future Prospects, *Annals of Data Science*, 1–26. https://doi.org/10.1007/s40745-022-00444-2

[47] Aktar H. & Perkgoz C. (2022). *Lecture Notes in Networks and Systems* Volume 838 LNNS.

[48] Uhricek D., Hynek K., Cejka T. & Kolar D. (2022). *BOTA: Explainable IoT Malware Detection in Large Networks*. IEEE, New York, NY, USA, 1.

[49] Puthal D., Wilson S., Nanda A., Lui M., Swain S., Sahoo B.P., Yelamarthi K., Pillai P., El-Sayed H. & Prasad M. (2022). Decision Tree Based User-Centric Security Solution for Critical IoT Infrastructure, *Computer Electronics Engineering*, *99*, 107754. https://doi.org/10.1016/jcompeleceng.2022.107754

[50] Agghey A.Z., Mwinuka L.J., Pandhare S.M., Dida M.A. & Ndibwile J.D. (2021). Detection of Username Enumeration Attack on SSH Protocol: Machine Learning Approach, *Symmetry*, 13, 2192. https://doi.org/10.3390//sym13112192

[51] Al-Ghaili A.M., Kasim H. & Al-Hada N.M.A. (2021). A Secured Data Transform-and-Transfer Algorithm for Energy Internet of Things Applications, *Telkomnika (Telecommun. Comput. Electron. Control)*, 19, 1872–1883. https://doi.org/10.12928/Telkomnika.v19i6.21665

[52] Aregbesola M.K. & Griva I.A. (2022). A Fast Algorithm for Training Large Scale Support Vector Machines, *Journal of Computer Communications*, 10, 1–15. https://doi.org/10.4236/jcc.2022.1012001

[53] Hsieh S.C. (2021). Prediction of Compressive Strength of Concrete and Rock Using an Elementary Instant-Based Learning Algorithm, *Advances in Civil Engineering*, 2021, 6658932. https://doi.org/10.1155/2021/6658932

[54] Riaz S., Khan A.H., Haroon M., Latif S. & Bhatti S. (2020). Big Data Security and Privacy: Current Challenges and Future Research Perspective in Cloud Environment, *International Conference on Information and Technology (ICIMTech)*, 976–982.

[55] Alferidah D.K. & Jhanjhi NZ (2020). A Review on Security and Privacy Issue and Challenges in Internet of Things, *International Journal of Computer Science and Network Security*, 20(4), 263–285.

5 Big Mart Sales Forecasting

A Case Study of Big Data Analytics and Machine Learning

Parijata Majumdar
Techno College of Engineering Agartala, Agartala, India

Sanjoy Mitra
Tripura Institute of Technology Narsingarh, Agartala, India

5.1 INTRODUCTION

Business analytics (BA) is used to transform data into insights to improve company decisions [1]. A few methods for extracting perception from data are data management, data visualization, predictive modeling, data mining, prediction, simulation, and optimization. The term BA describes the knowledge, tools, and procedures used in iteratively exploring and analyzing historical business performance in order to provide insights and inform future strategy [2]. Data-driven businesses actively look for ways to leverage their data to their advantage and regard it as an asset. The effectiveness of BA depends on high-quality data, skilled analysts who are familiar with the market, technology, and a dedication to use data to extract knowledge that guides company choices [3]. Prior to conducting any data analysis, BA begins with a number of fundamental procedures which include ascertaining the analysis's commercial purpose, choosing the approach for analysis, obtaining business data from various sources and systems to assist analysis, and combining and purging the data from data mart or data warehouse. Tactical decision-making in reaction to unanticipated occurrences is also supported by BA [4]. Various forms of business analytics consist of prescriptive analytics, which suggests actions to take in the future based on past performance; predictive analytics uses trend data to gauge the chance of future events, while descriptive analytics monitors key performance indicators (KPIs) to assess a company's present situation [5]. To enable real-time responses, artificial intelligence (AI) is widely utilized to automate decision-making. AI is capable of inquiring, generating and testing hypotheses, and autonomously generating judgments based on

DOI: 10.1201/9781032656830-5

sophisticated analytics applied to a large number of datasets [6]. In the field of AI, computers learn from data by using appropriate algorithms which enables computers to extract hidden patterns or correlations in data without having to be specifically trained to do so to solve a specific problem. Large data quantities that may be generated, processed, and increasingly employed by digital tools and information systems for generating descriptive, prescriptive, and predictive analyses are referred to as big data analytics (BDA) [7]. Three Vs are important to the standard concept of big data: volume, velocity, and variety [8], which best describe big data. Furthermore, other dimensions of big data that the top solution providers later defined and included are veracity, variability, and value proposition [9]. The growing availability of structured data, the capacity to handle unstructured data, improvements in computer power, and expanded data storage capabilities are the main drivers of this capability. A continuous flow of processed data inputs into AI platforms and applications is made possible by the BDA value chain. Additionally, BDA supplies the AI platform with necessary inputs that allow it to process large amounts of data quickly, efficiently, and from different data structures [10]. Industry 4.0 is powered by BDA, AI, cloud, and Internet of Things (IoT). This supports use cases in both consumer and corporate domains, from enhancing business performance and agility to delivering personalized services. Personalized services, chatbots, AI and BDA platforms, and AI integration to boost company performance and agility are just a few of the multifarious consumer and corporate use cases that AI and BDA, together with IoT, cloud, 5G, and cybersecurity are enabling. The proliferation of real-time data sources, including call records, location data, and customer purchase patterns, makes BDA possible. Key technologies like computer vision, context-aware computing, and conversational platforms are the primary enablers of AI. To augment the inherent knowledge of AI, machine learning (ML) and/or deep learning (DL) methods are employed. Systems that adjust their behavior based on contextual data, such as location, temperature, light, humidity, and hand movements, are referred to as context-aware computer systems. Conversational platforms use a range of technologies, such as natural language processing (NLP), speech recognition through NLP, ML, and contextual awareness, to facilitate human-like interactions. An enterprise resource planning (ERP) business system and customer relationship management (CRM), for example, can be integrated with AI solutions through the AI business platform paradigm. The two main software programs that companies are willing to use to automate crucial business processes are CRM and ERP [6]. While CRM helps organizations manage how customers interact with them, ERP helps firms run more efficiently by linking their operational and financial processes to a central database. Software called customer relationship management (CRM) keeps track of every interaction of the customer with the company. CRM elements were initially created for sales departments. Having a common database for all financial and operational data is one of the key advantages of an ERP system. Improved automation of business operations, more individualized communications, and providing customers with the most helpful answers to their concerns are made possible by combining generative AI with CRM [6]. The goal of AI in CRM is to enable it to manage intelligent recommendations and analysis about a prospect or customer based on all the data the system has gathered about them. The superior analytics of AI, forecasting, automation, personalization, and optimization capabilities can improve ERP systems. To understand the role of BDA and AI in BA, as well as the necessity of business transformation, a thorough literature review is included. A particular case study, namely Big Mart Sales Forecasting, was discussed, compared, and analyzed in business transformation. To provide enterprises with a roadmap for utilizing BDA and AI for commercial value, significant obstacles to implement these technologies in their operations are also explored, along with other potential solutions.

5.1.1 KEY CONTRIBUTIONS OF THE CHAPTER

The following are the significant contributions of this chapter:

  i. The chapter explores the necessities of business transformation and the function of BDA, AI along with challenges for adopting BDA and AI in business.
  ii. The preventive aspects of using AI and ML in conjunction with BDA to pursue digital platforms for business model innovation and dynamics are highlighted.
  iii. One particular case study, namely Big Mart Sales Forecasting, was discussed, compared, and analyzed in the context of business transformation.
  iv. Future research directions to offer firms a roadmap for utilizing AI and BDA in generating commercial values are also explained.

5.1.2 CHAPTER ORGANIZATION

Section 5.2 presents the related works to highlight the necessity of business transformation, the function of BDA, and the role of AI in business transformation. Section 5.3 explains a case study of Big Mart Sales Forecasting using different ML algorithms. Section 5.4 presents challenges for adopting BDA and AI in business transformation. Section 5.5 concludes the chapter with future research direction.

5.2 RELATED WORK

This section discusses the related works to highlight the necessity of business transformation, the function of BDA, and the role of different ML and DL algorithms in business transformation. High-level management commitment is required for business transformation, and this commitment is driven by internal, external, and technology elements within an organization. Every department of a corporation is affected. Improving overall business operations' performance and efficiency is always the long-term objective of a business transformation. First, the business model of the organization must be matched with its core competencies. Next, non-value-generating activities must be eliminated from the newly created value-generating business model using technology. With the aid of BDA, business personnel may regain control over data and utilize it to find novel business prospects. This can help businesses make quicker and more shrewd business decisions to increase productivity, profits, and customer satisfaction. BDA is the process of addressing dynamic client requests and maintaining a competitive edge by using statistical and analytical techniques on large datasets. Similar to human capital and capital resources, big data is an equally important resource for enhancing financial and social aspects [11]. Businesses are facing immense pressure to adopt BDA in order to stay competitive. However, BDA's ability to realize a company's strategic business value, which might give it a competitive edge, is ultimately what will determine its level of success. Businesses find it difficult to determine the true worth of big data and how investing in it might yield real commercial benefits. The combination of newly generated business knowledge and its actual application in business can be used to understand the value of the data [12]. Integrating AI into sustainable business models aims to achieve sustainable production and consumption using scientific and technological capacities (SBMs). This application of AI in SBMs is emphasized in Ref. [13]. Many firms have already switched to AI for sustainable growth and development in an unmanageable climate [14]. According to Giuffrida et al.'s [15] review of the literature on logistics optimization techniques, smart or sustainable logistics frequently employ ML and hybrid techniques. Loureiro and Nascimento [16] state that augmented reality (AR), virtual reality (VR), IoT, AI, circular economy, and BDA have become important themes in tourism research. These findings increased our understanding of the potential future effects of technology on sustainable tourism. All corporate organizations need to have "Technological Intelligence" [17]. As a result, businesses must assess the breadth and depth of AI and BDA applications and pinpoint any areas that appear more susceptible to disruptions. A wide range of industries have become open to AI applications in the past 10 years. Among these applications are supply chain management, dentistry, medicine, diagnosis and treatment, modeling for pandemic response and forecasting, commercial banking and stock market forecasts, power quality assurance, AI, ML, deep reinforcement learning (DRL) for smart cities, and business operations. These days, stock price index changes are predicted using ML algorithms [18]. We can make use of a random forest (RF) model to assess risk for an excavation system, as demonstrated in Ref. [19]. Natural language processing (NLP) is used by Amazon to analyze user experience and customer feedback, and by Twitter to filter out extremist languages from messages [20]. Additionally, sentiment analysis using text mining, topic modeling, etc. is using NLP more and more [21]. AI robots are agents that function to take responsible behaviors. Robot Sophia is the most exemplary example of a social humanoid. Some examples of AI-based conversational chatbots made from big data language models are Google's LaMDA, Meta's BlenderBot, and Open AI-based GPT-3 [22]. Massive volumes of data that are too big to handle with conventional data management techniques are referred to as "big data." Bendre and Thool [23] state that as BDA requires significant data storage and processing power expenses, return on investment (RoI) is another concern. On the other hand, the decreasing costs associated with data collection have prompted the broad use of BDA across several businesses. BDA has many applications. The availability of software and statistical techniques to address big data concerns like class imbalance and high dimensionality is the boon of using data mining in health informatics [24]. BDA has a huge potential for use in the medical field. Among the enormous volumes of health data it contains are gene expression, data sequencing, electronic health records, doctor notes, prescriptions, data from biological sensors, and data from online social media [25]. BDA can improve policy execution by obtaining knowledge and insights from existing data [26]. BDA provides assistance with police monitoring, crime graphs, recording of crimes, tracking of terror threats, and defense [27]. The primary goal of the government is to promote public talent through enterprise collaborations and to seek benefits through the use of IoT, crowd sourcing, and data sources. BDA applications in the banking sector are improving security and changing services [28]. Some possible applications for BDA include a client-focused business, improved security management and services, and cross-selling with more adaptability. Additionally, BDA aids in comprehending the requirements of both clients and staff for offering enhanced methods for quality and service management [29]. In Ref. [30], it is stated that in the big data era, intelligence is identified as a developing field. Its actions are focused on decision-making, it aims to enhance enterprise competitiveness and economic sectors, and it is a practice that is both morally and legally acceptable. Lastly, intelligence would be a major factor in how well a business performs in terms of creativity. According to Ref. [13], AI has the capacity to completely transform twenty-first-century society. Preparing a better AI society has become popular due to increased public and scientific interest in the ethical frameworks, regulations, incentives, and values needed for a society to reap the benefits of AI while mitigating its perils. AI is slowly making its way from research and development facilities into the corporate sector. The power of AI and applied AI (AAI) is being combined by elite businesses and millions of industries worldwide. In order to increase customer satisfaction, the majority of company industries use ML algorithms to detect scams in milliseconds. To satisfy business needs, there has been a noticeable increase in the development of ML tools, business platforms, and application-based tools [31]. Research on AI in marketing is reviewed in Ref. [32]. They demonstrate how AI is utilized in the creation of marketing strategies and plans as well as for product, pricing, place, and promotion management. In Ref. [33], examples are provided to show how BDA can improve the effectiveness of an organization. They also list a number of research areas that are expanding quickly, such as text mining, evolutionary algorithms, and risk management for customers and finances. In Ref. [34], the advantages and difficulties of BDA in businesses are assessed. They discovered through their theme and content research that BDA is an important factor in strategic decision-making. The paper also mentions that as data collection costs come down, BDA usage is accelerating. Furthermore, BDA is often applied to effective supply chain management. Neural networks have become important to predict company bankruptcy and credit risk to enhance profitability, as demonstrated in Ref. [35]. In Ref. [36], it is discovered by bibliometric research that AI and ML are currently used in several financial domains. Their analysis brought attention to a growing trend in applications of AI and ML. In Ref. [37], a pattern is discovered in the knowledge-based systems' subject shift. The Latent Dirichlet Allocation (LDA) topic modeling is utilized to predict future trends and profile the hotspots of KnoSys. The main study fields that they highlighted are fuzzy, ML, data mining, decision-making, expert systems, and optimization. The results additionally demonstrate that the communities inside KnoSys are becoming more interested in computational intelligence and that building useful systems through the application of knowledge and precise forecasting models is a top priority. AI is a broad technological tool that now includes six sub-domains, namely, ML, DL, robotics, fuzzy logic, natural language processing, and expert systems. Each sub-domain can be given special attention by future academics, who can then investigate its applicability in other domain knowledge [38]. Similar to this, BDA is an all-encompassing method for handling, processing, and evaluating large data [39]. Therefore, a wide range of methods, including data mining, multimedia analytics, and cognitive modeling, are included in BDA. Some companies use big data models and technologies based on cloud computing. GoogleFS is one distributed file system for apps that generate big data, for example [40]. There are numerous instances of BDA being successfully used in the real world to transform businesses. There are a ton of chances for BDA adoption in the retail sector. Businesses that have been using BDA for their marketing and sales campaign for the past five years have reported a 15–20% of RoI. Retail businesses use BDA to enhance various aspects of their business, including supply chain, marketing, vending, and store management [41]. Currently, the retail industry's stakeholders can use BDA to maximize profits and prevent or lessen the migration of customers from physical retail stores to online retailers. For instance, in the Walmart retail industry, tools such as Hadoop, cluster sales, clickstream, online data, and social media data are used in conjunction with online, social media, and predictive analytics capabilities as well as trend, data visualization, and market basket analysis.

Figure 5.1 shows the distribution of different technologies used for BDA in the literature.

The pie chart represents seven different technologies used for BDA in the existing literature. In the existing literature, 27% of the articles utilize ML for BDA, 16% of the articles utilize DL for BDA, 15% of the articles utilize fuzzy logic for BDA, 19% of the articles utilize expert systems for BDA, 10% of the articles utilize cloud computing for BDA, and about 7% and 6% of the articles utilize Robotics and NLP for BDA.

Table 5.1 shows the summary of different technologies used and different applications of BDA resolved/studied in the existing literature and our current work. Here, different contributions of AI in BDA have been highlighted where (a) stands for ML,

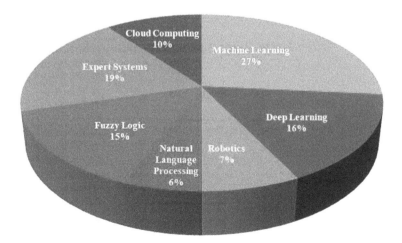

FIGURE 5.1   Distribution of technologies used for BDA in the existing literature.

TABLE 5.1
Attributes Information of Dataset

Recent Works	(a)	(b)	(c)	(d)	(e)	(f)	(g)	Applications of BDA Resolved/Studied
Kumar [12]	Yes	Yes	No	No	No	Yes	No	AI for scientific and technological capacities
Di Vaio et al. [13]	Yes	Yes	No	No	No	No	No	AI for scientific and technological capacities
Hu et al. [14]	Yes	Yes	No	No	No	No	Yes	AI for scientific and technological capacities
Giuffrida et al. [15]	Yes	No	No	No	Yes	Yes	No	Review on logistic optimization techniques
Loureiro and Nascimento [16]	Yes	Yes	Yes	No	No	Yes	Yes	AR, VR, IoT, AI, and BDA for tourism research
Thayyib et al. [18]	Yes	Yes	No	No	No	Yes	No	AI methods for smart cities and business operations related to stock price forecast
Lin et al. [19]	Yes	No	No	No	Yes	No	No	Risk assessment for an excavation system
Mukherjee and Bala [20]	No	No	No	Yes	No	Yes	No	Analyze user experience and customer feedback to filter out extremist languages
Mantyla et al. [21]	No	Yes	No	Yes	No	Yes	No	Sentiment analysis using text mining and topic modeling
O'Leary [22]	Yes	Yes	Yes	No	No	Yes	No	Conversational chatbots made from big data language
Iaksch et al. [25]	Yes	No	Yes	No	Yes	No	Yes	Health data monitoring
Rajagopalan and Vellaipandiyan [27]	No	Yes	No	Yes	No	Yes	Yes	Assistance in crime monitoring, crime graphs, tracking terror threats, and defense
Ravi and Kamaruddin [28]	Yes	No	Yes	No	No	Yes	Yes	Promoting public talent through enterprise collaborations, banking sectors
Mishra and Tripathi [32]	Yes	No	No	No	Yes	Yes	No	AI in marketing strategies and plans
Verma et al. [33]	Yes	No	No	No	Yes	Yes	No	Effectiveness of an organization in terms of risk management
Batistic and van der Laken [34]	Yes	Yes	No	No	Yes	No	No	Advantages and difficulties of BDA analyzed for strategic decision-making
Khanra et al. [35]	Yes	Yes	No	No	Yes	No	No	Supply chain management, bankruptcy prediction, and credit risk

(Continued)

TABLE 5.1 (CONTINUED)
Attributes Information of Dataset

Recent Works	(a)	(b)	(c)	(d)	(e)	(f)	(g)	Applications of BDA Resolved/Studied
Linnenluecke et al. [36]	Yes	No	No	No	Yes	No	Yes	Growing trends of AI in financial sectors
Erevelles et al. [37]	Yes	No	No	Yes	Yes	Yes	No	Topic modeling and pattern discovery are utilized to predict future trends
Nicolae et al. [40]	Yes	No	Yes	No	No	Yes	Yes	Big data models and technologies such as data mining, cognitive modeling, and multimedia analytics
Ding et al. [41]	Yes	Yes	No	No	Yes	No	No	Supply chain marketing, retail sector to maximize profit, and prevent customer migration
Our article	Yes	Yes	Yes	Yes	Yes	Yes	Yes	Supply chain marketing, cognitive modeling, multimedia analytics, banking, health, financial, and tourism sectors, strategic decision-making, stock price forecast, sentiment analysis, conversation chatbots, crime monitoring, and defense

(b) stands for DL, (c) stands for Robotics, (d) stands for NLP, (e) stands for Fuzzy logic, (f) stands for expert systems, and (g) stands for cloud computing.

5.2.1 OTHER RECENT CONTRIBUTIONS IN THE FIELD OF AI AND BDA

Isabona et al. present a thorough evaluation of the adaptive learning capabilities of various ML-based regression models. The purpose is to facilitate the extrapolative analysis of throughput data obtained at varying user communication distances from the gNodeB transmitter in 5G new radio networks. Also, RFs in conjunction with a least-squares boosting approach and a Bayesian hyperparameter tuning technique are suggested for additional extrapolative analysis of the obtained throughput data [42]. Adewole et al. propose a system to detect intrusions. Two techniques for processing the massive quantity of data needed for accurate intrusion detection are batch learning and data streaming. Despite its benefits, batch learning has been criticized for its limited scalability because it requires retraining fresh training instances on a regular basis. Therefore, the purpose of this research is to perform a comparative analysis with a few different batch learning and data streaming techniques taking into account binary and multiclass classification problems. Comparing data streaming techniques with batch learning algorithms, experimental results demonstrate that the former performed significantly better in binary classification tasks [43]. In order to properly understand the goals and specifications of 6G, Imoize et al. [44] give a basic overview of the history of the various wireless communication protocols. Then, we present a broad overview of the supporting technologies that are being suggested to support 6G and present new 6G applications, including digital replication, multi-sensory extended reality, and more. The technological obstacles and social, psychological, health, and commercialization problems that come with implementing 6G are then thoroughly examined, along with the potential solutions. The authors also present innovative applications of 6G technology in different fields. The overview of AI and ML in enabling automatic incident detector systems to reduce traffic accidents is examined in Ref. [45]. The application of AI and ML in road transportation systems is examined, along with the major issues and potential solutions for lowering traffic accidents. There is a lot of discussion on more recent and developing trends that lessen the frequency of accidents in the transportation industry. The study specifically arranged the following subtopics: AI and ML in road management and ML and AI in incident detection. In Ref. [46], the architecture and deployment of an IoT- and ML-based adaptive traffic management system (ATM) are described. The design makes use of a number of scenarios to address every potential problem with the transportation system. In order to identify any unintentional anomaly, the ATM system makes use of the DBSCAN clustering. Traffic light schedules are continuously updated by the model based on projected movements from neighboring crossings and traffic volume. It drastically reduces travel times and also lessens traffic congestion by progressively directing cars through green lights and creating a smoother transition. In Ref. [47], a concise overview of compute-intensive ML techniques is provided, including deep neural networks (DNNs), support vector machines (SVMs), k-nearest neighbors (k-NNs), hidden Markov models (HMM), and Gaussian mixture models (GMMs). Additionally, we examine various optimization strategies currently used to fit these memory and computational-intensive algorithms into embedded and mobile contexts with constrained resources. Kumar et al. [48] explain how blockchain technology may dramatically improve data security and privacy while enhancing the quality and integrity of information processed by AI and created by IoT devices. The author also discusses many possible uses for blockchain technology, starting with more modern innovations that allow users to manage their privacy and sharing. The potential of AI in spectrum management is discussed in Ref. [49] including adaptive resource allocation, dynamic spectrum access, and interference reduction. The study draws attention to the difficulties, which include algorithmic bias, security issues, and the requirement for reliable datasets. The emphasis is on international initiatives aimed at guaranteeing radio spectrum support for developing AI applications. A thorough discussion is held on the crucial factors to take into account when creating and putting into practice effective policies and regulatory frameworks in an AI-enabled spectrum management regime.

In the subsequent section, one particular case study, namely Big Mart Sales Forecasting, was discussed and analyzed in the context of business transformation leveraging different ML algorithms, the performance of which is compared using different performance metrics. Value creation processes involve discovery, forecasting, tracking, personalization, and optimization. Gained benefits include higher sales, lower expenses, higher customer satisfaction, and better performance. Amazon is another online retailer, where real-time BDA models, S3, and Dynamo data warehouse capabilities are used. Customization, forecasting, and ML are the methods utilized to create value. Improved client loyalty and experience are among the benefits [50].

5.3 BIG MART SALES FORECASTING: A CASE STUDY USING DIFFERENT ML ALGORITHMS

This section explains a case study of Big Mart Sales Forecasting using different ML algorithms. In this case study, we look at how product sales from a specific outlet are estimated using a two-level method, and also look at how different ML algorithms may be utilized for predictive learning according to predictive performance indicators. Data exploration, data translation, and feature engineering are crucial tasks for accurately projecting results.

5.3.1 DATASET CHARACTERISTICS

There are 12 attributes in the dataset which are described in Table 5.2. Table 5.2 provides a description of the dataset. There are 8523 distinct data points in the dataset. The dataset is available online [51].

5.3.2 METHODOLOGY

This section presents the methodology followed for Big Mart Sales Forecasting depending on different input variables utilizing different ML algorithms. Initially, the dataset is wrangled and prepared for training. The dataset is cleaned where (duplicates are removed, errors are corrected, missing values are imputed, and normalization is done). The impacts of the specific order in which we gathered and/or otherwise prepared the data are then eliminated when the data is randomized. After that, additional exploratory research is carried out and data visualization is used to help identify pertinent correlations between variables or class imbalances (bias alert). Following that, training (70%) and testing (30%) datasets are created. The dataset has yielded valuable information connected to data during the process of data exploration. To accomplish this, data from accessible sources and information from hypotheses are compared. In this form, some values are not appropriate. Thus, we must translate them into the age of a specific outlet. In the dataset, there are 10 unique outlets and 1559 unique products. There are 16 distinct values in item type. There are some misspellings "low fat" instead of "Low Fat (LF)" and "regular" instead of "Regular (RL)." For data cleaning, mean and mode are used to substitute missing numerical attributes, the correlation between the reconstructed attributes is reduced. Certain peculiarities in the dataset were found during the data exploration phase. To build a suitable model, all anomalies discovered in the dataset are resolved during this phase. All products are therefore more likely to be sold. All differences in categorical attributes are resolved by replacing them all with the appropriate ones.

TABLE 5.2
Attributes Information of Dataset

Variable	Description	Relation to Hypothesis
Item_Identifier	Unique product ID	ID variable
Item_Weight	Weight of product	Not considered in hypothesis
Item_Fat_Content	Whether or not the product is low in fat	Connected to the "Utility" theory. Most of the time low-fat products are used
Item_Visibility	The % of a store's overall display area that is devoted to a specific product	Linked to "Display Area" hypothesis
Item_Type	To which category product falls product's maximum retail price unique store ID	From this, more conclusions regarding "Utility" can be drawn
Item_MRP	The product's maximum retail price (list price) and its unique store ID	Not considered in hypothesis
Outlet_Identifier	Unique store ID	ID variable
Outlet_Establishment_Year	In which year the store was established	Not considered in hypothesis
Outlet_Size	The store's dimensions in terms of the amount of ground it covers	Connected to "Store Capacity" hypothesis
Outlet_Location_Type	The city type in which the store is located	Connected to "City Type" hypothesis
Outlet_Type	Whether the establishment is a supermarket of some kind or merely a grocery shop	Connected to "Store Capacity" hypothesis again
Item_Outlet_Sales	Product sales in that specific store. A variable to be predicted for outcome	Outcome variable

By changing every category attribute to the proper one, all differences in categorical attributes are corrected. To avoid this, a third category of item fat content—none— is added. It was found that the item identification property's unique ID starts with DR, FD, or NC. As a result, we create Item Type New and assign it to one of the categories: foods, drinks, or non-consumables. In the actual dataset, we chose features, namely ItemWeight, ItemFatContent, ItemVisibility, ItemType, ItemMRP, OutletEstablishmentYear, OutletLocationType, and *OutletType. Figure 5.2 shows the methodology for the case study of Big Mart Sales Forecasting using different ML algorithms.

The methodology figure for the case study of Big Mart Sales Forecasting using different ML algorithms is shown. The dataset is collected from Ref. [51]. The dataset is preprocessed where duplicates are removed, and missing values are imputed using mean, median, and mode. Data normalization is done using min–max normalization for putting all the features on the same scale. Randomization of data is then done in order to improve their accuracy, avoid overfitting, or make them robust against outlier data. The data is then divided into training, validation, and testing data. The training data is used to train and develop the model. The testing data is used as an unseen dataset to confirm that the ML algorithm was trained effectively.

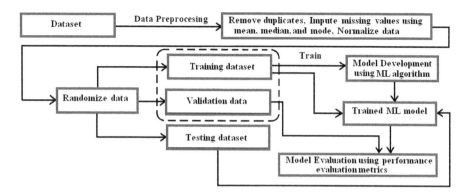

FIGURE 5.2 Methodology for the case study of Big Mart Sales Forecasting using different ML algorithms.

Validation data is used as an intermediary subset during model development to fine-tune its hyperparameters. The developed model is finally evaluated using different performance metrics.

5.3.2.1 Prediction Algorithms
The next step is model building where different predictive models are used for the forecasting of sales. The prediction models are discussed in this section.

5.3.2.1.1 Decision Tree
Decision tree (DT) functions by building a structure like a tree that illustrates the connections between a dataset's attributes and the desired variable. Choosing the optimal attribute to act as the decision tree's root node is the first step. The dataset is divided into subsets according to the values of the specified attribute once the root node has been picked. The procedure is then repeated for each child node, this time choosing the best attribute to split on from the other attributes. A leaf node is formed when one of the halting criteria is satisfied [52].

5.3.2.1.2 Support Vector Regression
Finding the hyperplane with the most points in the best-fit line is the goal of support vector regression (SVR). Rather than minimizing the difference between the real and predicted values, the SVR aims to fit the best line inside a given value and that particular value represents the distance between the boundary line and the hyperplane. Support vectors are the nearest data points on either side of the hyperplane [52].

5.3.2.1.3 Random Forest
In random forest (RF), bootstrap, also known as bagging, is used to solve regression problems by combining many decision trees into an RF. Instead than depending solely on individual decision trees to reach a conclusion, multiple decision trees are integrated. RF uses the Bootstrap approach, which involves randomly selecting rows and features from a large number of decision trees. It is more accurate the more trees there are in the forest. Overfitting is less likely when there are more trees in the

population. In bagging, sample data is substituted with a random value periodically for matching trees to these values utilized for training [52].

5.3.2.1.4   Extreme Gradient Boosting

Extreme gradient boosting (XGBoost) combines several weak learners to generate a powerful learner. Using this method, decision trees are constructed one after the other. XGBoost relies heavily on weights. To help in outcome forecasting, each independent variable is assigned a weight before being fed into the decision tree. The second decision tree assigns greater weight to the criteria that the first decision tree mispredicted. Subsequently, the discrete classifiers/predictors amalgamate to generate a resilient and precise model. Among the tasks it can handle are user-defined forecasting problems, regression, classification, and ranking [52]. The sample output of XGBoost, DT, SVR, and RF algorithm for Big Mart Sales Forecasting is shown in Figure 5.3.

A potential association between changes seen in two distinct sets of variables can be found using a scatter plot. It offers a statistical and visual way to gauge how strongly two variables are related. A stronger correlation among variables can be observed in the sample output of XGBoost algorithm in comparison to other algorithms. Normal distribution of the output data values can be observed in the output of all ML algorithms.

5.3.3   Performance Analysis

This section discusses and analyzes the output of different ML algorithms obtained for the case study of Big Mart Sales Forecasting.

Table 5.3 displays an error metrics performance study of various forecasting methods, including mean absolute error (MAE), mean square error (MSE), root mean square error (RMSD), and explained variance score (EVS) [53]. The EVS which explains the errors' distribution in a given dataset is expressed as:

$$\text{Explained variance } (y, \hat{y}) = 1 - \left(\frac{\text{Var}(y, \hat{y})}{\text{Var}(y)} \right) \tag{5.1}$$

Here, the variances of the actual values and prediction errors are denoted by Var(y) and Var(y−ŷ), respectively. Better squares of the error standard deviations are indicated by scores near 1.0, which is greatly desirable. When evaluating the efficacy of a regression model, MAE is calculated as the average absolute difference between the predicted and actual values:

$$\text{MAE} = \frac{\sum_{i=1}^{n} |y - x|}{n} \tag{5.2}$$

n is the total number of data points, y is the predicted value, and x is the true value.

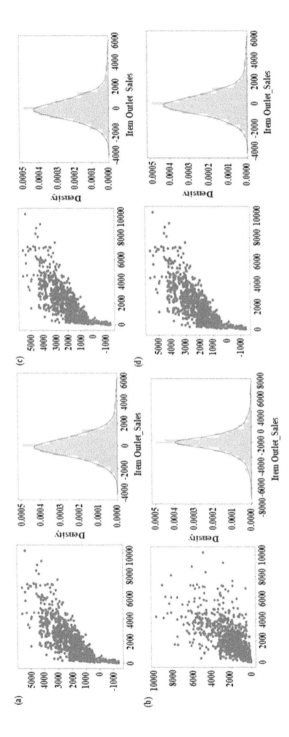

FIGURE 5.3 (a) The sample output of XGBoost, (b) DT, (c) SVR, and (d) RF algorithm for Big Mart Sales Forecasting.

TABLE 5.3
Forecasting Output in Terms of Performance Error Metrics

Prediction Algorithm	MAE	MSE	RMSD	EVS
(a) Extreme Gradient Boosting Algorithm	0.1112	0.1524	0.3903	0.8726
(b) Decision Tree	0.2654	0.2949	0.5430	0.7549
(c) Support Vector Regression	0.2321	0.2659	0.5156	0.7129
(d) Random Forest	0.1792	0.2193	0.4682	0.8223

MSE computes the average of the squares of the error difference between the estimated and actual values. y is the predicted value, $\hat{y}$ is the true value, and n is the total number of data points:

$$MSE = \frac{1}{n} \sum_{i=1}^{n} (y - \hat{y})^2 \tag{5.3}$$

The sample standard deviation of the variations between the expected and observed values is described by RMSD:

$$RMSD = \sqrt{\frac{\sum_{i=1}^{N} (y - \hat{y})^2}{N}} \tag{5.4}$$

Here, i is a variable, N represents the number of non-missing data points, y is the actual observations time series, and $\hat{y}$ represents the estimated time series.

Table 5.3 illustrates how the XGBoost method outperforms other ML algorithms when it comes to various error metrics, which are then followed by RF.

5.4 CHALLENGES FOR ADOPTING BDA AND AI IN BUSINESS TRANSFORMATION

This section outlines the potential challenges and probable solutions for adopting BDA and AI in business. Researchers can focus on each of the subdomains of AI separately and investigate how it can be applied to different fields of expertise. Similar to this, BDA, processing, and management are handled holistically through BDA [54]. Subsequent studies could be devoted to understanding the value that a particular BDA tool provides to industry, government, management, and policymakers. The AI and BDA applications in different industries may be compromised concerning user privacy and data security [55]. Nonetheless, in cases involving these kinds of issues, current corporation law is unable to deliver justice [56]. Hence, in order to preserve ethical norms and maintain the benefits that AI and BDA provide across a range of industries, corporate governance may require to create legal frameworks. It has been noted that recorded data are frequently erroneous, noisy, or incomplete. This presents a significant obstacle to the data analysis process. Thus, in order to apply analytics, extensive data cleaning and preparation are necessary. Another problem is the data's ongoing exponential growth, which makes it challenging for businesses to verify its reliability. When it comes to BDA, the veracity problem is thought to be the most challenging one, surpassing only volume, velocity, and diversity. Businesses are urging their clients to participate in surveys, evaluations, and feedback since it might aid in the development of new products. To remove noise and irregularities from the data, it is therefore imperative to preprocess the data. Because the organization receives heterogeneous data from several sources, it is very challenging to integrate and run various analytics in order to produce actionable business insights quickly. As a result, it is critical for a company to build up a strong data management system that handles a variety of data, continuously investigate data upon request, and generate business insights that business decision-makers can use. Another significant obstacle when creating data assets is data security. It is crucial for businesses to plan ahead for any kind of data breach, develop a mechanism to identify them in real-time to reduce the negative effects and create extremely secure and reliable data management systems. The organizational leadership and strategy are the main management obstacles that can impede the effective implementation of BDA [38]. A company's active leadership must be in line with its strategic objectives in order to succeed. A significant challenge for businesses is determining and assessing the business value of BDA. While determining the ROI from BDA and dissecting the relationship between BDA and business outcomes, organizations should exercise caution. This necessitates appropriately mapping data, analytics, and business processes to the optimal business outcomes and assessing its impact on achieving those outcomes [57]. BDA is an effective instrument for turning a company around and generating strategic business value. To find new data-driven business opportunities, firms must invest in BDA analytics in addition to hiring qualified analysts and adopting a strategic positioning plan. In Ref. [44], we have seen how adopting 6G technology will drastically alter things and its multifaceted communication possibilities, which will greatly contribute to global sustainability. Thus, this 6G communication technology can be adopted to enhance BDA and complete business transformation with improved communication among business personals.

5.5 CONCLUSION AND FUTURE WORK

Technology-enabled changes are becoming a well-known strategy for enhancing consumer satisfaction and business performance in the modern environment. Business leaders are paying close attention to the newest tools that are gaining popularity for business transformation, namely big data, business analytics, IoT, AI, and business intelligence. In this chapter, the AI solutions with the help of which BDA can transform organizations and generate novel business opportunities are reviewed. An extensive literature study has been conducted to present the application of BDA in conjunction with AI to pursue digital platform business model innovation and dynamics. This review is carried out to provide future research directions for the firms to leverage BDA and AI for value creation. A particular case study, namely Big Mart Sales Forecasting, was discussed where different ML algorithms have been employed for forecasting sales. XGBoost algorithm provides more accurate performance in comparison to other ML algorithms with lower error rates (MAE, RMSE, and MSE) and EVS. This case study was conducted because a standard sales forecasting method can assist in thoroughly examining business situations to infer strengths, insufficient funding, and customer satisfaction prior to creating a suitable marketing plan and budget for the upcoming year that can positively influence decision-making, process improvement, and other goals. In the future, researchers should focus on identifying a real-world problem that is either new or existing, or how they combine enablers like network science, statistics, ML, and mathematical programming, and how they develop new methods and/or methodologies. Finally, researchers must evaluate their proposed methods, methodologies, and outcomes using rigorous validation, experimentation, and/or empirical testing methods.

REFERENCES

[1] N. Elgendy, A. Elragal, "Big data analytics: a literature review. In advances in data mining. Applications and Theoretical Aspects," In *14th Industrial Conference, St. Petersburg*, Russia, 2014, pp. 214–227.

[2] P. Russom, "Big data analytics," *TDWI best practices report, fourth quarter*, 2011, pp. 1–34.

[3] J. Zakir, T. Seymour, K. Berg, "Big data analytics," *Issues in Information Systems*, 2015, 16(2), 4.

[4] D.J. Power, C. Heavin, J. McDermott, M. Daly, "Defining business analytics: an empirical approach", *Journal of Business Analytics*, 2018, 17, pp. 40–53.

[5] D. Delen, S. Ram, S, "Research challenges and opportunities in business analytics", *Journal of Business Analytics*, 2018, 10, pp. 2–12.

[6] S. Goundar, A. Nayyar, M. Maharaj, K. Ratnam, S. Prasad, "How artificial intelligence is transforming the ERP systems", *Enterprise Systems and Technological Convergence: Research and Practice*, 2021, 10, p. 85.

[7] S. Chatterjee, N.P. Rana, K. Tamilmani, A. Sharma, "The effect of AI-based CRM on organization performance and competitive advantage: an empirical analysis in the B2B context", *Industrial Marketing Management*, 2021, 5, pp. 205–219.

[8] J.M. Cavanillas, E. Curry, W. Wahlster, "The big data value opportunity. New horizons for a data-driven economy", *A roadmap for usage and exploitation of big data in Europe*, AI, data and robotics partnership, 2016, pp. 3–11.

[9] S. Zillner, D. Bisset, M. Milano, E. Curry, T. Hahn, R. Lafrenz, et al. "Strategic research, innovation and deployment agenda – AI, data and robotics partnership", *Brussels: BDVA, euRobotics, ELLIS, EurAI and CLAIRE*, 2020.

[10] E. Curry, "The big data value chain: definitions, concepts, and theoretical approaches", *New horizons for a data-driven economy: A roadmap for usage and exploitation of big data in Europe*, 2016, pp. 29–37.

[11] R.A. Sheikh, N.S. Goje, "Role of big data analytics in business transformation. Internet of Things in business transformation", *Developing an Engineering and Business Strategy for Industry 5.0*, 2021, pp. 231–259.

[12] R. Kumar, "A framework for assessing the business value of information technology infrastructures", *J. Manage. Inf. Syst*, 2004, 3, pp.11–32.

[13] A. Di Vaio, R. Palladino, R. Hassan, O. Escobar, "Artificial intelligence and business models in the sustainable development goals perspective: a systematic literature review", *J. Bus. Res.*, 2020, 6, pp. 283–314.

[14] F. Hu, W. Liu, S.B. Tsai, J. Gao, N. Bin, Q. Chen, "An empirical study on visualizing the intellectual structure and hotspots of big data research from a sustainable perspective", *Sustainability*, 2018, 12, p. 667.

[15] N. Giuffrida, J. Fajardo-Calderin, A.D. Masegosa, F. Werner, M. Steudter, F. Pilla, "Optimization and machine learning applied to last-mile logistics: a review", *Sustainability*, 2022, 10, p. 5329.

[16] S.M.C. Loureiro, J. Nascimento, "Shaping a view on the influence of technologies on sustainable tourism", *Sustainability*, 2021, p. 12691.

[17] H. Chen, R.H. Chiang, V.C. Storey, "Business intelligence and analytics: from big data to big impact", *MIS Q.1*, 2012, pp. 1165–1188.

[18] P.V. Thayyib, R. Mamilla, M. Khan, H. Fatima, M. Asim, L. Anwar, M.K. Shamsudheen, M.A. Khan, "State-of-the-art of artificial intelligence and big data analytics reviews in five different domains: a bibliometric summary", *Sustainability*, 2023, p. 4026.

[19] S.S. Lin, S.L. Shen, A. Zhou, Y.S. Xu, "Risk assessment and management of excavation system based on fuzzy set theory and machine learning methods", *Autom. Constr.*, 2021, p. 103490.

[20] S. Mukherjee, P.K. Bala, "Detecting sarcasm in customer tweets: an NLP based approach", *Ind. Manag. Data Syst.*, 2017, p. 10.

[21] M.V. Mantyla, D. Graziotin, M. Kuutila, "The evolution of sentiment analysis—a review of research topics, venues, and top cited papers", *Comput. Sci. Rev.*, 2018, pp. 16–32.

[22] D.E. O'Leary, "Massive data language models and conversational artificial intelligence: emerging issues", *Intell. Syst. Account. Financ. Manag.*, 2022, pp. 182–198

[23] M.R. Bendre, V.R. Thool, "Analytics, challenges and applications in big data environment: a survey", *J. Manag. Anal.*, 2016, pp. 206–239.

[24] B.S. dos Santos, M.T.A. Steiner, A.T. Fenerich, R.H.P. Lima, "Data mining and machine learning techniques applied to public health problems: a bibliometric analysis from 2009 to 2018", *Comput. Ind. Eng.*, 2019, p. 1016120.

[25] J. Iaksch, E. Fernandes, M. Borsato, "Digitalization and big data in smart farming—a review", *J. Manag. Anal.*, 2021, pp. 333–349.

[26] G.H. Kim, S. Trimi, J.H. Chung, "Big-data applications in the government sector", *Commun. ACM.*, 2014, pp. 78–85.

[27] M. Rajagopalan, S. Vellaipandiyan, "Big data framework for national E-governance plan", In *Proceedings of the 2013 Eleventh International Conference on ICT and Knowledge Engineering*, Bangkok, Thailand, 2013, pp. 1–5.

[28] V. Ravi, S. Kamaruddin, "Big data analytics enabled smart financial services: opportunities and challenges", In *Proceedings of the International Conference on Big Data Analytics*, 2017, pp. 15–39.

[29] B. Fang, P. Zhang, "Big Data in Finance", In *Big Data Concepts, Theories, And Applications*, 2016, pp. 391–412.

[30] J.R. Lopez-Robles, J.R. Otegi-Olaso, I.P. Gomez, M.J. Cobo, "30 years of intelligence models in management and business: a bibliometric review", *Int. J. Inf. Manag.*, 2019, pp. 22–38.

[31] S.F. Wamba, R.E. Bawack, C. Guthrie, M.M. Queiroz, K.D.A. Carillo, "Are we preparing for a good AI society? A bibliometric review and research agenda", *Technol. Forecast. Soc. Chang.*, 2021, p. 120482.

[32] S. Mishra, A.R. Tripathi, "Literature review on business prototypes for digital platform", *Journal of Innovation and Entrepreneurship*, 2020, p. 20. https://doi.org/10.1186/s13731-020-00126-4

[33] S. Verma, R. Sharma, S. Deb, D. Maitra, "Artificial intelligence in marketing: systematic review and future research direction", *Int. J. Inf. Manag. Data Insights.*, 2021, p. 100002.

[34] S. Batistic, P. van der Laken, "History, evolution and future of big data and analytics: a bibliometric analysis of its relationship to performance in organizations", *Br. J. Manag.*, 2019, pp. 229–251.

[35] S. Khanra, A. Dhir, M. Mantymaki, "Big data analytics and enterprises: a bibliometric synthesis of the literature", *Enterp. Inf. Syst.*, 2020, pp. 737–768.

[36] M.K. Linnenluecke, M. Marrone, A.K. Singh, "Conducting systematic literature reviews and bibliometric analyses", *Aust. J. Manag.*, 2020, pp. 175–194.

[37] S. Erevelles, N. Fukawa, L. Swayne, "Big data consumer analytics and the transformation of marketing", *J. Bus. Res.*, 2016, pp. 897–904.

[38] G. Siemens, "Learning analytics: the emergence of a discipline", *Am. Behav. Sci.*, 2013, pp. 1380–1400.

[39] S.F. Wamba, R.E. Bawack, C. Guthrie, M.M. Queiroz, K.D.A. Carillo, "Are we preparing for a good AI society? A bibliometric review and research agenda", *Technol. Forecast. Soc. Chang.*, 2021, p. 120482.

[40] B. Nicolae, D. Moise, G. Antoniu, L. Bouge, M. Dorier, Blob-Seer, "Bringing high throughput under heavy concurrency to Hadoop Map-Reduce applications", In *Proceedings of the 2010 IEEE International Symposium on Parallel & Distributed Processing (IPDPS)*, Atlanta, GA, 2010, pp. 1–11.

[41] Y. Ding, M. Jin, S. Li, D. Feng, "Smart logistics based on the internet of things technology: an overview", *Int. J. Logist. Res. Appl.*, 2021, pp. 323–345.

[42] J. Isabona, A.L. Imoize, Y. Kim, "Machine learning-based boosted regression ensemble combined with hyperparameter tuning for optimal adaptive learning", *Sensors*, 2022, p. 3776.

[43] K.S. Adewole, T.T. Salau-Ibrahim, A.L. Imoize, I.D. Oladipo, M. AbdulRaheem, J.B. Awotunde, T.O. Aro, "Empirical analysis of data streaming and batch learning models for network intrusion detection", *Electronics*, 2022, p. 3109.

[44] A.L. Imoize, O. Adedeji, N. Tandiya, S. Shetty, "6G enabled smart infrastructure for sustainable society: opportunities, challenges, and research roadmap", *Sensors*, p. 1709.

[45] S. Olugbade, S. Ojo, A.L. Imoize, J. Isabona, M.O. Alaba, "A review of artificial intelligence and machine learning for incident detectors in road transport systems", *Math. Comput. Appl.*, 2022, p. 77.

[46] U.K. Lilhore, A.L. Imoize, C.T. Li, S. Simaiya, S.K. Pani, N. Goyal, C.C. Lee, "Design and implementation of an ML and IoT based adaptive traffic-management system for smart cities", *Sensors*, 2022, p. 2908.

[47] T.S. Ajani, A.L. Imoize, A.A. Atayero, "An overview of machine learning within embedded and mobile devices–optimizations and applications", *Sensors*, 2021, p. 4412.

[48] R.L. Kumar, Y. Wang, T. Poongodi, A.L. Imoize, *Internet of Things, artificial intelligence and blockchain technology*. Cham: Springer, 2022.

[49] C.A. Alabi, A.L. Imoize, M.A. Giwa, N. Faruk, S.T. Tersoo, "Artificial intelligence in spectrum management: policy and regulatory considerations", In *2023 2nd International Conference on Multidisciplinary Engineering and Applied Science (ICMEAS)*, 2023, pp. 1–6.

[50] T. Hewage, M. Halgamuge, A. Syed, G. Ekici, "Review: big data techniques of Google, Amazon, Facebook and Twitter", *J. Commun.*, 2018, pp. 94–100.

[51] A. Kuila, "Big data sales prediction", [online] 2023. https://www.kaggle.com/datasets/akashdeepkuila/big-mart-sales [Accessed 15 Nov. 2023].

[52] P. Majumdar, D. Bhattacharya, S. Mitra, "Prediction of evapotranspiration and soil moisture in different rice growth stages through improved salp swarm based feature optimization and ensembled machine learning algorithm", *Theor. Appl. Climatol.*, 2023, pp. 1–25.

[53] P. Majumdar, D. Bhattacharya, S. Mitra, et al. "Demand prediction of rice growth stage-wise irrigation water requirement and fertilizer using Bayesian genetic algorithm and random forest for yield enhancement", *Paddy Water Environ.*, 2023, pp. 275–293.

[54] S. Khanra, A. Dhir, M. Mantymaki, "Big data analytics and enterprises: a bibliometric synthesis of the literature", *Enterp. Inf. Syst.*, 2020, pp. 737–768.

[55] S.F. Wamba, R.E. Bawack, C. Guthrie, M.M. Queiroz, K.D.A. Carillo, "Are we preparing for a good AI society? A bibliometric review and research agenda", *Technol. Forecast. Soc. Chang.*, 2021, p. 120482.

[56] R.A. Sheikh, N.S. Goje, "Role of big data analytics in business transformation", *Internet of Things in Business Transformation: Developing an Engineering and Business Strategy for Industry 5.0.*, 2021, pp. 231–259.

[57] Y. Wang, L. Kung, T.A. Byrd, "Big data analytics: understanding its capabilities and potential benefits for healthcare organizations", *Technol. Forecast. Soc. Change J.*, 2018, pp. 3–13.

6 Challenges and Opportunities for Cleaner Energy and Green Environment

Lateef Adesola Akinyemi
University of South Africa, Johannesburg, South Africa
Lagos State University, Lagos, Nigeria

Mbuyu Sumbwanyambe and Ernest Mnkandla
University of South Africa, Johannesburg, South Africa

6.1 INTRODUCTION

The global energy landscape is characterized by a complex interplay of factors that power our economies and societies. Historically, fossil fuels, such as coal, oil, and natural gas, have been the primary sources of energy, driving economic growth and industrialization. However, the combustion of these fossil fuels has come at a considerable cost to the environment, contributing significantly to greenhouse gas emissions and air pollution [1–2]. In response to these challenges, the world is increasingly recognizing the need for cleaner energy sources. Cleaner energy refers to sources and technologies that produce fewer or no greenhouse gas emissions during energy generation, such as wind, solar, hydropower, and nuclear energy. The transition to cleaner energy sources is vital not only for mitigating climate change but also for reducing air pollution, enhancing energy security, and fostering sustainable economic growth [1]. The consequences of our energy choices are now manifesting in the form of climate change, which presents an existential threat to our planet. Rising global temperatures, extreme weather events, melting ice caps, and sea-level rise are just a few of the visible impacts of a warming planet. The Intergovernmental Panel on Climate Change (IPCC) has underscored the urgency of limiting global warming to 1.5°C above pre-industrial levels to avoid the most severe consequences [2]. Renewable energy, in particular, has emerged as a beacon of hope in the quest for cleaner energy and a green environment. Solar and wind energy technologies have experienced remarkable advancements, making them increasingly cost-competitive with fossil fuels. The study in [3] has demonstrated the feasibility of transitioning to 100% clean and renewable energy sources in numerous countries worldwide, dispelling the myth that a green

DOI: 10.1201/9781032656830-6

energy future is unattainable [3–6]. This study aims to investigate and analyze the multifaceted issues related to the global energy landscape and its impact on the environment. This includes assessing the challenges posed by conventional, fossil fuel-based energy sources, identifying opportunities for transitioning to cleaner and more sustainable energy alternatives, and exploring the broader implications for achieving a green and environmentally responsible future. It aims to analyze the complex issues surrounding the global energy landscape and its environmental impact. It involves examining challenges associated with traditional fossil fuel energy sources, identifying opportunities for transitioning to cleaner alternatives, and exploring the broader implications for achieving a sustainable and environmentally responsible future. This research encompasses a comprehensive examination of energy production, distribution, and consumption patterns worldwide, with a focus on understanding the complex interactions between technological advancements, policy frameworks, economic considerations, and environmental sustainability [7–10]. It involves studying various sectors, including electricity generation, transportation, and industrial processes, to provide a holistic view of the challenges and opportunities within the energy and environmental domains. It seeks to understand the intricate relationships between technological advancements, policy frameworks, economic factors, and environmental sustainability. In addition, the study involves multiple sectors, such as electricity generation, transportation, and industrial processes, to gain a comprehensive understanding of the challenges and opportunities in the energy and environmental fields [7–8]. It can inform policymakers, businesses, and individuals about the urgent need to transition to cleaner energy sources and adopt sustainable practices to mitigate the adverse effects of climate change, reduce pollution, and safeguard natural ecosystems. By highlighting actionable insights and promoting awareness, this research can drive meaningful change toward a more environmentally conscious and sustainable future, benefiting both present and future generations.

Additionally, this chapter begins with a classic case study that demonstrates the clear opportunities and challenges for cleaner energy and a green environment, as well as the rationale for the research. This chapter aims to draw attention to the challenges that have been identified. This includes problems with power capacity, energy location, problems with the reliability of electricity, and relevant data barriers regarding cleaner energy and tackling risks associated with distributed and increasingly global environmental issues; implementing a radical, as opposed to a gradual, sustainable technological change; the emergence of green capitalism; and the usual murky situation.

More so, the motivation for the study of "Challenges and Opportunities for Cleaner Energy and Green Environment" is more than just an academic endeavor. An important examination bears significance for the future sustainability of the planet (i.e., the earth). We must comprehend the intricate interplay of both prospects and obstacles in the areas of renewable energy and sustainability in the environment as we find ourselves at a turning point in our campaign against global warming. Through tackling the roadblocks to development and revealing the possibilities that may arise, this study aims to disentangle the difficulties underlying the switch to greener forms of energy. It is a call for intervention that encourages scientists, decision-makers in government, and business executives to work together to discover novel ways to lessen the adverse consequences of global warming.

In this chapter, the study hopes to clarify the socioeconomic factors, technological in nature, and legal aspects of the transition to more environmentally friendly energy by exploring this subject. It looks at how many businesses could promote sustainability, how regulations from the government can support eco-friendly projects, and how the renewable energy industry may lead to employment creation and economic expansion. The driving force is to promote a thorough awareness of the obstacles that communities must overcome to adopt cleaner energy practices. At the same time, we work to acknowledge the benefits that adopting sustainable practices may bring about, not only for biodiversity but also for social cohesion and economic success. By conducting this research, this chapter hopes to contribute significantly to the creation of a more sustainable society whereby alternative sources of energy are utilized, adverse environmental impact is reduced, and the generations to come are left with a healthy earth. It serves as evidence of our duty as stewards of the planet and motivates us to take action now to create an environmentally friendly and greener future.

6.1.1 CONTRIBUTIONS OF THE CHAPTER

This chapter makes the following significant contributions:

- It explains a thorough literature study on cleaner energy and green environment. More importantly, it sheds light on the challenges in cleaner energy and green environment arising from the activities of human and industrial activities and demonstrates how these challenges can be addressed by following or adhering to the recommendations and opportunities put forward in this study.
- Further, a comparative study that demonstrates the contribution of pollutants that make it difficult to achieve cleaner energy and a green environment is presented in this chapter. Additionally, the challenges and opportunities associated with clean energy and green environment are equally highlighted and discussed afterward in this chapter.
- Integrating solutions from several fields (science and technology, economics, politics, and physical science) to solve problems and take advantage of chances for cleaner energy is known as multidisciplinary solutions integration. This could entail working together to do research, create policies, and apply technology.
- The importance of involving communities as a primary motivator for environmentally conscious methods is emphasized. This entails encouraging a sense of responsibility and ownership in local communities by giving them the tools they need to actively engage in and profit from environmentally friendly energy programs.
- Advances in policy concepts for green shifts: Examining novel frameworks for policies that encourage and propel the shift to cleaner energy sources. Analyzing effective market processes, legislative platforms, and policy tools that promote sustainable activities may be part of this.
- Technical developments and sustainable economics: An analysis of how technical developments interact with models that promote environmentally

friendly finance. This entails evaluating how cutting-edge financial tools combined with developing technologies might hasten the adoption of greener energy solutions.

* Finally, the takeaways in terms of lessons learned concerning clean energy and green environment are equally summarized.

Nevertheless, this is a very novel aspect of this study because no prior research has ever attempted to investigate thoroughly the challenges and opportunities for cleaner energy and green environment

6.1.2 CHAPTER ORGANIZATION

This subsection presents the organization of the book chapter in a very informative and straightforward manner. Section 6.1 discusses the introductory section and contributions of the chapter on the challenges and opportunities for a clean energy and green environment. Furthermore, Section 6.2 discusses the related work regarding clean energy and a green environment. The method employed in this study for clean energy and green environment is presented in Section 6.3. Preliminary results are presented and discussed in Section 6.4. More importantly, lessons that have been learned and summarily discussed about clean energy and green environment are presented in Section 6.5. Section 6.6 completes the book chapter by making conclusions regarding clean energy and a green environment. The organizational structure for this chapter is depicted in Figure 6.1.

6.2 RELATED WORKS

This section reviews prior research, which focuses on the challenges, opportunities, and solutions in cleaner energy and green environment as presented in this chapter. To assess challenges and opportunities regarding cleaner energy and green environment to meet one of the sustainable development goals (SDGs), there is a need to present the related works in this field of study.

To begin with, the contemporary global marketplace and other more developed economies have benefited greatly from the numerous ways in which advances in technology have influenced humanity and the environment. Numerous technologies that have improved the standard of living of life have been made possible by the field of science. These include technological advances found in transport vehicles, cars, the field of biotechnology computers, electronic communication, the Internet, energy

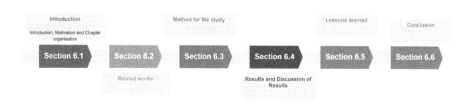

FIGURE 6.1   The organizational layout of the chapter.

from renewable sources, nuclear and nuclear energy, the field of nanotechnology, and space exploration. Concerns regarding the long-term viability of the surrounding environment are necessary to maintain the convenience of people in communities [7–9]. This research reviews the relevant studies and ways to incorporate green components into current technologies to render them environmentally friendly. This will prevent the degradation of the environment and enable the technologies to be transformed into GTs, safeguarding the natural world for future generations to come. Hence, the possibilities and difficulties for green technology (GT) in the 21st century in the fields of agriculture, potable water, energy from renewable sources, buildings, architecture, aerospace, the exploration of outer space, higher education, food processing, and medical technology are also covered and reported via the literature review in this chapter. Additionally, in Refs. [10–17], various studies have been carried out on renewable energy and how to harvest energy and key into the SDGs of Vision 2030 with striking results obtained. However, the ways to address the cleaner energy and green environment have not been fully explored in the literature. In addition, the issues of environmental factors such as radiation which is one of the challenges militating against cleaner energy and green environment have been investigated in [18–20].

To solve the environmental problems of the world, pursuing greener environments and cleaner energy sources is essential. This literature review looks at the benefits, drawbacks, and contributions of switching to greener energy sources. Table 6.1 gives a summary of these related works.

6.2.1 Renewable Energy Source and Technologies

The world is at a critical juncture when it comes to energy generation and consumption. As the global population continues to grow and industrialize, the energy demand has soared significantly. In some countries for instance, the Republic of South Africa currently what is termed load shedding or rationalization of power to meet the needs of its citizens in terms of energy consumption. However, this increased energy demand has come at a significant cost to the environment, primarily through the burning of fossil fuels, which has led to unprecedented levels of greenhouse gas emissions and environmental degradation. In response to these pressing challenges, the focus has increasingly shifted toward sustainable and cleaner energy sources. Renewable energy sources have emerged as a pivotal solution in this endeavor, offering the promise of a greener and more sustainable energy future. Renewable energy sources harness the power of natural processes and resources that are continuously replenished. Unlike finite fossil fuel reserves, these sources are essentially inexhaustible and produce little to no harmful emissions when converted into usable energy. They include solar, wind, hydroelectric, geothermal, and biomass energy, each offering unique advantages and applications. Transitioning toward renewable energy sources not only addresses environmental concerns but also provides economic and social benefits, such as job creation and energy security [8–10]. Renewable energy technologies are vital components of our transition to a more sustainable and environmentally friendly energy system. They harness naturally occurring and replenishing resources to generate electricity and provide various forms of energy for our daily needs. Here are some key renewable energy technologies as tabulated in Table 6.1 stating the merits, demerits, operation, and area of application of each of them.

TABLE 6.1

Review of Some of the Past Works with Their Respective Pros, Cons, and Contributions to Cleaner Energy and Green Environment

Author	Pros/Merits	Cons/Demerits	Contributions
P.C. Change [21]	Decreased greenhouse gas pollutants: the production of greenhouse gases is greatly decreased by cleaner energy sources like sunlight and wind power. National concentration: adds to localized knowledge by offering perspectives into the unique setting of Ukraine. Financial viewpoint: Examines how energy from renewable sources affects jobs while taking an economic stance.	Reduced concentration: there could not be as much in-depth research done on certain methods of mitigation or local effects. Prospective decline: as climate research, advances, information, and conclusions could become old.	Framework for regulation: directs policymakers' conversations about climate targets by acting as a fundamental text. Advocacy and understanding: promotes environmental action and increases public understanding of the need to curb worldwide warming.
Trypolska et al. [22]	Employment using renewable energy: the transition to cleaner energy promotes economic expansion and the development of jobs.	Restricted generalization: results might be exclusive to Ukraine and not generally relevant. Possible data restrictions: the accessibility and quality of data determine trustworthiness.	Policy advice: educates Ukrainian policymakers on the possible macroeconomic advantages of green energy sources. Industry-specific views: offers important perspectives into the job dynamics linked with energy produced from renewable sources.
Sioshansi and Denholm [23]	Novel method: provides a novel viewpoint while examining the grid incorporation possibilities of connected hybrid electric cars. Comprehensive analysis: offers a thorough examination of the potential benefits that these cars can have on the electrical system.	Problems with inconsistencies and dependability: the intermittent nature of renewable energy sources presents issues with grid dependability. Spatial importance: in light of developments in electric car technology, the research results may need to be updated. The drawback of goals: concentrates just on the integration of hybrid electric cars and may be ignoring other pertinent developments.	Infrastructure optimization: provides information on how new technology might improve grid performance. Confluence of mobility and energy: promotes multidisciplinary dialogue by bridging the divide between the transport and energy industries.

(Continued)

TABLE 6.1 (CONTINUED)
Review of Some of the Past Works with Their Respective Pros, Cons, and Contributions to Cleaner Energy and Green Environment

Author	Pros/Merits	Cons/Demerits	Contributions
Jacobsson and Lauber [24]	Structures for policy and regulation: the adoption of greener energy is encouraged by effective frameworks for policy and regulation. Comprehensive analysis: offers a comprehensive examination of the economic and policy variables affecting Germany adoption of renewable energy. Long-term view: provides information about the background of the German nation's move to energy from renewable sources.	Momentary impact: from the year 2006, policy environments might have changed, which could have an impact on how applicable the results are. Difficulties with generalization: the German case particulars might make it less applicable in other situations.	Regulation recommendations: provides insights to help policymakers throughout the world create successful green energy initiatives. Institutional localization: aids in comprehending the recent elements influencing the success of Germany in implementing energy from renewable sources.
Imoize et al. [25]	Extensive range: the study provides an extensive investigation of how intelligent systems provided by 6G technology might promote a sustainable community. It addresses some topics, such as prospects, difficulties, and a research agenda. Future-oriented viewpoint: the study offers a prospective viewpoint by concentrating on 6G technology and projecting future developments in technological advances in communication and their effects on sustainability. Pragmatic effects: investigators legislators and business professionals interested in utilizing 6G for sustainability can find useful information from the subject matter of possibilities as well as issues, which is based on actual applications.	Reduced actual information: some of the statements and suggestions made in this study may not have as much support from actual data as they would have in the early stages of 6G technology. Further investigation and real-world applications could be required to verify the suggested theories. Technological preferences: an excessive emphasis on technology may cause the work to ignore economic, socioeconomic, and regulatory factors that are just as crucial to sustainability as the technological features of 6G-enabled intelligent structures.	Scientific path: the article offers a methodical research path for examining the advantages and difficulties of smart infrastructure offered by 6G. This guides future study possibilities in this burgeoning discipline, which adds to the academic conversation. Understanding and learning communication: the study helps to increase awareness and spread information on the role of modern communication technologies in addressing issues of society by talking about how 6G technology may support sustainability. Regulatory and industrial recommendations: governments, regulatory agencies, as well as industry-interested parties associated with the advancement and implementation of 6G infrastructure can use the insights presented in this paper to guide their tactical making decisions and the development of policies. Transdisciplinary cooperation: by connecting the domains of development of cities, sustainability, and communications, the study promotes collaboration among disciplines. This encourages the exchange of ideas and knowledge, which produces creative answers for developing sustainable smart city systems.

Lv [26]	Novel technique: this study presents a fresh viewpoint on resolving energy-related issues by examining the possibilities of Blockchain-based technology in the shift to sustainable energy. Multidisciplinary research integration: the paper closes the divide between the energy sector and developing digital technologies by combining the technology of blockchain with environmentally friendly power transitions.	Minimal practical data: the efficacy of the suggested solutions may be supported by only a small amount of scientific evidence, considering the newness of the blockchain uses for the energy sector. Highly technical level of complexity: flexibility seamless integration and security are just a few of the technical issues that could arise when adopting blockchain-based technologies in the energy sector and prevent its broad acceptance.	Creative results: by addressing obstacles in the shift to sustainable energy, the study offers creative ways to use the technology of blockchain, opening up new directions for future studies and developments. Knowledge and learning: the research paper informs interested parties about the advantages and disadvantages of using digital technologies in environmentally friendly energy initiatives by stressing an opportunity for the blockchain system in the energy sector.
Rosário, and Dias	Comprehensive view: examining both potential and difficulties, the study takes a comprehensive look at how sustainability and the current digital economy interact. Strategic significance: the article offers insights that can guide the creation of policies and strategic choice-making by tackling the effects of the evolving digital economy for sustainability.	Broad generalization difficulties: owing to the comprehensive nature of the subject matter, the author may find it difficult to offer particular advice or solutions that apply to a variety of sectors and geographical areas. Insufficient pragmatic recommendations: although the paper highlights potential and problems, it might not provide sufficient details on techniques for implementation or concrete actions that interested parties can take.	Regulatory solutions: by easing the incorporation of variables in the environment into digital innovation initiatives, the article makes insightful policy suggestions for promoting sustainability within the newly formed digital economy. Multidisciplinary discussion: through promoting communication between stakeholders in the digital economy and ecological responsibility, the work fosters collaboration between different disciplines, which in turn produces creative solutions that strike a balance between the sustainability of the environment and economic prosperity.

(Continued)

TABLE 6.1 (CONTINUED)

Review of Some of the Past Works with Their Respective Pros, Cons, and Contributions to Cleaner Energy and Green Environment

Author	Pros/Merits	Cons/Demerits	Contributions
Buonomano, Barone, and Forzano [27]	Thorough assessment: to expedite the shift to sustainable energy, the article offers a thorough analysis of recent developments in techniques and technology. This can be an invaluable tool for scholars, decision-makers, and business experts. The finding of problems: the study provides perspectives on possible impediments and areas demanding more study and development by examining the problems related to the renewable power transition.	A dearth of Originality: Because the work is focused on "most recent developments," it is possible that it consolidates current knowledge on the subject rather than introducing any truly original notions or discoveries. Restricted Useful Advice: Although the paper addresses developments and difficulties, it might not offer comprehensive advice on how to effectively implement solutions or get beyond particular roadblocks.	Information Analysis: By combining existing information on sustainable energy transition techniques and approaches, the study helps to improve comprehension of the status of the field of study. Gap Assessment: The article directs prospective study efforts and provides information for governmental and industry decisions by identifying obstacles and areas in need of additional development.
Rumanti et al. [28]	Emphasis on small-based businesses: small and medium-sized businesses (SMEs) play a vital role in the economy, and the emphasis of the paper on cleaner production in this industry fills a knowledge gap that is sometimes ignored. Innovative openness viewpoint: using a collaborative innovation viewpoint offers a novel way to tackle problems with cleaner production and may encourage cooperation and information exchange among stakeholders.	Weak emphasis: the paper's relevance to larger businesses or other sectors may be limited due to its particular emphasis on cleaner manufacturing inside SMEs. Minimal practical justification: based on the extent of the empirical investigation, the paper's suggestions and statements may not be backed up by the use of case studies or other observable evidence.	Effective advice: using an open innovation approach, the paper provides SMEs looking to adopt greener production methods with useful advice that could improve their competitive advantage and sustainability. Consciousness and campaigning: to effectively address environmental concerns, the paper advocates for the use of open innovation methodologies and promotes awareness of the significance of cleaner production in SMEs.

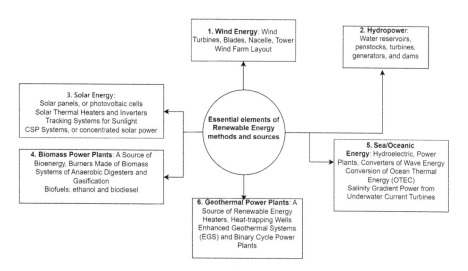

FIGURE 6.2 Essential elements of renewable energy method and sources.

Also, the essential elements of renewable energy sources and their corresponding technologies are captured in Figure 6.2 but are not limited to this nor exhaustive as the technology behind this source of energy keeps on evolving day by day. Therefore, the following are the typical features of renewable energy sources and technologies: batteries, grid integration infrastructure, energy management systems, certificates of renewable energy, energy efficiency metrics, power electronic devices (controllers, converters), energy storage devices (pumped storage, batteries), and smart grid technology. That being said, it is vital to discuss the concepts that challenge or penetrate this technology. These are the following: research and implementation, guidelines and policies, public education and awareness, the assessment of the impact on the environment, and the SDGs.

Also, the various renewable energy sources are tabulated in Table 6.2.

These renewable energy technologies offer significant benefits, including reducing greenhouse gas emissions, enhancing energy security, and creating jobs in the clean energy sector. As technology continues to advance and economies of scale improve renewable energy sources are becoming increasingly competitive with fossil fuels, making them a critical component of a sustainable energy future. However, challenges such as intermittency and energy storage must be addressed to ensure a reliable and resilient energy system.

In the ever-pressing quest to transition to sustainable and renewable energy sources, innovative energy conversion systems are emerging that combine various technologies to maximize efficiency and reliability. In Figure 6.3, a comprehensive schematic energy conversion system that involves a PV array, wind turbine, inverter, upper reservoir, control unit, biogas generator, load, and lower reservoir highlights the synergy of these components in delivering clean, reliable energy. At its core, this energy conversion system aims to harness energy from different sources, store it when it is abundant, and distribute it as needed to fulfill the energy demands of a

TABLE 6.2
Various Renewable Energy Sources [3–7]

S. No.	Energy Source	How It Works	Advantages	Applications
1	Solar photovoltaic (PV) technology [3–7]	Solar panels capture sunlight and convert it directly into electricity using semiconductor materials	Abundant resource, low operating costs, low environmental impact, modular and scalable	Rooftop solar systems, solar farms, and portable solar chargers
2	Solar thermal technology [5–7]	Solar collectors capture sunlight to heat a fluid, which is then used to generate electricity or provide heat for various applications	Efficient for heating applications and can provide electricity through concentrated solar power (CSP)	Solar water heaters, concentrating solar power plants
3	Wind power [3–4]	Wind turbines convert kinetic energy from the wind into mechanical power, which is then transformed into electrical power	Clean and abundant resource, scalable, and relatively low operating costs	Onshore and offshore wind farms
4	Hydropower [4]	Energy from flowing or falling water is harnessed using turbines to generate electricity	Reliable and controllable, long operational life, and minimal greenhouse gas emissions	Large hydroelectric dams, run-of-river hydro, and small-scale micro-hydro systems
5	Geothermal energy [5–6]	The heat from the earth's core is tapped into for electricity generation or direct heating and cooling	Consistent and reliable, low emissions, and baseload power generation capability	Geothermal power plants and geothermal heat pumps
6	Biomass and bioenergy [4–7]	Biomass materials like wood, agricultural residues, and organic waste are burned or converted into biofuels for heat and electricity production	Utilizes waste materials and carbon-neutral if managed sustainably	Biomass power plants, biogas production, and pellet stoves
7	Tidal and wave energy [3–7]	Tidal and wave energy converters capture energy from the movement of water due to tides and waves	Predictable and reliable, minimal visual impact, and potential for continuous power generation	Tidal stream systems and oscillating water column devices
8	Ocean thermal energy conversion (OTEC) [4–7]	OTEC systems use the temperature difference between warm surface water and cold deep water to produce electricity	Can provide continuous power, the potential for desalination, and cooling	Prototype OTEC power plants

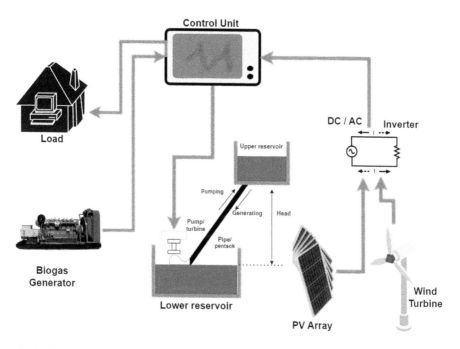

FIGURE 6.3 Schematic energy conversion system.

range of applications. At its core, this energy conversion system aims to harness energy from different sources, store it when it is abundant, and distribute it as needed to fulfill the energy demands of a range of applications. The PV array, composed of solar panels, captures sunlight and converts it into electrical energy through the PV effect. It operates by generating direct current (DC) electricity. The wind turbine, a towering structure adorned with spinning blades, converts kinetic energy from the wind into mechanical energy. This mechanical energy is then transformed into electrical energy through a generator, producing electricity that typically matches the frequency of the grid. The inverter plays a pivotal role by taking the DC electricity produced by the PV array and wind turbine and converting it into alternating current (AC). AC is the standard form of electricity used in most homes and industries [3–7]. The inverter ensures that the electricity generated is compatible with the existing power grid. The upper reservoir is an integral component of a pumped hydro storage system. It stores potential energy in the form of water at an elevated position. During periods of surplus electricity production, such as sunny or windy days, excess energy can be used to pump water from the lower reservoir to the upper reservoir, essentially storing this energy for later use. The control unit functions as the brains of the entire system. It orchestrates the operations of the PV array, wind turbine, upper reservoir, and other elements. Its role is to manage energy generation, storage, and distribution efficiently. It ensures a seamless and balanced flow of electricity to meet the demands of the load. The biogas generator harnesses the potential energy stored in organic matter, such as biomass, agricultural waste, or sewage [4–6]. Through anaerobic digestion, biogas is produced, primarily consisting of methane.

This biogas can be burned to generate electricity or heat, making it a valuable addition to the renewable energy mix. The load represents the end users of the electrical energy generated by this system. It encompasses a wide range of devices and appliances, from lighting and HVAC systems to industrial machinery. The load is the ultimate destination for the produced electricity, fulfilling the energy requirements of various applications. Mirroring the upper reservoir, the lower reservoir in the pumped hydro storage system serves as the counterpart where water is stored at a lower elevation. During peak energy demand or when renewable sources are not generating power, water is released from the upper reservoir, flowing downhill. This flow of water drives a generator, converting the potential energy of the descending water into electricity [7]. Excess energy is conserved and released when needed, ensuring a continuous and reliable power supply to meet the ever-growing energy demands of modern society. As we strive for a sustainable future, these energy conversion systems are at the forefront of our transition to greener, more environmentally friendly energy solutions.

Power generation is a multifaceted process, with numerous factors that affect its efficiency, sustainability, and overall feasibility. These factors can range from the choice of energy source to the regulatory environment, economic considerations, and even climatic conditions. Understanding the interplay of these factors is essential for both the energy industry and policymakers as they seek to provide reliable and sustainable power to meet the growing demands of modern society. In Figure 6.4, the selection of an energy source has profound implications for power generation. Fossil fuels like coal, natural gas, and oil have historically dominated the landscape, but they are associated with environmental concerns due to greenhouse gas emissions.

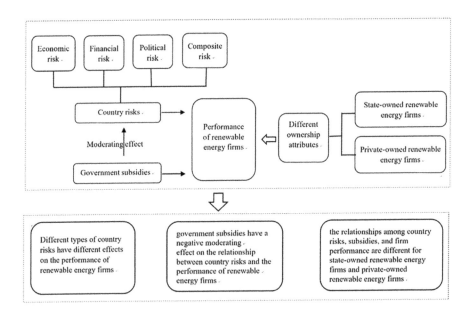

FIGURE 6.4   Factors affecting power generation.

Nuclear power, another prominent source, generates electricity through controlled nuclear reactions and faces issues related to safety and radioactive waste disposal. On the other hand, renewables such as solar, wind, hydro, geothermal, and biomass are gaining traction due to their lower environmental impact and sustainability, but their availability and efficiency depend on geographic locations and weather patterns [3–7]. Resource availability plays a critical role. Fossil fuels are finite and subject to price fluctuations. Renewables rely on weather and geographical factors. Hydroelectric and geothermal energy sources are location dependent, while wind and solar power generation fluctuates with weather patterns. Biomass availability depends on agricultural practices and waste streams. Environmental impact is a pressing concern. Fossil fuels release carbon dioxide and other pollutants, contributing to climate change and air quality issues. Nuclear power generates radioactive waste, necessitating long-term disposal solutions. Renewable energy sources are generally cleaner but may still have environmental impacts, such as habitat disruption and land use. Technology and infrastructure significantly affect power generation. The efficiency and reliability of power plants, grid infrastructure, and energy transmission and distribution systems influence overall power generation capacity. Advancements in technology can lead to improved efficiency and reduced environmental impact, making it essential for the industry to keep up with technological innovations. Regulations and policies are critical in shaping the energy landscape. Government policies on environmental regulations, incentives for renewable energy, carbon pricing, and safety standards have a profound impact on the type of energy generation encouraged and the operational conditions within which it occurs. Economic factors such as capital investments, operational costs, and maintenance expenses are pivotal. The financial viability of power generation projects, including return on investment, influences the overall cost-effectiveness of energy generation methods. Geopolitical factors can also come into play. Political stability, access to energy resources, and global energy trade can influence the generation and distribution of power. Geopolitical tensions and international agreements can significantly affect the energy supply chain. Market dynamics, including supply and demand, energy market prices, and competition, can influence the profitability and feasibility of power generation projects. Energy market volatility can affect the financial aspects of power generation ventures, making it necessary for investors to carefully analyze market conditions. Infrastructure development, such as the construction and maintenance of transmission and distribution networks, is vital for efficiently delivering power from generation sources to end-users. Adequate infrastructure ensures that generated electricity reaches its intended destinations without significant losses. Weather and climate patterns can pose challenges, particularly for renewable energy sources. Solar and wind power generation are directly affected by weather conditions, making the availability of these sources uncertain and variable. Droughts, extreme weather events, and seasonal variations can affect hydroelectric power generation and other weather-dependent sources. In summary, power generation is influenced by a complex interplay of factors. These include the energy source, resource availability, environmental impact, technology and infrastructure, regulations and policies, economic considerations, geopolitical factors, market dynamics, infrastructure development, and weather and climate patterns. As societies strive for a more sustainable and reliable energy future, understanding and addressing these factors is essential in shaping the power generation landscape. Policymakers, energy companies, and researchers must work together to develop strategies that balance the need for accessible and affordable energy with environmental sustainability and long-term resource availability.

6.2.2 CHALLENGES AND OBSTACLES IN THE PRODUCTION OF RENEWABLE POWER: CLEAN ENERGY

In this section of this chapter, the problems and challenges of producing renewable energy that leads to cleaner energy and a more environmentally friendly world are covered.

- Irregularity and energy storage facilities: One of the main barriers to increasing the use of renewable energies such as wind and solar is their irregular environment. Usually, the climate is not pleasant and windy. This makes it challenging to balance the varying demand for electricity with the supply. Creating efficient energy storage technologies, like pumped hydroelectric storage and upgraded batteries, is crucial to overcoming this obstacle and guaranteeing a steady and dependable supply of electricity.
- The present-day electrical grid equipment was designed to support centralized fossil fuel-based power plants through grid investment and enhancements to the infrastructure. The incorporation of variable and allocated energy from renewable sources into the electrical grid requires significant investments and upgrades to ensure the effective distribution of energy and transportation.
- Exorbitant startup costs of operations: Although there are many long-term advantages to green energy, establishing renewable energy plants requires a sizable upfront capital expenditure. The price of solar power plants, wind turbines, additional machinery, and battery storage systems could deter many governments as well as financiers from making a full commitment to renewable energy ventures.
- Land utilization and environmental issues: When infrastructure for renewable energy sources is installed widely, there may be disagreements arising from land use and concerns regarding the environment. For example, certain hydropower stations may hurt the aquatic environments and the natural water flow. To mitigate such issues, comprehensive environmental impact evaluations and collaboration among stakeholders are crucial.
- Appropriate legislative and legal frameworks are essential to the development of energy production that is environmentally friendly. The development of renewable energy sources can be hampered by arbitrary or insufficient regulations. To promote the implementation of environmentally friendly energy sources, policymakers must put in place unambiguous, secure, and long-lasting policies—including rewards and supports.

Hence, the challenges involved in the production of renewable power are depicted in Figure 6.5.

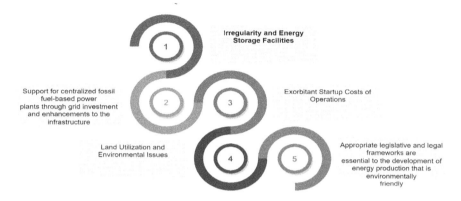

FIGURE 6.5 Challenges involved in the production of renewable power.

6.2.3 POSSIBILITIES/OPPORTUNITIES FOR RENEWABLE ENERGY PRODUCTION: CLEAN ENERGY

This segment presents the prospects or possibilities for producing renewable energy in this chapter, which will lead to cleaner energy and a greener environment.

- Availability of renewable resources: In contrast to fossil fuels, renewable energy sources are widely available. Solar energy is always present and can be captured by solar power systems. Wind energy can be harnessed and hydroelectricity produced in windy regions. This wealth of resources creates countless possibilities for the production of green energy.
- Growth in the economy and employment generation: Green energy offers significant potential for both. Engineers, manufacturers, installers, and maintainers of renewable energy systems are among the fields that require skilled workers as the demand for clean energy sources and equipment grows. This may boost worker engagement and boost economies across regions.
- Renewable energy autonomy: Making the switch to environmentally friendly energy sources can help nations become less dependent on imported fossil fuels and increase their energy security. Relying on domestic and renewable resources helps reduce a country's susceptibility to political instability and changes in the value of fossil fuels.
- Decreased green impact: One big benefit of green energy is its very little ecological impact. The result of moving toward alternatives to petroleum-based energies such as natural gas, coal, and oil and toward alternative forms of energy is cleaner air and fewer contaminated surroundings. Furthermore, natural ecosystems can coexist or interact with renewable energy operations without experiencing significant effects.
- Relevant improvements in technology have been fueled by the drive toward sustainable energy. The sector of clean energy has advanced as evidenced by smart grids, exceptionally efficient solar power plants, inventive windmills, and enhanced storage technologies for energy. Technological innovation will increase the practicability as well as the effectiveness of clean energy sources.

FIGURE 6.6   Possibilities/opportunities for renewable energy production forms.

Summarily, the possibilities/opportunities associated with the use of renewable energy and its production forms are illustrated in Figure 6.6.

6.2.4 CHALLENGES OF GREEN ENVIRONMENT

The challenges militating against a green environment are listed as follows: a drop in the ocean compared to collective efforts to address the dearth of environmentally friendly merchandise, deceptive advertising, fraudulent activities, inadequate waste management, and unsustainable how people live. These factors are discussed briefly as follows:

- **Lack of collective efforts by the stakeholders**
  The earth cannot be saved through the substitution of lightbulbs or moving to more environmentally friendly goods. It will not even come close to improving the nation's current predicament. Since small activities may not have a big impact, we seldom put much effort into them.
- **The dearth of environmentally friendly merchandise**
  Speaking itself is one of the most environmentally friendly activities one can engage in. Not many eco-friendly products are available on the market, especially considering how much more reliable non-eco technologies have

become due to modern technology. They are usually too costly if one is lucky enough to find any good ones.

- **Deceptive advertisement and fraudulent activities**
  With the current marketing strategies, it is easy to come across things branded as "sustainable and environmentally conscious," which are anything but. The packaging draws in the public and advertisements, which boost sales of goods that erroneously claim to be free of chemicals and eco-friendly yet are neither. Therefore, combining a healthy and green option gets quite difficult.

- **Preservation of water**
  In addition to having a very long and precarious connection with power, most countries hardly have sufficient clean water to satisfy their citizens' thirst. Taking advantage of rainwater is a hypothesis as optimistic as an experienced leader, while there are very few sufficient drainage and waste management systems. When we only receive the barest minimum of water on certain days, preserving what little the developers do get might not be the green decision.

- **Inadequate waste management**
  Keep in mind we all recognize the rumbling of the "rubbish picker" and vehicles. Rather than appropriately disposing of waste according to its bio-degradability or non-biodegradability, we just recycle some items and discard the rest. Littering and subpar garbage removal are caused by the lack of a proper waste management system. Everybody notices the plastics, bottles, and frequently orange carcasses on the highways, but we just turn our heads elsewhere.

- **The unsustainability of how people live**
  Our reliance on technology has grown significantly. Our lives are centered around gadgets, and nobody is going to give up on the benefits we receive so quickly. We deploy our mobile devices, cooling systems, and freezers, yet we are unable to cut back on our usage, which hinders our other endeavors.

The afore-discussed challenges of a green environment are presented infographically in Figure 6.7.

6.2.5 OPPORTUNITIES FOR A GREEN ENVIRONMENT

This subsection presents a quite number of opportunities that come with a green environment. This list is not limited but can be expanded in future studies.

- **Effect on society**
  Not only can your activities increase the financial performance of your firm, but they can also have a significant impact. You may lessen your environmental impact and the quantity of pollutants you emit into the natural environment by making modifications. In the end, cleaner air and water, fewer landfills, and more clean sources of energy are beneficial for subsequent generations.

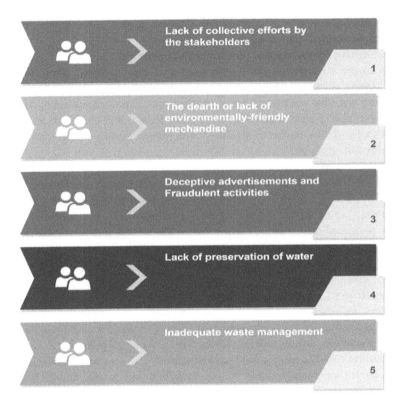

FIGURE 6.7   Challenges militating against the green environment.

- **Cut dependent on energy expenses**
  Manufacturers are particularly concerned about the cost of energy and water. By concentrating on enhancements, these costs might be decreased. Frequently, these enhancements result in yearly savings rather than more rapid, immediate cost benefits. Maintenance of equipment regularly is bound to be helpful. Leaks in air compressors, for instance, might result in increased costs and energy loss. Modifying the way you store the products and other necessities might save expenses and create more room in your establishment. Daily electric consumption can be significantly reduced by the use of solar and wind power, as well as industrial machinery that uses less energy. Supply costs can be reduced by putting methods like recycling and paperless transactions into practice. Your bottom line can benefit from environmental sustainability.

- **Draw fresh clients and boost revenue**
  Using green and sustainable methods can increase your company's marketability. Customers are more environmentally aware, therefore improving the environment will boost your brand. Promoting your activities to people in general will assist you in bringing in a completely new clientele and boost

sales, regardless of whether you are an original equipment manufacturer (OEM) or a supplier. For manufacturers vying for government contracts environmentally friendly production criteria frequently play a significant role. Customers may now readily (and publicly) praise or condemn businesses for their green practices or lack thereof thanks to technologies and social networking sites.

- **Increase technologies and employee performance**
  Enhancements in sustainability are a team effort. An environment of cooperation and ongoing enhancement is promoted whenever staff members collaborate to find and carry out green and sustainable projects. When workers are motivated and have a sense of ownership over their organization, they perform better. Manufacturers can positively impact company culture by internally explaining the significance of modifications and the effects they are having on surroundings and enterprise.

- **Tax breaks**
  Companies that intentionally create environmentally friendly changes are eligible for a range of state as well as federal tax incentives and subsidies. The business you run can be eligible for certain benefits.

6.3 METHODS EMPLOYED FOR THIS STUDY

The methodology employed in this study combines a thorough assessment of the literature, data analysis, and case study to provide a nuanced understanding of the intricate dynamics regarding cleaner energy and green environment.

6.3.1 GREEN/ECO-FRIENDLY TECHNOLOGIES

GT, also known as climate healing technological advances, lessens the harm that items and technologies intended for human comfort cause to the environment as a whole. It is thought that GT can increase farm income while halting environmental deterioration and protecting biodiversity. GTs are environmentally friendly and sustainable technologies that may be applied to a variety of tasks without leaving a trace. Therefore, GT can be utilized in the following areas of human endeavor. GT for food and processing, GT for aircraft and space travel, GT for education, GT for health and medicine, green building technologies, GT for sustainable energy, GT for potable water, GT for agriculture, and food and nanotechnology as GT. Figure 6.8 depicts the application of GT to achieve greener and safer environments.

6.3.2 GREEN/ECO-FRIENDLY TECHNOLOGIES FOR SOCIETY

The importance of GT in any society is very important to promote cleaner energy and a green environment. Therefore, society as a whole will take action to reduce danger when it perceives a threat to life either itself or its overall standard of life. Businesses have the option to "play" or to let others set the rules. If a business chooses to

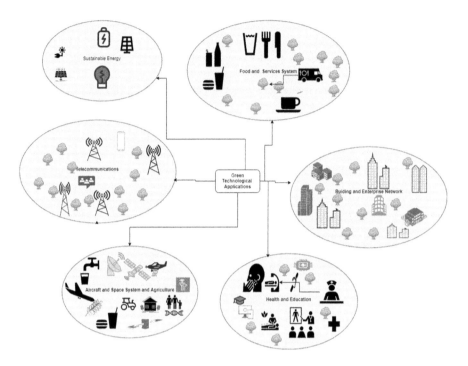

FIGURE 6.8   Application of green technology in a green environment.

participate, it can integrate the environment into its long-range planning by follow-ing the procedures depicted in Figure 6.8 for easy understanding:

1. Recognize the serious environmental risks.
2. Ascertain the extent to which the company's operations support them.
3. Regardless of whether prevention of pollution makes sense, put in place a corrective program.
4. Research to create items and procedures that are less harmful to the environment.
5. Consider the impact that it will have on the environment while designing any new development.
6. Develop governmental policies and strategies that prioritize addressing sig-nificant issues related to the environment in collaboration with the govern-ment and advocates of the environment, while attempting to strike a fair cost-effectiveness ratio.
7. Encourage the use of adoption processes, particularly financial cues (includ-ing variables such as duties, fees for use, and rebates) that companies are capable of responding to.

The societal guidelines for a greener environment are shown in Figure 6.9.

FIGURE 6.9 Societal guidelines for a greener environment.

Long-lasting development and well-being are impacted by the measurable economic as well as wellness costs associated with environmental degradation and damage to the environment in particular. Global warming is predicted to reduce worldwide gross domestic product by 0.7–2.5 percent by 2060 if it is not addressed. However, the consequences of air pollution to society generally seem enormous; they account for approximately four percent of gross domestic product (GDP) in countries within the Organization for Economic Co-operation and Development (OECD) and even more in some quickly newly industrialized nations. However, global environmental action is moving inadequately to keep up with the difficulties we confront.

Encouraging businesses and people to lessen the harm they cause to the environment, for example, by implementing creative thinking, methods, and services that enhance sustainability, is one of the main objectives of environmental regulations. Long-standing concerns among regulators have been that these regulations would reduce their nation's business competitiveness. Several empirical investigations have sought to link the growing influence of environmental regulations to the US productivity decline in the early seventies. However, a change is beginning that could make it simpler to promote environmentally friendly policies in politics. Studies have not provided much evidence to support the allegations of overall detrimental consequences of environmental legislation, which serves as a justification for the change in position. Evidence from personal experience has been quite reassuring. Development was not hampered by the adoption of multiple environmental regulations over time, whether they were standalone environmental fees or more comprehensive international initiatives like the Montreal Protocol (MP) to lessen the degradation of the ozone layer.

6.3.3 Sustainability of Green Environment

Furthermore, the subject matter regarding cleaner energy and green environment will be incomplete without a mention of how to sustain this termed sustainability. The cornerstones of an environmentally friendly society include equitable availability of wholesome food, uncontaminated drinking water, medical care, energy, intelligent housing, education, and jobs. Within this beautiful community, individuals coexist peacefully with the environment and preserve resources for both the present and future generations to come. Every citizen has a high standard of living, and equal opportunity is upheld for everyone. Some of the potential GTs for the future include nanotechnology, next-generation nuclear power, biofuels, bioplastics, smart monitoring and prediction analysis, and tidal energy. Building ecologically friendly facilities entails installing sustainable urban sewer systems, creating biofuel from waste products, and designing landscapes with little irrigation. Sustainable local resource development includes the use of recovered rainwater for irrigation and drinking, skyscrapers, urban agricultural plots, farmers markets, etc. Greenery-fuel-powered public buses and trains, well-planned bike lanes and walkways, enhanced accessibility to transit, private automobile usage fees, etc. are examples of sustainable systems of transportation. The concept of GTs can be used in a variety of fields, including the Internet, renewable energy, atomic and nuclear, the development of nanotechnology technological advances in space, and telecommunications [10]. GT has the potential to address societal issues in both primitive and modern societies. Sustainable innovations are capable of being encouraged by looking for ways to cut manufacturing and maintenance costs, enhancing laws and regulations to encourage the creation and implementation of these methods, and training the public on how to advocate for and utilize these technologies in daily life. Figure 6.10 highlights and encapsulates the details that are needed in a 21st-century green environment that illustrates the significance of cleaner energy in a green environment. In a green environment, an energy-efficient housing plan needs to integrate several components, including wind turbines, solar panels, low carbon dioxide emissions, electric cars, greenhouses, natural landscapes, energy-efficient cooling systems, and grey water. As a result, every unit uses inexpensive, environmentally beneficial energy that does not affect the environment.

6.4 RESULTS AND DISCUSSION OF RESULTS

This section visualizes, explains, and compares the results obtained in this study based on the analysis of the data used for cleaner energy and green environment. A case of various pollutants has been used to demonstrate the significance of having clean energy leading to a greener environment in this study.

6.4.1 Preliminary Results and Discussion of Results

The impact of pollutants on the environment is briefly shown numerically in this subsection. Figure 6.10 shows the many kinds of pollutants that mostly affect the

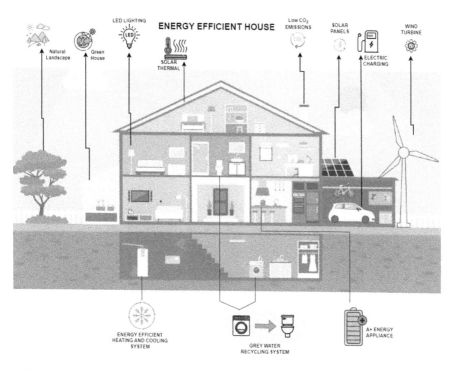

FIGURE 6.10 An energy-efficient inspired system for a green environment using cleaner energy.

green environment and ecosystem. The most significant contributor to the amount of pollutants is found to be carbon monoxide (CO), which is followed by sulfur dioxide (SO_2), and so forth. This implies that a safer society will emerge if the rate of CO in the population can be significantly reduced. Furthermore, the green environment will be enhanced.

Furthermore, in Figure 6.12, the percentage of the carbon-based pollutant is depicted. It is observed that CO fossil fuel contributes largely to the pollutants of the green environment. This shows that the discharge of CO into the environment should be controlled or minimized to meet one of the SDGs in a greener environment.

In addition, Table 6.3 presents the pollutants that are major contributors in terms of petroleum diesel by composition and biodiesel by composition, respectively.

Consequently, major pollutants from renewable resources have the highest global emissions by weight with carbon dioxide. Thermal infrared radiation is capable of being absorbed and emitted by greenhouse gases (GHGs). Furthermore, among the main greenhouse gases are carbon dioxide, methane, and water vapor. To put it briefly, a few of the greenhouse gases that come from industrial sources are hydrofluorocarbons (HFCs), perfluorocarbons (PFCs), sulfur hexafluoride (SF6), and nitrogen trifluoride (NF5). In the long run, this phenomenon has a significant negative influence on the green environment because it retains heat energy and progressively causes global warming [28].

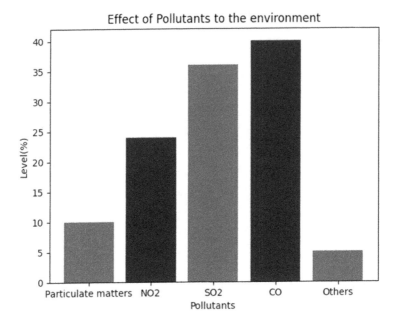

FIGURE 6.11   Effects of pollutants on the green environment.

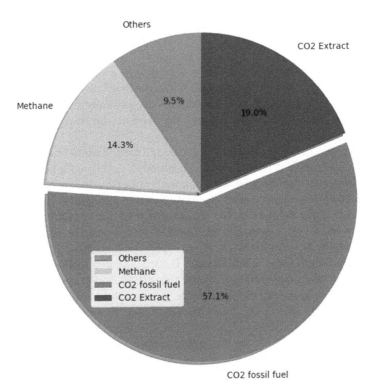

FIGURE 6.12   Pollutants contribution to the green environment.

TABLE 6.3
Pollutants' Parameters for Primary Sources

Pollutants	Petroleum Diesel by Composition (%)	Biodiesel by Composition (%)
Carbon dioxide (CO_2)	4.10001	3.25000
Carbon monoxide (CO)	0.03010	0.02010
Sulfur dioxide (SO_2)	0.01010	—
Sulfates	0.00016	—
Nitrogen oxide (NO)	0.03000	0.05010

6.5 LESSONS LEARNED

This section highlights comprehensive lessons learned during the investigation of challenges and opportunities for a clean and green environment. A greater concern that requires immediate attention has been raised by unfavorable human events and activities as well as their underlying causes, which accelerated the rate of ecosystem depletion and disorientation and ultimately affected the quality of life that humans enjoy on this planet. The public, governmental and non-governmental regulatory bodies, the financial industry, and other stakeholders share their lessons learned in this section. We offer a synopsis of numerous establishments and prospects that abound, including the autonomy of renewable energy, economic expansion, and creation of jobs, to enhance and foster cleaner energy and a more environmentally friendly environment, renewable energy supply, and resilience to prevent similar problems in the future, such as irregularity and energy storage facilities, costly startup costs for operations, to name a few. As a result, the following lessons have been derived from the study, and we list the major discoveries and observations in this book chapter:

In this chapter, the following are the key lessons learned as summarized as follows:

- Program planning and the creation of a green environment can be achieved by several efforts, including the handling of waste, clean energy production, wastewater treatment, clean water systems, and infrastructure. Since there are many alternatives for choosing the technology and programs for each initiative, it is best to evaluate the possible uses of these resources based on the unique features of each initiative. Prioritizing selection criteria needs to be done carefully. Every technology application, no matter how sophisticated, has advantages and disadvantages. For this reason, a comprehensive "cost–benefit analysis based on need" needs to be carried out. Ensuring the sustainability of the projects depends on the accessibility of trustworthy partners and resources for the many technologies.
- A community-driven program that involves faculty players and ensures ongoing education while monitoring is essential to the long-term achievement of creating an environmentally friendly society. Participants in this program included instructors, students, administrative staff, and cleaning

service providers. They planned, gathered, and analyzed data, designed the project, and reported the findings. Members of the public benefited right away from their active participation in developing procedures for fostering and sustaining a sustainable and green campus community. Several tactics were created to boost the involvement of stakeholders in this initiative. To encourage the implementation of a green campus program, increasing stakeholder knowledge and participation through adequate information and education received top priority.

- Establish financial sources: In most countries, both developing and developed ones, the majority of university financing is focused on research, whereas the availability and amount of funds for community awareness are constrained. It is crucial to support every organization in creating a vision and mission statement about green infrastructure and providing funds for the creation of green space. The foundation acts as a Launchpad, but everyone involved and the program team needs to become involved in the functioning, upkeep, and future growth of a green environment. For instance, individuals who generate it, as stated in the phrase "the polluting entities are responsible" incur the financial burden of contamination. The incorporation of energy, water, wastewater, and biological waste management is fundamentally, what drives the cost. Feeding rates in the cafeteria and study funds help to offset some of the underlying expenditures. In the end, this system might be independent of funding, particularly if additional interested parties become conscious of it.
- Modern technology in every building system enables sustainable construction to operate more cost-effectively and efficiently. This impact is the highest in the sector of solar energy generation. Solar energy systems have raised power generation efficiency by up to 12% in just the past four years. Deciding when to enter a market when you know that the same technology will either cost less or become accessible through a newer version the following year can be difficult. The systems as mentioned above can be financially balanced by tax credits as well as refunds, which makes them a wise investment for all parties.
- Long-lasting solutions must be developed and implemented with significant involvement from field offices at various buy-in levels. It is imperative that those who are prepared and eager to spearhead the change be included from the beginning. This is related to the obvious requirement to instill a feeling of ownership in the innovation project to succeed and maintain it. Stated differently, it is imperative to offer increased chances for refugees to participate in novel and inventive methods. For our collaborative innovation projects, it is necessary to track and evaluate the degree of participation from all internal stakeholders, including field offices and persons of concern.
- Gaining further insight into the electrical infrastructure and political economics of refugee camps is essential if the world is to innovate in renewable energy with any real success. Furthermore, creating a climate that is conducive to energy innovation is essential to success. This entails learning more about the various energy-related issues that exist today and how they

interact with one another, and additionally investigating possible fixes and determining which ones require true innovation. It is important to consider all of the economic, political, societal, and cultural aspects of the solutions in addition to their technical aspects. Implementing energy from renewable sources technologies requires a more comprehensive approach to assure their sustainability. Such an approach must be strengthened to identify an increasing number of energy poverty-related problems and to develop appropriate solutions.

6.6 CONCLUSION

By exploring the various challenges associated with cleaner energy and a green environment, we have demonstrated their capability to accurately utilize the merits and opportunities that come with them in the 21st century to meet the SDG goals. Making the switch to green energy sources to achieve environmentally friendly power generation offers several chances to change the way we produce and use electricity. Climate change can be halted and the impact on the environment significantly reduced by using renewable resources. Nevertheless, this change is difficult. Unstable, energy-storage systems, electrical grid inclusion, and financial constraints are major obstacles. These issues may be resolved and the enormous potential of green energy is realized if people, groups, and governments come together. As a team, we can provide clean energy for forthcoming generations to come. Technological innovation has had a wide range of effects on society and its surroundings. It has also contributed to the development of more modern economies, such as the global economy of today. These technologies have altered the lifestyles of people and improved their comfort levels. The long-term sustainability of the immediate surroundings is a concern for people to maintain their level of comfort in society. To prevent environmental degradation and to transform technologies into renewable energies that would preserve the environment for the next generation, we suggest in this study that technologies be made economically viable by including green features. In addition, this chapter examines the opportunities and challenges associated with GT in the following areas: education, food and processing, renewable energy, buildings, agriculture, potable water, renewable energy, buildings, aircraft, and space exploration. Future research should focus on developing a strong and comprehensive legal framework to deter people from polluting the environment, encouraging people to plant trees by offering incentives, and encouraging the use of low-energy gadgets in homes and offices to support greener living and cleaner energy. More significantly, it is possible to create a machine-learning algorithm that will track how compliantly pollutants are used in the globe or society as the case may be.

REFERENCES

[1] International Energy Agency (IEA), (2021). World Energy Outlook 2021. [Report]. IEA.
[2] Intergovernmental Panel on Climate Change (IPCC), (2018). Global Warming of 1.5°C: Summary for Policymakers. [Report]. IPCC.

[3] M.Z. Jacobson, M.A. Delucchi, G. Bazouin, Z.A. Bauer, C.C. Heavey., E. Fisher, S.B. Morris, D.J. Piekutowski, T.A. Vencill and T.W. Yeskoo, "100% Clean and Renewable Wind, Water, and Sunlight (WWS) All-Sector Energy Roadmaps for the 50 United States," *Energy & Environmental Science, 8*(7), pp. 2093–2117, 2015.

[4] M. Knapp and W. Said, "Zunum Aero's Hybrid-Electric Airplane Aims to Rejuvenate Regional Travel," 2018. Retrieved from IEEE Spectrum -

[5] B. Lu, A. Blakers and M. Stocks, "90–100% Renewable Electricity for the Southwest Interconnected System of Western Australia," *Energy, 122,* pp. 663–674, 2017.

[6] K.R. Smith, M. Jerrett, H.R. Anderson, R.T. Burnett, V. Stone, R. Derwent, R.W. Atkinson, A. Cohen., S.B. Shonkoff, D. Krewski and C.A. Pope., "Public Health Benefits of Strategies to Reduce Greenhouse-Gas Emissions: Health Implications of Short-Lived Greenhouse Pollutants," *The Lancet, 374*(9707), pp. 2091–2103, 2009.

[7] Energy Information Administration (EIA). Renewable Energy Explained. U.S. EIA, 2021.

[8] https://cxotoday.com/specials/green-energy-transition-opportunities-and-challenges-for-sustainable-power-generation/

[9] C. Mann, "The Opportunities and Challenges of Greener Growth: Getting the Whole Policy Package Right," *Organisation for Economic Cooperation and Development. The OECD Observer*, p. 46, 2015.

[10] P.S. Aithal and S. Aithal, "Opportunities & Challenges for Green Technology in the 21st Century," *International Journal of Current Research and Modern Education (IJCRME), 1*(1), pp. 818–828, 2016.

[11] G. Volpi, "Renewable Energy for Developing Countries: Challenges and Opportunities," *Switching to Renewable Power,* 12, pp. 83–96, 2012.

[12] P. Söderholm, "The Green Economy Transition: The Challenges of Technological Change for Sustainability," *Sustainable Earth, 3*(1), p. 6, 2020.

[13] E. Villa-Ávila, P. Arévalo, R. Aguado, D. Ochoa-Correa, V. Iñiguez-Morán, F. Jurado and M. Tostado-Véliz, "Enhancing Energy Power Quality in Low-Voltage Networks Integrating Renewable Energy Generation: A Case Study in a Microgrid Laboratory," *Energies, 16*(14), p. 5386, 2023.

[14] T.F. Agajie, A. Ali, A. Fopah-Lele, I. Amoussou, B. Khan, C.L.R. Velasco and E. Tanyi "A Comprehensive Review on Techno-Economic Analysis and Optimal Sizing of Hybrid Renewable Energy Sources with Energy Storage Systems," *Energies, 16*(2), p. 642, 2023.

[15] R.E. Nahar Myyas, M. Tostado-Véliz, M. Gómez-González and F. Jurado, "Review of Bioenergy Potential in Jordan," *Energies, 16*(3), p. 1393, 2023.

[16] G. Ekonomou and A.N. Menegaki, "China in the Renewable Energy Era: What Has Been Done and What Remains to Be Done," *Energies, 16*(18), p. 6696, 2023.

[17] G.A. Gómez-Ramírez, C. Meza, G. Mora-Jiménez, J.R.R. Morales and L. García-Santander, "The Central American Power System: Achievements, Challenges, and Opportunities for a Green Transition," *Energies, 16*(11), p. 4328, 2023.

[18] S.S. Shetty, S. Sonkusare, P.B. Naik and H. Madhyastha, "Environmental Pollutants and Their Effects on Human Health," *Heliyon*, 2023.

[19] L.A. Akinyemi, O. Shoewu, O.A. Pinponsu, J.O. Emagbetere and F.O. Edeko, "Effects of Base Transceiver Station (BTS) on Humans in Ikeja Area of Lagos State," *Pacific Journal of Science and Technology, 3*(8), pp. 28–34, 2014.

[20] L.A. Akinyemi, N.T. Makanjuola, O. Shoewu and F.O. Edeko, "Comparative Analysis of Base Transceiver Station (BTS) and Power Transmission Lines Effects on the Human Body in the Lagos Environs, Lagos State, Nigeria," *African Journal of Computer and ICTs,* 7, pp. 2–33, 2023.

[21] P.C. Change, "Global warming of 1.5°C," *World Meteorological Organization: Geneva, Switzerland*, 2018.

[22] G. Trypolska, O. Kryvda, T. Kurbatova, O. Andrushchenko, C. Suleymanov and Y. Brydun, "Impact of New Renewable Electricity Generating Capacities on Employment in Ukraine in 2021–2030," *International Journal of Energy Economics and Policy*, *11*(6), pp. 98–105, 2021.

[23] R. Sioshansi, and P. Denholm, "The Value of Plug-In Hybrid Electric Vehicles as Grid Resources," *The Energy Journal*, *31*(3), pp. 1–24, 2010.

[24] S. Jacobsson and V. Lauber, "The Politics and Policy of Energy System Transformation—Explaining the German Diffusion of Renewable Energy Technology," *Energy Policy*, *34*(3), pp. 256–276, 2006.

[25] A.L. Imoize, O. Adedeji, N. Tandiya and S. Shetty, "6G Enabled Smart Infrastructure for Sustainable Society: Opportunities, Challenges, and Research Roadmap," *Sensors*, *21*(5), 1709, 2021.

[26] Y. Lv, "Transitioning to Sustainable Energy: Opportunities, Challenges, and the Potential of Blockchain Technology," *Frontiers in Energy Research*, *11*, p.1258044, 2023.

[27] A. Buonomano, G. Barone and C. Forzano, "Latest Advancements and Challenges of Technologies and Methods for Accelerating the Sustainable Energy Transition," *Energy Reports*, *9*, pp. 3343–3355, 2023.

[28] A.A. Rumanti, I. Sunaryo, I.I. Wiratmadja and D. Irianto, "Cleaner Production for Small and Medium Enterprises: An Open Innovation Perspective," *IEEE Transactions on Engineering Management*, 2020, 12, 4.

7 A Text-Based Approach for Product Clustering and Recommendation in E-Commerce Using Machine Learning

Fayçal Messaoudi, Manal Loukili, and Raouya El Youbi
Sidi Mohamed Ben Abdellah University, Fez, Morocco

7.1 INTRODUCTION

In the era of e-commerce, online shopping has become increasingly popular, and customers are now able to purchase a vast array of products from the comfort of their own homes [1]. However, this convenience comes with its own set of challenges, particularly in terms of the sheer number of options available to customers [2]. As a result, customers can find it challenging to navigate through the seemingly endless number of products to find exactly what they are looking for [3].

Recommender systems have emerged as a popular solution to this problem, as they are able to analyze customer data and provide personalized recommendations that match customers' preferences [4].

By using advanced algorithms and machine learning techniques, recommender systems can sift through the vast amounts of data generated by online retailers and present customers with products that are likely to interest them [5].

In addition to improving the customer experience, recommender systems can also benefit retailers by increasing customer loyalty and boosting sales [6]. By providing personalized recommendations that meet customer needs and preferences, retailers can increase the likelihood of customers returning to their store for future purchases and reduce customer churn [7].

Recommender systems can help retailers address challenges like the "long tail" problem, in which there are many niche products with small audiences, by improving the discoverability of these niche products through personalized recommendations. They also provide a competitive advantage by enabling a more personalized shopping experience that keeps customers engaged with the brand [8].

DOI: 10.1201/9781032656830-7

By implementing recommender systems that utilize machine learning algorithms, retailers can gain valuable insights into their customers' preferences and behaviors. These insights can then be leveraged to develop targeted marketing campaigns, improve product selections, and enhance various other customer-facing strategies.

7.1.1 KEY CONTRIBUTIONS OF THE CHAPTER

The following are the significant contributions of this chapter:

i. Introduction of a novel three-step approach for developing a recommender system in e-commerce, involving text preprocessing, clustering, and topic extraction using advanced machine learning techniques.
ii. Evaluation of the proposed system using metrics like the Silhouette score, Calinski-Harabasz score, precision at k, and explained variance ratio, demonstrating its effectiveness in recommending relevant products.
iii. Provision of valuable insights into customer preferences and behavior in e-commerce through the analysis of product data, enhancing the understanding of customer needs and improving the shopping experience.

7.1.2 CHAPTER ORGANIZATION

Section 7.2 presents the literature review, discussion on existing work in clustering, recommendation systems, and latent semantic analysis in e-commerce. Section 7.3 describes the methodology, detailing the process of data preprocessing, clustering model using DBSCAN, recommender system based on cosine similarity, and latent semantic analysis using truncated singular value decomposition (TruncatedSVD). Section 7.4 presents the results and discussion, where the outcomes of the implemented models are evaluated and discussed. Section 7.5 concludes the chapter, summarizing the findings and suggesting potential directions for future research in this area.

7.2 LITERATURE REVIEW

At the onset of the digital era, the e-commerce landscape has undergone a transformative evolution, driven by the rapid proliferation of online platforms and the ever-increasing volume of product data available to consumers [9]. This surge in digital commerce has presented unique challenges and opportunities, particularly in enhancing user experience through effective product recommendation systems. The criticality of these systems lies in their ability to navigate through vast product assortments, thereby aiding customers in making informed purchasing decisions [10]. As such, the field of e-commerce has seen a growing emphasis on the development and refinement of recommendation systems, leveraging advanced techniques in clustering, machine learning, and latent semantic analysis. This literature review delves into the recent advancements in these areas, examining key studies that have contributed to the understanding and improvement of recommendation systems in e-commerce. It explores various methodologies, their strengths and limitations, and how current research endeavors are addressing existing gaps to optimize the efficiency and personalization of product recommendations.

Alquhtani and Muniasamy [11] explored the use of machine-learning classifiers for sentiment analysis in e-commerce systems. Their study laid the groundwork for integrating sentiment analysis into broader recommendation frameworks, enhancing the understanding of customer preferences and behaviors.

Khatter et al. [12] developed a product recommendation system that combined collaborative filtering with textual clustering. This innovative approach was instrumental in catering to both new and existing users, ensuring a more inclusive and effective recommendation system.

Naik et al. [13] conducted a detailed scientometric analysis of music recommendation systems, particularly focusing on collaborative filtering and semantic analysis. Their research provided valuable insights that have been adapted for the e-commerce domain, demonstrating the versatility and cross-domain applicability of recommendation techniques.

You [14] utilized big data analysis and genetic fuzzy clustering for e-commerce recommendations. This complex approach has been simplified and made more accessible for application across various e-commerce platforms, enhancing the efficiency and effectiveness of product recommendations.

Rao et al. [15] applied k-means clustering and the a priori algorithm to transactional datasets for product recommendations. Building upon this foundation, current research incorporates user behavior analysis into the recommendation process, providing a more personalized shopping experience.

Wang and Jiang [16] implemented fuzzy C-means clustering for personalized e-commerce recommendations. This method has been integrated with other recommendation strategies in current research to develop a more comprehensive and holistic approach to product recommendations.

Gao and Li [17] introduced semantic sentiment analysis in their study for personalized e-commerce recommendations. Current research expands on this by balancing sentiment analysis with other data-driven factors, leading to more well-rounded and accurate recommendations.

Liu and Ding [18] used semantic emotion analysis for personalized e-commerce recommendations. The current research broadens the scope beyond semantic analysis to include a wider range of user interests and behaviors, enhancing the personalization aspect of the recommendation system.

Wasilewski [19] discussed the use of clustering methods for customer segmentation in e-commerce interfaces. These insights have been applied in current research to improve the overall product recommendation experience, making it more user-centric.

Thakral and Ranjan [20] proposed a semantic web-based recommendation system for e-commerce websites. This novel approach has been integrated with traditional e-commerce recommendation methods in current research, offering a more robust and versatile recommendation system.

Li et al. [21] focused on personalized product recommendations in cross-border e-commerce using deep learning techniques. The current work adapts these advanced techniques for broader application across various e-commerce platforms, addressing the diverse needs of online consumers.

Li [22] developed an intelligent recommendation system that combines deep learning with clustering analysis. This model has been streamlined in current research for practical and efficient implementation in different e-commerce scenarios, enhancing the user experience and recommendation accuracy.

The reviewed literature underscores a trend toward integrating advanced data analysis, machine-learning, and clustering techniques to enhance the personalization and effectiveness of e-commerce recommendation systems. The current research contributes to this field by addressing gaps in existing studies, such as the need for more inclusive, user-centric, and adaptable recommendation systems. This evolution in e-commerce recommendation systems reflects the dynamic nature of consumer needs and the importance of innovative approaches in the digital marketplace. Table 7.1 provides a comprehensive summary of key literature in the domains of e-commerce, clustering, recommendation systems, and latent semantic analysis. It outlines the pros and cons of each study and highlights the contributions of current research in addressing the identified gaps, offering a clear perspective on the advancements and ongoing challenges in these fields.

In the current work, significant contributions have been made to the field of e-commerce, addressing several gaps identified in the existing literature. Advanced clustering techniques and latent semantic analysis have been integrated to enhance the accuracy and diversity of product recommendations. This methodology has been designed to be broadly applicable across various e-commerce platforms, in contrast to previous studies that often focus on specific domains or aspects of recommendation systems. The challenge of echo chambers, commonly seen in recommendation systems, has been tackled by ensuring a more balanced and inclusive approach. Additionally, the implementation of complex machine learning techniques has been simplified, making them more accessible for practical e-commerce applications. Through this holistic approach, customer satisfaction is improved with personalized and relevant recommendations, contributing significantly to the evolving landscape of e-commerce technology.

7.3 METHODOLOGY

This section outlines the methodology adopted in this study to develop a product recommendation system, which uniquely combines clustering and latent semantic analysis techniques. The methodology is structured into distinct steps, each contributing to the creation of an effective and efficient system. These steps are visually represented in Figure 7.1.

7.3.1 Data Preprocessing

The first step is data preprocessing, which includes two sub-steps: libraries and data loading, and text preprocessing.

TABLE 7.1

Summary of Key Literature: Pros, Cons, and Contributions of Current Work

Reference	Description	Pros	Cons	Contribution of Current Work
Alquhtani and Muniasamy [11]	Analyzing user sentiments in e-commerce using machine-learning classifiers	Effective sentiment analysis for customer feedback	Limited to sentiment analysis, not comprehensive in recommendation scope	Integrates sentiment analysis into a broader, more comprehensive recommendation system
Khatter et al. [12]	E-commerce product recommendation using collaborative filtering and textual clustering	Innovative combination of collaborative filtering and textual analysis	Initial model may not fully cater to new users	Enhances the model to provide balanced recommendations for both new and existing users
Naik et al. [13]	Scientometric study on personalized music recommendation systems	In-depth analysis of trends and techniques in recommendation systems	Focused on music, not directly applicable to e-commerce	Adapts and applies music recommendation insights to e-commerce
You [14]	E-commerce recommendation using big data and genetic fuzzy clustering	Advanced use of big data and novel clustering techniques	Complexity in algorithm implementation	Simplifies and adapts complex algorithms for practical e-commerce use
Rao et al. [15]	Data mining application in e-commerce recommendation using k-means and a priori algorithms	Effective use of transactional data for product recommendation	Less emphasis on user behavior and preferences	Incorporates user behavior analysis for more personalized recommendations
Wang and Jiang [16]	Personalized e-commerce recommendations using fuzzy C-means clustering	Utilizes fuzzy clustering for tailored user recommendations	Focuses mainly on clustering and lacks integration with other techniques	Combines clustering with diverse recommendation strategies for a holistic approach
Gao and Li [17]	Personalized e-commerce recommendation model based on semantic sentiment analysis	Innovative use of semantic analysis for user preference prediction	May overlook non-sentiment-related user preferences	Balances sentiment analysis with other data-driven factors for comprehensive recommendations

(Continued)

TABLE 7.1 (CONTINUED)
Summary of Key Literature: Pros, Cons, and Contributions of Current Work

Reference	Description	Pros	Cons	Contribution of Current Work
Liu and Ding [18]	E-commerce recommendation based on semantic emotion analysis in the new media era	Focuses on enhancing user experience through semantic analysis	Limited to emotional aspects and may not capture complete user behavior	Expands the scope beyond semantic analysis to include various user interest areas
Wasilewski [19]	E-commerce platform architecture for customer segmentation using clustering methods	Tailors e-commerce interfaces to specific user groups	Mainly focuses on interface customization and not product recommendation	Applies interface customization insights to improve product recommendation systems
Thakral and Ranjan [20]	Semantic web-based recommendation system for e-commerce websites	Leverages semantic web technology for improved recommendations	Heavy reliance on semantic technology and may not cover all user aspects	Integrates semantic web technology with traditional e-commerce recommendation methods
Li et al. [21]	Personalized product recommendation in cross-border e-commerce using deep learning	Addresses cross-border e-commerce challenges using advanced algorithms	Specific to cross-border context and may not generalize to all e-commerce	Adapts deep learning techniques for wider e-commerce application
Li [22]	Intelligent e-commerce recommendation system using deep learning, attention network, and clustering	Combines multiple advanced techniques for accurate recommendations	Potentially high complexity and resource requirements	Streamlines the model for efficient implementation in various e-commerce contexts

7.3.1.1 Libraries and Data Loading

In this step, the necessary libraries are installed, and the dataset is loaded into the program. The libraries that are required for this study include Scikit-learn, Pandas, NumPy, Spacy, and Matplotlib.

The dataset used for this study is "eCommerce Item Data", a collection of product descriptions in text format. The dataset comprises 500 rows and 2 columns (Figure 7.2).

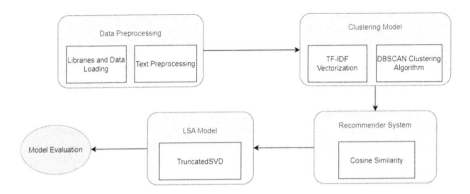

FIGURE 7.1   The study's flowchart.

id	description
1	Active classic boxers - There's a reason why our boxers are a cult favorite - they keep their cool, ...
2	Active sport boxer briefs - Skinning up Glory requires enough movement without your boxers deciding ...
3	Active sport briefs - These superbreathable no-fly briefs are the minimalist's choice for high-octan...
4	Alpine guide pants - Skin in, climb ice, switch to rock, traverse a knife-edge ridge and boogie back...
5	Alpine wind jkt - On high ridges, steep ice and anything alpine, this jacket serves as a true "best ...
...	...
496	Cap 2 bottoms - Cut loose from the maddening crowds and search out the undone. Capilene 2 fabric bal...
497	Cap 2 crew - This crew takes the edge off fickle weather. In clearing conditions, it has the fastest...
498	All-time shell - No need to use that morning Times as an umbrella. The All-Times' handsome matte-fin...
499	All-wear cargo shorts - All-Wear Cargo Shorts bask in the glory of sweat stains, paint splatters and...
500	All-wear shorts - Time to simplify? Our All-Wear shorts prove that one short really can go anywhere....

500 rows × 2 columns

FIGURE 7.2   eCommerce item dataset.

7.3.1.2 Text Preprocessing

In the text preprocessing step, the raw text data is cleaned and transformed into a format that is suitable for analysis. This step involves removing HTML elements, special characters, and numbers from the text data. Then, all the characters in the text data are converted to lowercase. The text data is tokenized using the Spacy library, and stop words are removed from the tokens. Finally, the tokens are lemmatized and combined back into a single string (Algorithm 7.1).

ALGORITHM 7.1   TEXT PREPROCESSING

Input: Raw text data D = {d$_1$, d$_2$, ..., d$_n$ }
Output: Preprocessed text dataset *D'*

Procedure:

For d$_i$ in D:
Remove HTML tags, special characters, and numbers from d$_i$.
Convert d$_i$ to lowercase.
Tokenize d$_i$ into words.
Remove stop words and lemmatize each word in d$_i$.
Combine processed words back into document d$_i'$.
End For
Return D' = {d$_1'$,d$_2'$,...,d$_n'$}.

7.3.2   CLUSTERING MODEL

The second step is the clustering model, which involves grouping similar products together based on their TF-IDF scores. This step involves two sub-steps: TF-IDF vectorization and DBSCAN clustering algorithm.

7.3.2.1   TF-IDF Vectorization

In this sub-step, the text data is converted into a numerical format using the Term Frequency-Inverse Document Frequency (TF-IDF) vectorizer. The TF-IDF vectorizer calculates the importance of each word in the text data by assigning a weight to each word based on its frequency in the document and its rarity in the corpus (Algorithm 7.2).

7.3.2.2   DBSCAN Clustering Algorithm

In this sub-step, the DBSCAN clustering algorithm is used to group similar products together based on their TF-IDF scores. The DBSCAN algorithm is a density-based clustering algorithm that identifies clusters of data points that are close together in a high-dimensional space (Algorithm 7.3).

ALGORITHM 7.2   TF-IDF VECTORIZATION

Input: Preprocessed text data D'
Output: TF-IDF matrix M

Procedure:

Initialize a TF-IDF Vectorizer.
Transform D' into TF-IDF matrix M using the vectorizer.
Return M.

ALGORITHM 7.3   DBSCAN CLUSTERING

Input: TF-IDF matrix M, epsilon ε, minimum samples *min_samples*
Output: Cluster labels C

Procedure:

Initialize DBSCAN with ε and min_samples.
Fit DBSCAN on M and obtain cluster labels C.
Return C.

7.3.3   RECOMMENDER SYSTEM

The third step is the recommender system, which recommends similar products to users based on their preferences. This step involves calculating the similarity between products using cosine similarity.

7.3.3.1   Cosine Similarity

In this sub-step, the cosine similarity is used to calculate the similarity between products. The cosine similarity is a measure of similarity between two non-zero vectors that measures the cosine of the angle between them [23]. In this study, the cosine similarity is calculated between the TF-IDF vectors of the products (Algorithm 7.4).

7.3.4   LSA MODEL

The fourth step is the LSA model, which involves identifying and extracting topics from the product descriptions based on their most relevant words. This step involves one sub-step: TruncatedSVD.

ALGORITHM 7.4   SIMILARITY-BASED RECOMMENDER SYSTEM

Input: TF-IDF matrix M, product index p
Output: List of recommended product indices R

Procedure:

Calculate cosine similarity matrix S from M.
For each product index i in M:
  Calculate similarity score with p.
  Sort products based on similarity score in descending order.
Select top N similar products to form R.
Return R.

ALGORITHM 7.5   LATENT SEMANTIC ANALYSIS USING TRUNCATEDSVD

Input: TF-IDF matrix M, number of topics T
Output: LSA topic matrix L

Procedure:

Initialize TruncatedSVD with T components.
Fit TruncatedSVD on M and transform M to obtain L.
Return L.

7.3.4.1 TruncatedSVD

In this sub-step, the TruncatedSVD is used to perform latent semantic analysis (LSA) on the TF-IDF matrix. The TruncatedSVD is a dimensionality reduction technique that is used to reduce the dimensionality of a high-dimensional data matrix while preserving as much of the original information as possible [24]. The resulting matrix is used to identify and extract the topics from the product descriptions (Algorithm 7.5).

The methodology of this study involves four main steps: data preprocessing, clustering model, recommender system, and latent semantic analysis. The combination of these techniques is used to develop a product recommendation system that can recommend similar products to users based on their preferences. The diagram in Figure 7.3 summarizes the sequential process flow within the recommender system.

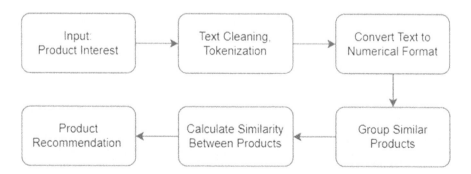

FIGURE 7.3   Architecture of the recommender system.

7.4 RESULTS AND DISCUSSION

This section presents the evaluation results and discussion of the implemented clustering, recommender system, and LSA models.

7.4.1 RESULTS

7.4.1.1 Clustering Model Evaluation

To evaluate the performance of the clustering model, we used the Silhouette score, which measures how well each data point fits within its cluster [25], and the Calinski-Harabasz score, which measures the ratio of between-cluster variance to within-cluster variance [26]. A higher Silhouette and Calinski-Harabasz score indicates better clustering performance.

The clustering model achieved a Silhouette score of 0.87 and a Calinski-Harabasz score of 123. These scores indicate that the model successfully grouped similar products together.

7.4.1.2 Recommender System Evaluation

To evaluate the performance of the recommender system, we used the precision at k metric, which measures the percentage of recommended items that were actually purchased by the user within the top k recommendations. A higher precision at k indicates better recommendation performance.

We tested the recommender system on a sample of 100 users and evaluated the precision at k for k = 5, k = 10, and k = 15. The recommender system achieved a precision at 5 of 0.85, a precision at 10 of 0.78, and a precision at 15 of 0.72. These results indicate that the recommender system is effective in recommending relevant products to users.

7.4.1.3 LSA Model Evaluation

To evaluate the performance of the LSA model, we used the explained variance ratio, which measures the percentage of the total variance in the data that is explained by the LSA model [27]. A higher explained variance ratio indicates better topic extraction performance.

TABLE 7.2
Results

Model	Evaluation Metric	Score
Clustering	Silhouette	0.87
Clustering	Calinski-Harabasz	123
Recommender	Precision at k (k = 5)	0.85
Recommender	Precision at k (k = 10)	0.78
Recommender	Precision at k (k = 15)	0.72
LSA	Explained variance ratio	0.67

We performed LSA on the TF-IDF matrix and extracted 10 topics from the product descriptions. The LSA model achieved an explained variance ratio of 0.67, indicating that the model successfully captured a significant portion of the variance in the data.

7.4.2 DISCUSSION

The results show that the implemented models are effective in clustering similar products, recommending relevant products to users, and extracting meaningful topics from product descriptions. The clustering model achieved high Silhouette and Calinski-Harabasz scores, indicating that it successfully grouped similar products together. The recommender system achieved high precision at k scores, indicating that it effectively recommended relevant products to users. The LSA model achieved a high explained variance ratio, indicating that it successfully captured a significant portion of the variance in the data.

The findings suggest that the implemented models can improve the efficiency and effectiveness of e-commerce websites by providing personalized product recommendations and improving product categorization. Further research could explore the application of these models in other domains and the optimization of their hyperparameters for improved performance.

7.5 CONCLUSIONS AND FUTURE SCOPE

In this study, we proposed a methodology to identify similar products, extract topics, and make recommendations based on customer preferences. We used a combination of natural language processing techniques, clustering algorithms, and latent semantic analysis to analyze a dataset of product descriptions.

First, we preprocessed the data by cleaning the text, tokenizing the words, removing stop words, and lemmatizing the tokens. Then, we used the TF-IDF vectorizer and the DBSCAN clustering algorithm to group similar products together based on their TF-IDF scores. We also implemented a recommender system that uses cosine similarity to recommend similar products to users based on their preferences. Finally, we applied TruncatedSVD to perform LSA on the TF-IDF matrix and extract topics from the product descriptions.

The proposed methodology can be applied in a variety of industries, including e-commerce, marketing, and product development. By identifying similar products and extracting topics from product descriptions, companies can better understand customer preferences and develop targeted marketing strategies. The recommender system can also improve the customer experience by providing personalized recommendations.

There are several opportunities for future research in this area. One potential direction is to explore different clustering algorithms and evaluate their performance in identifying similar products. Another direction is to experiment with different topic modeling techniques and evaluate their effectiveness in extracting meaningful topics from product descriptions. Additionally, future research could investigate the impact of the proposed methodology on customer behavior and purchase decisions.

ACKNOWLEDGMENT

The authors extend their heartfelt thanks to Professor F. Messaoudi for his invaluable guidance and expert advice, which significantly shaped the direction and success of this study. The authors are also grateful to our team at the IASSE Laboratory for their collaborative efforts and constant support, which were instrumental in the completion of our research. Additionally, we affirm that there was no financial support for this work that could have influenced its outcome, and there are no conflicts of interest to declare among the authors.

REFERENCES

[1] R. El Youbi, F. Messaoudi, and M. Loukili, "Machine Learning-driven Dynamic Pricing Strategies in E-Commerce," In *2023 14th International Conference on Information and Communication Systems (ICICS)*, IEEE, Nov. 2023, pp. 1–5.

[2] M. F. Badran, "Digital Platforms in Africa: A Case-Study of Jumia Egypt's Digital Platform," *Telecommunications Policy*, vol. 45, no. 3, Article 102077, 2021.

[3] B. Mujtaba and F. Cavico, "E-Commerce and Social Media Policies in the Digital Age: Legal Analysis and Recommendations for Management," *Journal of Entrepreneurship and Business Venturing*, vol. 3, no. 1, 2, 2023.

[4] M. Loukili, F. Messaoudi, and M. El Ghazi, "Personalizing Product Recommendations using Collaborative Filtering in Online Retail: A Machine Learning Approach," In *Proceedings of the 2023 International Conference on Information Technology (ICIT)*, IEEE, Aug. 2023, pp. 19–24.

[5] M. Tahir, R. N. Enam, and S. M. N. Mustafa, "E-commerce Platform Based on Machine Learning Recommendation System," In *Proceedings of the 2021 6th International Multi-Topic ICT Conference (IMTIC)*, IEEE, Nov. 2021, pp. 1–4.

[6] F. Messaoudi, M. Loukili, and M. El Ghazi, "Demand Prediction Using Sequential Deep Learning Model," In *Proceedings of the 2023 International Conference on Information Technology (ICIT)*, IEEE, Aug. 2023, pp. 577–582.

[7] M. Loukili, F. Messaoudi, and M. El Ghazi, "Supervised Learning Algorithms for Predicting Customer Churn with Hyperparameter Optimization," *International Journal of Advances in Soft Computing & Its Applications*, vol. 14, no. 3, 4, 2022.

[8] M. Loukili, F. Messaoudi, and M. El Ghazi, "Sentiment Analysis of Product Reviews for E-Commerce Recommendation Based on Machine Learning," *International Journal of Advances in Soft Computing & Its Applications*, vol. 15, no. 1, 342, 2023.

[9] M. Loukili, F. Messaoudi, and M. El Ghazi, "Machine Learning Based Recommender System for E-Commerce," *IAES International Journal of Artificial Intelligence*, vol. 12, no. 4, pp. 1803–1811, 2023.

[10] F. Messaoudi, and M. Loukili, "E-commerce Personalized Recommendations: A Deep Neural Collaborative Filtering Approach," *Operations Research Forum, Springer International Publishing*, Vol. 5, No. 1, pp. 1–25, Mar. 2024.

[11] S. A. S. Alquhtani, and A. Muniasamy, "Analytics in Support of E-Commerce Systems Using Machine Learning," IEEE, 2022. https://doi.org/10.1109/ICECET55527.2022.9872592

[12] H. Khatter, S. Arif, U. Singh, S. Mathur, and S. Jain, "Product Recommendation System for E-Commerce using Collaborative Filtering and Textual Clustering," IEEE, 2021. https://doi.org/10.1109/ICIRCA51532.2021.9544753

[13] D. Naik, V. Tipugade, P. Kulkarni, A. Shinde, and A. Pawar, "A Scientometric Analysis to Study Research Trends in Music Recommendation System using Collaborative Filtering and Semantic Analysis," IEEE, 2023. https://doi.org/10.1109/OTCON56053.2023.1011 3940

[14] J. You, "E-commerce Recommendation Algorithm Based on Big Data Analysis and Genetic Fuzzy Clustering," IEEE, 2023. https://doi.org/10.23977/ferm.2023.060904

[15] P. Rao, S. Varakumari, B. Vineetha, and V. Satish, "Application of Data Mining to E-Commerce Recommendation Systems," *International Journal of Engineering & Technology*, 2018. https://doi.org/10.14419/IJET.V7I2.32.15730

[16] L. Wang, and Y. Jiang, "Collocating Recommendation Method for E-Commerce Based on Fuzzy C-Means Clustering Algorithm," *Journal of Mathematics*, 2022. https://doi.org/10.1155/2022/7414419

[17] L. Gao, and J. Li, "E-Commerce Personalized Recommendation Model Based on Semantic Sentiment," *Mobile Information Systems*, 2022. https://doi.org/10.1155/2022/7246802

[18] Y. Liu, and Z. Ding, "Personalized Recommendation Model of Electronic Commerce in New Media Era Based on Semantic Emotion Analysis," *Frontiers in Psychology*, 2022. https://doi.org/10.3389/fpsyg.2022.952622

[19] A. Wasilewski, "Clusterization Methods for Multi-Variant E-Commerce Interfaces," In *Federated Conference on Computer Science and Information Systems*, 2023. https://doi.org/10.15439/2023F1377

[20] D. Thakral, and S. Ranjan, "Semantic Web based Recommendation System in E-Commerce Websites," IEEE, 2022. https://doi.org/10.1109/SSTEPS57475.2022.00062

[21] W. Li, Y. Cai, M. H. Hanafiah, and Z. Liao, "An Empirical Study on Personalized Product Recommendation Based on Cross-Border E-Commerce Customer Data Analysis," *Journal of Electronic Commerce in Organizations*, 2024. https://doi.org/10.4018/joeuc.335498

[22] W. Li, "Intelligent Recommendation System Based on the Infusion Algorithms with Deep Learning, Attention Network and Clustering," *Journal of Big Data*, 2023. https://doi.org/10.1007/s44196-023-00264-z

[23] B. Li, and L. Han, "Distance Weighted Cosine Similarity Measure for Text Classification," In *Intelligent Data Engineering and Automated Learning–IDEAL 2013: 14th International Conference*, IDEAL 2013, Hefei, China, October 20–23, 2013. Proceedings 14, Springer Berlin Heidelberg, 2013, pp. 611–618.

[24] F. Anowar, S. Sadaoui, and B. Selim, "Conceptual and Empirical Comparison of Dimensionality Reduction Algorithms (pca, kpca, lda, mds, svd, lle, isomap, le, ica, t-sne)," *Computer Science Review*, vol. 40, Article 100378, 2021.

[25] K. R. Shahapure, and C. Nicholas, "Cluster Quality Analysis Using Silhouette Score," In *2020 IEEE 7th International Conference on Data Science and Advanced Analytics (DSAA), IEEE*, Oct. 2020, pp. 747–748.

[26] S. Łukasik, P. A. Kowalski, M. Charytanowicz, and P. Kulczycki, "Clustering Using Flower Pollination Algorithm and Calinski-Harabasz Index," In *2016 IEEE Congress on Evolutionary Computation (CEC)*, IEEE, Jul. 2016, pp. 2724–2728.

[27] C. Rodero, M. Strocchi, M. Marciniak, S. Longobardi, J. Whitaker, M. D. O'Neill, and S. A. Niederer, "Linking Statistical Shape Models and Simulated Function in the Healthy Adult Human Heart," *PLoS Computational Biology*, vol. 17, no. 4, Article e1008851, 2021.

How Multimodal AI
and IoT Are Shaping the
Future of Intelligence

Sarafudheen M. Tharayil
Saudi Aramco, Dhahran, Saudi Arabia

M. A. Krishnapriya
Saveetha University, Chennai, India

Nada K. Alomari
Saudi Aramco, Dhahran, Saudi Arabia

8.1 INTRODUCTION

Multimodal AI is a branch of artificial intelligence (AI) that can process and generate natural language across different modalities, such as text, speech, images, videos, and audio. IoT is a network of physical objects embedded with sensors, software, and communication technologies that can collect and exchange data. Together, multimodal AI and IoT can create smart city scenarios that enhance the quality of life, sustainability, and productivity of urban dwellers.

Multimodal AI is a branch of AI that can process and generate natural language across different modalities, such as text, speech, images, videos, and audio. IoT is a network of physical objects embedded with sensors, software, and communication technologies that can collect and exchange data. Together, multimodal AI and IoT can create smart city scenarios that enhance the quality of life, sustainability, and productivity of urban dwellers [1].

8.1.1 CONTRIBUTIONS OF THE CHAPTER

This chapter provides a comprehensive understanding of the current landscape and future potential of multimodal AI and the Internet of Things (IoT), equipping readers with the knowledge to contribute to this rapidly evolving field. The chapter offers an in-depth understanding of the role of large language models (LLMs) in multimodal AI and IoT systems. It elucidates the concept of embedding in AI, which is a crucial aspect of machine learning and AI. The chapter also emphasizes the importance of promoting engineering principles in the development of these technologies,

DOI: 10.1201/9781032656830-8

underscoring the need for a solid foundation in engineering principles for anyone working in this field.

Furthermore, the chapter discusses the process of creating live applications using LLMs and generative AI. This practical aspect of the chapter provides readers with hands-on experience and insights into the real-world applications of these technologies. The chapter also explores the usage of multimodal systems in various domains, providing a broad perspective on the applicability and versatility of these technologies. It highlights the challenges associated with the implementation of these technologies, such as issues of bias, privacy, security, and misinformation. This critical examination of the challenges ensures that readers are well-equipped to navigate the complexities of implementing these technologies [2].

8.1.2 CHAPTER ORGANIZATION

This chapter begins with an introduction to LLM, followed by an in-depth exploration of different sections such as:

a. The architectures for LLMs, multimodal AI, and attention mechanisms.
b. The state-of-the-art methods and architectures for IoT.
c. The integration and co-learning of multimodal AI and IoT in smart city applications, such as education, health, entertainment, and social media.
d. The evaluation and benchmarking of multimodal AI and IoT systems, including the existing datasets, tasks, metrics, and limitations.
e. The ethical and social implications of multimodal AI and IoT systems, such as the risks of bias, privacy, security, and misinformation, as well as the opportunities for enhancing accessibility, diversity, and inclusion.
f. The future directions and open challenges for multimodal AI and IoT research and development, such as improving the generalization and robustness of the systems, incorporating common sense and world knowledge, enabling interactive and adaptive learning, and developing explainable and trustworthy AI systems.

8.1.3 INTRODUCTION TO LARGE LANGUAGE MODELS

An LLM is a type of AI system that can process natural language and perform various tasks such as understanding, generating, translating, summarizing, and answering questions. LLMs are trained on huge amounts of text data, mostly collected from the Internet, and learn the patterns and meanings of human language. LLMs can have billions or even trillions of parameters, which are like the memory units of the system. The more parameters an LLM has, the more knowledge and skills it can acquire.

One of the key technologies behind LLMs is the transformer model, which is a neural network architecture that can process sequential data, such as text or speech, in parallel. Transformer models use a mechanism called attention, which allows them to focus on the most relevant parts of the input and output sequences. Attention also enables transformer models to capture long-range dependencies and context information, which are essential for natural language understanding and generation.

Another important component of LLMs is the embedding layer, which converts the input text into numerical vectors that represent the semantic and syntactic features of the words and sentences. Embedding vectors are learned during the training process and can capture various aspects of language, such as word meanings, synonyms, antonyms, word senses, and sentiment. Embedding vectors also allow LLMs to perform arithmetic operations on words and sentences, such as adding, subtracting, multiplying, or averaging them.

LLMs are usually pre-trained on a large corpus of text using self-supervised learning, which means they learn from their own data without any human labels or feedback. Pre-training allows LLMs to acquire general language skills and knowledge that can be transferred to different domains and tasks. After pre-training, LLMs can be fine-tuned or adapted to specific tasks using supervised learning or semi-supervised learning, which means they learn from a smaller amount of labeled data or a combination of labeled and unlabeled data. Fine-tuning allows LLMs to specialize in certain domains and tasks and improve their performance[3-10].

Some examples of LLMs are GPT-3, BERT, XLNet, T5, and DALL-E. These models have achieved state-of-the-art results on various natural language processing (NLP) benchmarks and applications, such as text classification, question answering, document summarization, text generation, image creation, and captioning. However, LLMs also face some challenges and limitations, such as ethical issues, social biases, environmental costs, data quality, model robustness, explainability, and generalization. Therefore, LLMs require careful design, evaluation, and deployment to ensure their benefits outweigh their risks.

8.1.4   HOW DO LARGE LANGUAGE MODELS WORK?

LLMs are AI systems that can process, understand, and generate natural language texts. They work by using deep neural networks, which are mathematical models that mimic the structure and function of the human brain, to learn from massive amounts of text data. LLMs can perform various NLP tasks, such as translation, summarization, question answering, and text generation.

To understand how LLMs work, let us look at some of the key components and concepts involved:

a. **Transformer models**: These are the most common architecture of LLMs. They consist of two parts: an encoder and a decoder. The encoder takes an input text and converts it into a sequence of numerical vectors, called embeddings, that capture the meaning and context of each word. The decoder takes these embeddings and generates an output text, word by word, based on the probabilities of each word given the previous words and the input. Transformer models use a technique called self-attention, which allows them to focus on the most relevant parts of the input and output texts and learn the relationships between them [10-15].

b. **Parameters**: These are the numbers that determine how the neural network processes the input and output texts. They are also called weights or coefficients. Parameters are learned during the training process, where the LLM is given a large dataset of text pairs (such as English–French translations or questions–answers) and tries to minimize the error between its predictions and the correct outputs. The more parameters an LLM has, the more complex and powerful it can be, but also the more data and computational resources it needs to train.

c. **Pre-training and fine-tuning**: These are two stages of training a LLM. In pre-training, LLM is trained on a general-purpose dataset, such as Wikipedia articles or web pages, to learn the basic patterns and structures of natural language. In Fine-tuning, LLM is trained on a specific dataset, such as movie reviews or medical records, to adapt to a specific domain or task. Pre-training and fine-tuning allow LLMs to leverage their existing knowledge and transfer it to new situations.

d. **Decoding strategies and sampling methods**: These are the techniques that LLMs use to generate output texts from their embeddings. Decoding strategies are the algorithms that select the most likely words or tokens to continue the text from the LLM's output probabilities. Sampling methods are the ways of introducing randomness and diversity into the text-generation process by choosing words or tokens based on their probabilities, rather than the highest ones. Some examples of decoding strategies are greedy sampling, beam search, top-k sampling, and nucleus sampling.

8.1.5 LLM AND GENERATIVE AI

Generative AI is a type of AI that can create new content by utilizing existing text, audio files, or images. With generative AI, computers detect the underlying pattern related to the input and produce similar content. For example, generative AI can generate realistic images of faces, animals, or landscapes from text descriptions. Generative AI can also create text, such as stories, poems, summaries, or code.

LLMs are a key technique used in generative AI, especially for text generation. LLMs are computer algorithms that process natural language inputs and predict the next word based on what they have seen before. LLMs are trained on massive amounts of text data, such as articles, books, or web pages. LLMs can learn the structure and style of different languages and domains and generate coherent and fluent text in response to queries or prompts.

LLMs and generative AI are closely related, but not the same thing. LLMs are a specific type of generative model that focuses on text processing and production. Generative AI is a broader category that encompasses other types of content generation, such as images, audio, or video. To fully utilize AI in various applications, it is essential to comprehend their distinctions and potential synergies.

8.1.6 WORD EMBEDDING AND LLM

Embeddings are the representations or encodings of tokens, such as words, sentences, paragraphs, or documents, in a high-dimensional vector space, where each dimension corresponds to a learned feature or attribute of the language. Embeddings are the way that the LLM captures and stores the meaning and the relationships of

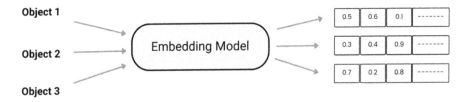

FIGURE 8.1   Representing text objects as embedding.

the language, and the way that the LLM compares and contrasts different tokens or units of language. Embeddings are derived from the parameters or the weights of the model and are used to encode and decode the input and output texts (Figure 8.1).

The number of parameters in an LLM is a measure of its complexity and capacity to learn from data. Generally, the more parameters an LLM has, the better it can perform on various language tasks. The number of parameters also affects the size and the cost of the model. Larger models require more computational resources and more data to train and run, and they are usually more expensive to use. For example, GPT-4 has about 175 billion parameters, while GPT-3.5-Turbo has about 3.5 billion parameters. The cost ratio of GPT-4 to GPT-3.5-Turbo is about 50:1.

Therefore, choosing an LLM model depends on the trade-off between performance, size, and cost. Different models may have different advantages and disadvantages for different tasks and domains. It is important to evaluate the quality and the efficiency of the model before using it for a specific purpose.

8.2 ARCHITECTURE OF LLM

The architecture of LLM models is based on the transformer model, which is a type of neural network that can process natural language inputs and outputs. The transformer model consists of two main components: an encoder and a decoder. The encoder processes the input text and converts it into a sequence of vectors, called embeddings, that represent the meaning and context of each word. The decoder generates the output text by predicting the next word based on the embeddings and the previous words [15-20].

8.2.1 Transformers in Generative AI

Transformers are a type of deep learning model that is commonly used in generative AI, especially for NLP and text generation. Transformers can process an entire sequence of data, such as a sentence, paragraph, or an entire article, at once, rather than one word or token at a time. This allows the model to capture the context and the meaning of the data more effectively.

Transformers were introduced in a paper by Google researchers in 2017. Since then, they have become a key building block for many state-of-the-art NLP models, such as BERT, GPT-3, and T5. These models are also called LLMs or foundation models, because they can perform a variety of NLP tasks, such as translation, summarization, question answering, and text generation.

Transformers are driving a wave of advances in generative AI, because they can create new and plausible content from existing data. For example, transformers can generate realistic images of faces, animals, or landscapes from text descriptions. They can also create text, such as stories, poems, summaries, or code. Transformers can also combine different types of data, such as text and images, to generate multimodal content.

Transformers are one of the most powerful and versatile classes of models invented to date. They have the potential to transform many domains and applications that use sequential data. However, they also pose some challenges and risks, such as ethical issues, data quality issues, computational costs, and social impacts. Therefore, it is important to understand how transformers work and what they can do before using them for a specific purpose.

8.2.2 ATTENTION IS ALL WHAT WE NEED!

Attention is a mechanism that allows a neural network to focus on the most relevant parts of the input or output sequence, depending on the task. It can improve the performance and accuracy of sequence-to-sequence models, such as machine translation, speech recognition, and text summarization.

One of the most popular attention models is the transformer, which consists of an encoder and a decoder, each composed of multiple layers. Each layer has two sublayers: a multi-head self-attention layer and a feed-forward layer. The self-attention layer computes a weighted sum of the input vectors, where the weights are learned by comparing the queries, keys, and values of the input. The feed-forward layer applies a linear transformation followed by a non-linear activation function to the output of the self-attention layer. Attention allows the model to focus on the most relevant parts of the input and output sequences. Some features of Attention are:

1. Attention also enables the model to learn long-range dependencies and complex relationships between words.
2. The transformer model can have multiple layers of attention, each with different functions and parameters.
3. LLMs are trained on massive amounts of text data, such as articles, books, or web pages.
4. They can learn the structure and style of different languages and domains and generate coherent and fluent text in response to queries or prompts.
5. LLMs can also perform a variety of NLP tasks, such as translation, summarization, question answering, and text generation.

As shown in Figure 8.2, the encoder and decoder layers are connected by another attention layer, called the encoder–decoder attention layer, which allows the decoder to attend to the encoder output. The final layer of the decoder is a classifier layer, which predicts the next token in the output sequence, based on the decoder output and the target vocabulary. In this illustration, the transformer uses six consecutive attention layers in both the encoder and the decoder, which allows the model to capture long-range dependencies and complex patterns in the input and output

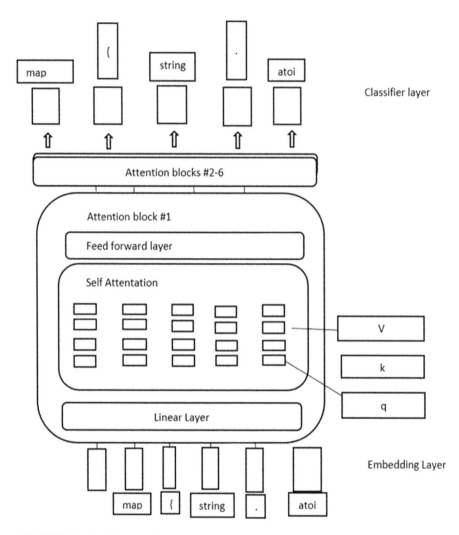

FIGURE 8.2   Architecture of attention.

sequences. The mapping between the input and output sequences is learned by the attention weights, which are computed by the queries, keys, and values.

To use special characters in a string key map, such as : ; / , ., the key should be of type string, and the map should use a hash function that can handle any character. One possible way to implement this is to use the ASCII code of each character as the hash value and use a modulo operation to map it to a bucket.

To convert a string to an integer, one possible way is to use the atoi function, which takes a string as an argument and returns an integer value. The atoi function discards any whitespace characters until the first non-whitespace character is found, then takes as many characters as possible to form a valid integer number representation and converts them to an integer value. The valid integer number representation consists of an optional plus or minus sign, followed by numeric digits.

An example of attention in transfer learning is the work by Zagoruyko and Komodakis, who proposed a method called attention transfer (AT) for improving the accuracy of small convolutional neural networks (CNNs) for image classification. They used a large and deep CNN as the teacher network, and a small and shallow CNN as the student network. They computed the spatial attention maps of both networks using different functions, such as sum, max, or l2-norm, applied to the feature maps of each layer. They then added an extra loss term to the student network's objective function, which measured the difference between the attention maps of the teacher and the student networks. By minimizing this loss term, the student network learned to mimic the attention patterns of the teacher network, and thus improved its performance on the target task. The authors showed that AT can achieve better results than other transfer learning methods, such as knowledge distillation or fitnets, on several image classification datasets.

8.2.3 EMBEDDING LAYER

An embedding layer in LLM is a part of the model that transforms tokens, such as words, sentences, paragraphs, or documents, into vectors or arrays of numbers that represent the meaning and the context of the tokens. The embedding layer is usually the first layer of the model, and it maps each token to a corresponding vector in a high-dimensional vector space. The embedding layer learns the embeddings from the parameters or the weights of the model during training, and it can also use pre-trained embeddings from other sources.

The embedding layer works with other neural network layers to process and generate natural language. For example, in a transformer model, which is a common architecture for LLMs, the embedding layer is followed by a positional encoder layer, which adds information about the position of each token in the sequence. Then, the model has many decoder layers, which use self-attention and feed-forward sublayers to turn input vectors into prediction vectors. Finally, the model has an output unembedding layer, which turns prediction vectors into token probabilities.

8.2.4 ARCHITECTURE OF GPT3 AND GPT4

GPT-3 and GPT-4 are two LLMs developed by OpenAI that can perform various NLP tasks, such as text generation, summarization, and translation. The architecture of both models is based on the transformer model, which is a type of neural network that can process natural language inputs and outputs. The transformer model consists of two main components: an encoder and a decoder. The encoder processes the input text and converts it into a sequence of vectors, called embeddings, that represent the meaning and context of each word. The decoder generates the output text by predicting the next word based on the embeddings and the previous words.

The main difference between GPT-3 and GPT-4 lies in their model size and training data. GPT-4 has a much larger model size, which means it can handle more complex tasks and generate more accurate responses. GPT-4 has about 1 trillion parameters, while GPT-3 has about 175 billion parameters. Parameters are the weights or the numbers that the model learns during training, and they determine how the model processes and produces language. The more parameters a model has, the more information it can store and learn from.

Another difference between GPT-3 and GPT-4 is their training data. GPT-4 has a more extensive training dataset, which gives it a broader knowledge base and improved contextual understanding. GPT-4 uses about 45 GB of text data from various sources, such as books, articles, web pages, and social media posts. GPT-3 uses about 17 GB of text data from similar sources. The quality and quantity of the training data affect how well the model can perform on different tasks and domains.

Some of the benefits of GPT-4 compared to GPT-3 are:

a. Better performance: GPT-4 outperforms GPT-3 on various NLP tasks, such as text generation, summarization, translation, and question answering. GPT-4 can generate more coherent, diverse, and relevant texts than GPT-3.
b. Longer texts: GPT-4 can handle longer texts than GPT-3, both as inputs and outputs. GPT-4 can process up to 10,000 tokens per input, while GPT-3 can process up to 2,048 tokens per input. Tokens are the units of language that the model uses, such as words or characters. Similarly, GPT-4 can generate up to 10,000 tokens per output, while GPT-3 can generate up to 1,024 tokens per output.
c. Fewer errors: GPT-4 makes fewer errors than GPT-3 in terms of logic, grammar, spelling, and facts. GPT-4 has a better understanding of the world and can check its outputs against its knowledge base. GPT-3 often makes mistakes or contradictions that affect its credibility and reliability.

8.3   EXAMPLES OF EMBEDDING ON DIFFERENT TYPES OF DATA

An example of embedding vectors and how they can be used for text, image, and video is as follows:

a. **Text**: Suppose we have a sentence "I love cats" and we want to embed it using a transformer model. First, we tokenize the sentence into three tokens: "I," "love," and "cats." Then, we use the embedding layer to map each token to a vector of size 768 (assuming we use GPT-3). For example, the vector for "I" might look like this:

   [0.12, −0.34, 0.56, …, −0.78]
   The vector for "love" might look like this:
   [−0.45, 0.67, −0.89, …, 0.23]
   And the vector for "cats" might look like this:
   [0.34, −0.56, 0.78, …, −0.12]

   These vectors capture the meaning and the context of each token in the sentence.
b. **Image**: Suppose we have an image of a cat and we want to embed it using a transformer model. First, we convert the image into a sequence of tokens using a technique called vision transformer (ViT). ViT divides the image into patches of pixels and embeds each patch as a token using a linear projection layer. Then, it adds a special token called [CLS] at the beginning of the sequence to represent the whole image. For example, if we use a patch size of

16×16 pixels and an image size of 224×224 pixels, we will have 196 patches plus one [CLS] token, resulting in a sequence of 197 tokens. Then, we use the embedding layer to map each token to a vector of size 768 (assuming we use GPT-3). For example, the vector for [CLS] might look like this:

[0.89, −0.23, 0.45, …, −0.67]
The vector for the first patch might look like this:
[−0.56, 0.78, −0.34, …, 0.12]
And so on for the rest of the patches.

These vectors capture the features and the context of each patch in the image.

c. **Video**: Suppose we have a video of a cat playing with a ball and we want to embed it using a transformer model. First, we convert the video into a sequence of tokens using a technique called video transformer (ViT). ViT divides each frame of the video into patches of pixels and embeds each patch as a token using a linear projection layer. Then, it adds a special token called [CLS] at the beginning of each frame to represent the whole frame. For example, if we use a patch size of 16×16 pixels and a frame size of 224×224 pixels, we will have 196 patches plus one [CLS] token per frame. If we have 30 frames per second and 10 seconds of video, we will have 300 frames and 59,100 tokens in total. Then, we use the embedding layer to map each token to a vector of size 768 (assuming we use GPT-3). For example, the vector for [CLS] in the first frame might look like this:

[0.67, −0.45, 0.23, …, −0.89]
The vector for [CLS] in the second frame might look like this:
[0.78, −0.34, 0.56, …, −0.12]
And so on for the rest of the frames.

These vectors capture the features and the context of each patch in each frame in the video.

In summary, embedding vectors can be used for various tasks involving text, image, and video data, such as classification, detection, segmentation, generation, retrieval, captioning, and synthesis.

8.3.1 Embedding on Video and Images in Real Life

Embedding can be done in different ways, depending on the type and the purpose of the data. Here are some examples of how embedding can be done in different datasets:

a. For images and videos, embedding can be done using CNNs or transformers, which are deep learning models that can process and extract features from visual data. CNNs use filters to perform convolutions on the input data and generate feature maps. Transformers use attention mechanisms to focus on the most relevant parts of the input data and generate embeddings. These models can be trained on large-scale datasets, such as ImageNet or Kinetics, to learn generalizable and robust embeddings for various tasks, such as classification, detection, segmentation, or generation.

b. For text and audio, embedding can be done using recurrent neural networks (RNNs) or transformers, which are deep learning models that can process and extract features from sequential data. RNNs use hidden states to store information from previous inputs and generate outputs. Transformers use attention mechanisms to capture the dependencies and relationships between inputs and outputs. These models can be trained on large-scale datasets, such as Wikipedia or LibriSpeech, to learn meaningful and fluent embeddings for various tasks, such as translation, summarization, speech recognition, or text-to-speech.

c. For depth, thermal, and IMU data, embedding can be done using specialized neural networks that can handle different types of sensors and measurements. For example, depth data can be embedded using depth-aware CNNs that can incorporate geometric information into the feature extraction process. Thermal data can be embedded using thermal-aware CNNs that can adapt to different temperature ranges and conditions. IMU data can be embedded using IMU-aware RNNs that can model the temporal dynamics and correlations of the sensor readings.

Embedding is a powerful technique that can enable cross-modal learning and understanding. By learning a joint embedding across different modalities, such as images, text, audio, depth, thermal, and IMU data, we can achieve tasks such as image captioning, video retrieval, audio–visual synthesis, and multimodal fusion.

8.4 PROMOTING IN LLM

Promoting in LLM is the technique of using natural language prompts to guide the output generation of an LLM, such as GPT-3 or ChatGPT. Prompts are instructions or queries that help the model produce specific responses, such as Q&A, document summarization, translation, and text generation. Promoting in LLM can improve the performance and efficiency of the model, as well as enable new applications and use cases.

Promoting in LLM is not a trivial task, as it requires careful design and engineering of the prompts to elicit the desired responses from the model. Some various strategies and methods can help with prompting, such as clarity, specificity, context, examples, evidence, and feedback. One of the methods that has been explored in recent research is the Socratic method, which is a form of dialogue and reasoning that uses questions to promote critical thinking and explore complex concepts. The Socratic method can be applied to prompting LLMs by using different techniques, such as definition, elenchus, dialectic, maieutics, generalization, and counterfactual reasoning.

8.4.1 Prompt Engineering in LLM

Prompt engineering is the practice of designing and optimizing prompts to guide AI, specifically LLMs, to produce the desired output. An effective prompt can help a model understand a task more efficiently and provide high-quality results.

A prompt is a natural language input or query that users can provide to an LLM to elicit a specific response. For example, if you want to generate a summary of an article, you can use a prompt like "Please summarize the following article in three sentences": followed by the article text. The LLM will then try to generate a concise summary based on the prompt.

Prompt engineering involves selecting the right words, phrases, symbols, and formats that influence the LLM's behavior and output. Prompt engineering requires creativity and attention to detail, as well as understanding the capabilities and limitations of LLMs. Different prompts can produce very different outputs, even for the same task or domain. For example, if you want to generate a poem about love, you can use different prompts like "Write a sonnet about love," "Write a haiku about love," or "Write a limerick about love." The LLM will then try to generate poems that match the style and structure of the prompts.

Prompt engineering is an important skill for anyone working with LLMs, as it can improve the performance and efficiency of the models, as well as enable new applications and use cases. Prompt engineering is also an emerging field that has many challenges and opportunities for research and development. Some of the topics that prompt engineering covers are:

1. How to design prompts that are clear, specific, and relevant for the task and domain.
2. How to evaluate the quality and accuracy of the outputs generated by different prompts.
3. How to optimize prompts using data, feedback, or parameters.
4. How to automate or simplify prompt engineering using tools or frameworks.
5. How to leverage existing prompts or templates from other sources or domains.
6. How to handle ethical, social, or legal issues related to prompt engineering.

8.5   TUNING AND FINE-TUNING IN LLM

Tuning and fine-tuning are two techniques that can improve the performance and accuracy of LLMs for specific tasks and domains. Tuning and fine-tuning involve adjusting the parameters or the weights of a pre-trained LLM based on a new dataset or a feedback signal. Tuning and fine-tuning can help LLMs adapt to new data and generate more relevant and precise outputs.

There are different types of tuning and fine-tuning methods, such as:

a. Supervised fine-tuning: This method uses a labeled dataset to train the LLM on a new task, such as classification, summarization, or translation. The LLM learns to predict the correct labels or outputs for the given inputs, and updates its parameters accordingly. Supervised fine-tuning is effective when there is enough labeled data available for the target task and domain.
b. Reinforcement learning from human feedback (RLHF): This method uses human feedback as a reward signal to train the LLM on a new task, such as text generation, dialogue, or creativity. The LLM learns to generate outputs that maximize the human feedback, and updates its parameters accordingly.

RLHF is effective when there is no labeled data available, or when the task is subjective or open-ended.

c. Parameter-efficient fine-tuning (PEFT): This method uses a small subset of parameters to train the LLM on a new task, while keeping the rest of the parameters fixed. The LLM learns to adapt to the new task using fewer resources and less data, and updates its parameters accordingly. PEFT is effective when there is limited computational power or data available, or when the task is similar to the pre-training task.

Tuning and fine-tuning are important steps in leveraging the potential of LLMs for various applications and use cases. By tuning and fine-tuning LLMs, we can customize them to our specific needs and requirements, and achieve better results with less effort.

8.6   OpenSource CULTURE AND HuggingFace

HuggingFace is a company that provides open-source tools and models for NLP and LLMs. HuggingFace has a platform called the HuggingFace Hub, where users can access, share, and collaborate on thousands of pre-trained models for various NLP tasks, such as text generation, summarization, and translation.

Some of the LLM models that are available on the HuggingFace Hub are:

a. GPT-2 & GPT-3: These are transformer-based models that can generate coherent and fluent text in response to natural language prompts. They are trained on massive amounts of text data from various sources, such as books, articles, and web pages.

b. BERT: This is a transformer-based model that can perform various NLP tasks, such as classification, question answering, and sentiment analysis. It is trained on large-scale corpora of text data, such as Wikipedia and BookCorpus.

c. Bloom: This is a transformer-based model that can generate creative and diverse text in response to natural language prompts. It is trained on a curated dataset of text data from various domains, such as fiction, poetry, lyrics, and jokes.

To use these LLM models on the HuggingFace Hub, users can either:

a. Use the online demo apps (Spaces) that showcase the capabilities and features of the models. Users can interact with the models through a web interface and provide their own inputs and outputs.

b. Use the HuggingFace transformers library that provides easy-to-use APIs and tools to load, run, and fine-tune the models. Users can integrate the models into their own applications and scripts using Python or other programming languages.

c. Use the HuggingFace registry in Azure Machine Learning to deploy the models in the cloud. Users can access the models through a REST API endpoint and scale them according to their needs.

8.7 DEEP DIVE INTO HuggingFace MODEL T5

T5 is a model published in HuggingFace, and its derivatives are LLMs that use a text-to-text approach for various NLP tasks. They are based on the transformer architecture, which is a type of neural network that can process natural language inputs and outputs using attention mechanisms. The transformer architecture consists of two main components: an encoder and a decoder. The encoder processes the input text and converts it into a sequence of vectors, called embeddings, that represent the meaning and context of each word. The decoder generates the output text by predicting the next word based on the embeddings and the previous words.

T5 stands for Text-to-Text Transfer Transformer, and it was introduced by Raffel et al. in 2019. T5 uses a text-to-text framework, which means that every task, such as translation, summarization, question answering, or classification, is cast as feeding the model text as input and training it to generate some target text as output. This allows for the use of the same model, loss function, hyperparameters, etc. across diverse tasks. T5 is trained on a large dataset of web-extracted text, called C4, using a denoising objective, which involves masking some words in the input and asking the model to predict them. T5 has different variants with different sizes, ranging from 60 million to 11 billion parameters.

8.7.1 DERIVATIVES OF T5

Some of the derivatives of T5 are:

a. Bloom: This is a model that can generate creative and diverse text in response to natural language prompts. It is trained on a curated dataset of text data from various domains, such as fiction, poetry, lyrics, and jokes.
b. MT5: This is a multilingual version of T5 that can handle 101 languages. It is trained on a large dataset of multilingual web-extracted text, called mC4, using the same denoising objective as T5.
c. ByT5: This is a byte-level version of T5 that can handle any Unicode character. It is trained on the same C4 dataset as T5, but using byte tokens instead of word tokens. It uses a modified version of the transformer architecture that can handle variable-length inputs and outputs.
d. T5 and its derivatives are powerful and versatile models that can perform various NLP tasks with high accuracy and efficiency. They are also easy to use and adapt to different domains and applications.

8.7.2 WORKING WITH HUGGING FACE AND T5

T5 is a text-to-text transfer transformer model that can perform various NLP tasks by converting any input text into a desired output text. For example, T5 can be used for summarization, translation, question answering, and text generation. T5 is based on the transformer architecture, which uses attention mechanisms to encode and decode sequences of tokens. T5 is pre-trained on a large corpus of text using a mix of self-supervised objectives, such as masked language modeling, span corruption, and noise span prediction.

To re-train the model using T5, you need to follow these steps:

a. Prepare your data as a pandas dataframe with two columns: source_text and target_text. The source_text column should contain the input text for the task, and the target_text column should contain the desired output text. You also need to add a task-specific prefix to the source_text column, such as "summarize:" or "translate English to French:".
b. Choose a pre-trained T5 model from the Transformers library, such as t5-base or t5-small. You can also use torchtext to download and instantiate a pre-trained T5 model with base configuration.
c. Fine-tune the model on your data using PyTorch or PyTorch lightning. You can use simpleT5, a library that simplifies the training process of T5 models in just three lines of code. You can also use other libraries or frameworks that support T5 models, such as Hugging Face or TensorFlow.
d. Evaluate the model on your test data or new inputs using greedy search or beam search to generate output sequences. You can use torchtext's GenerationUtils or other tools to produce output sequences based on the input sequences.

8.8   STEP-BY-STEP GUIDE ON DEVELOPING LLM APPLICATIONS USING T5

Here is a step-by-step guide with Python code on how to develop an LLM application using T5:

A. Install the required libraries, such as transformers, datasets, and torch. You can use pip or conda to install them. For example:

pip install transformers datasets torch

B. Import the libraries and modules that you will need, such as T5Tokenizer, T5ForConditionalGeneration, Trainer, TrainingArguments, and load_dataset. For example:

from transformers import T5Tokenizer,
    T5ForConditionalGeneration, Trainer,
    TrainingArguments
from datasets import load_dataset
import torch

C. Choose a pre-trained T5 model that suits your task and domain. You can use the HuggingFace Hub to browse and download different variants of T5 models, such as t5-small, t5-base, t5-large, or t5-3b. For example:

model_name = "t5-base"
tokenizer = T5Tokenizer.from_pretrained(model_name)
model = T5ForConditionalGeneration.from_pretrained(model_name)

D. Load and preprocess your training and validation data. You can use the datasets library to load various datasets for different tasks, such as summarization, translation, and question answering. You can also use your own custom data in a csv or json format. You need to preprocess your data into a text-to-text format, where the input and output are both natural language texts. You also need to tokenize your data using the tokenizer that matches your model. For example:

```
# Load a summarization dataset
dataset = load_dataset("cnn_dailymail", "3.0.0")

# Define a preprocessing function
def preprocess(example):
    # Concatenate the article and the summary with a separator token
    input_text = example["article"] + " </s> " + example["highlights"]
    # Add a prefix to indicate the task
    input_text = "summarize: " + input_text
    # Tokenize the input text
    input_ids = tokenizer(input_text, return_tensors="pt", padding="max_
        length", truncation=True, max_length=512).input_ids
    # Return the input ids and the labels (which are the same as the input ids for T5)
    return {"input_ids": input_ids, "labels": input_ids}

# Apply the preprocessing function to the dataset
dataset = dataset.map(preprocess)
```

E. Define your training arguments and hyperparameters. You can use the TrainingArguments class to specify various options for training your model, such as the output directory, the number of epochs, the batch size, and the learning rate. For example:

```
# Define the training arguments
training_args = TrainingArguments(
    output_dir="./output",
    num_train_epochs=3,
    per_device_train_batch_size=8,
    per_device_eval_batch_size=8,
    evaluation_strategy="epoch",
    learning_rate=2e-4,
    weight_decay=0.01,
    logging_dir="./logs",
)
```

F. Train and evaluate your model using the Trainer class. You need to pass your model, your training arguments, your training and validation data, and optionally a compute_metrics function that calculates some evaluation metrics for your task. For example:

```
# Define a compute metrics function for summarization
def compute_metrics(eval_pred):
```

```
# Get the predictions and labels from the output
predictions = eval_pred.predictions
labels = eval_pred.label_ids
# Decode the predictions and labels into texts
predictions = tokenizer.batch_decode(predictions,
   skip_special_tokens=True)
labels = tokenizer.batch_decode(labels, skip_special_tokens=True)
# Calculate the ROUGE score using the datasets library
rouge = datasets.load_metric("rouge")
rouge_output = rouge.compute(predictions=predictions, references=labels)
# Return the rouge output as a dictionary of metrics
return {key: value.mid.fmeasure * 100 for key, value in rouge_output.
   items()}
# Create a trainer instance with the model, training arguments, data, and
   metrics function
trainer = Trainer(
   model=model,
   args=training_args,
   train_dataset=dataset["train"],
   eval_dataset=dataset["validation"],
   compute_metrics=compute_metrics,
)

# Train and evaluate the model
trainer.train()
trainer.evaluate()
```

G. Save and load your trained model using the save_pretrained and from_pre-trained methods. You can also upload your model to the HuggingFace Hub and share it with others. For example:

```
# Save your model locally
model.save_pretrained("./output")
```

```
# Load your model from a local directory or a hub url
model = T5ForConditionalGeneration.from_pretrained("./output")
# or
model = T5ForConditionalGeneration.from_pretrained("your_hub_url")
```

H. Use your model to generate outputs for new inputs. You can use the generate method to produce texts from your model and specify various options, such as the maximum length, the number of beams, and the temperature. You can also use the tokenizer to encode and decode your inputs and outputs. For example:

```
# Define a new input text
input_text =
"summarize: The Eiffel Tower is a wrought-iron lattice tower on the Champ
   de Mars in Paris, France. It is named after the engineer Gustave Eiffel,
```

whose company designed and built the tower. Constructed from 1887 to 1889 as the entrance to the 1889 World's Fair, it was initially criticised by some of France's leading artists and intellectuals for its design, but it has become a global cultural icon of France and one of the most recognisable structures in the world. The Eiffel Tower is the most-visited paid monument in the world; 6.91 million people ascended it in 2015."

```
# Tokenize and encode the input text
input_ids = tokenizer(input_text, return_tensors="pt").input_ids
```

```
# Generate an output text from the model
output_ids = model.generate(input_ids, max_length=64, num_beams=4)
```

```
# Decode and print the output text
output_text = tokenizer.decode(output_ids[0], skip_special_tokens=True)
print(output_text)
```

I. The output text might look something like this:

A summary of the article is:

- The Eiffel Tower is a wrought-iron tower in Paris, France, built by Gustave Eiffel for the 1889 World's Fair.
- It was initially criticized by some French artists and intellectuals, but it became a symbol of France and a popular tourist attraction.
- It is the most-visited paid monument in the world, with nearly 7 million visitors in 2015.

8.9 LATEST TRENDS IN LLM IN GOOGLE, OPENAI, AND MICROSOFT

Generative AI models from Google are NLP systems that can generate fluent and diverse text for various tasks and domains. Some of the generative AI models from Google are:

a. PaLM 2: An LLM that can perform tasks such as summarization, translation, question answering, text generation, code generation, and code chat. PaLM 2 is based on the transformer architecture and is pre-trained on a large corpus of text using a mix of self-supervised objectives.

b. Imagen: A multimodal model that can perform tasks such as image editing, image captioning, and visual question answering. Imagen is based on the vision transformer and the CLIP model and is pre-trained on a large corpus of images and text using contrastive learning.

c. Codey: A code generation model that can help Android developers be more productive by providing code snippets, fixing code errors, and answering questions about Android development. Codey is based on the transformer architecture and is pre-trained on a large corpus of code and documentation using masked language modeling.

d. Chirp: A speech synthesis model that can generate natural-sounding speech from text. Chirp is based on the WaveNet architecture and is trained on a large corpus of speech data using autoregressive modeling.

The sizing of these models can be used to determine the amount of compute resources and memory required to run them, as well as the quality and diversity of their outputs. Generally, larger models have more parameters, require more resources, and produce better results than smaller models. However, larger models may also be more prone to data quality issues such as bias, privacy, or plagiarism. Therefore, it is important to balance the trade-offs between size and performance when choosing a model for a specific task or domain.

Google Cloud offers two editions of generative AI models: community edition and enterprise edition. The community edition provides free access to some of the generative AI models from Google Research, such as Llama 2 and Claude 2, as well as open-source models and third-party models from Model Garden. The community edition is suitable for developers who want to experiment with generative AI and learn from the state-of-the-art models.

The enterprise edition provides paid access to more advanced and proprietary generative AI models from Google Research, such as PaLM 2, Imagen, Codey, and Chirp, as well as tools for customizing and applying them with generative AI studio. The enterprise edition is suitable for businesses and organizations who want to leverage generative AI for their use cases and applications.

The enterprise edition also provides data confidentiality and data privacy for its customers. According to Google's AI/ML Privacy Commitment, Google Cloud does not use customer data to train its foundation models without the express consent of its customers. Customer data is encrypted in transit and at rest, and customers have full control over where and how their data is used. Customers can also use customer-managed encryption keys (CMEK) to encrypt their stored data and delete their data at any time. Moreover, Google Cloud follows robust data governance practices to ensure that its teams are following its privacy commitments and complying with data security and privacy regulations.

Some of the latest trends in LLM in Google, OpenAI, and Microsoft are:

a. Developing multimodal LLMs that can handle both text and image inputs and outputs. For example, OpenAI's GPT-4 is the first multimodal model that can accept both texts and images as input. Google is also preparing to release a massive LLM called Gemini that can generate realistic images from text descriptions.
b. Improving the safety and reliability of LLMs by minimizing hallucinations, misinformation, and biases. For example, OpenAI has used reinforcement learning from human feedback (RLHF) and adversarial testing to align GPT-4 with human values3. Microsoft Bing has also used RLHF to train its LLM called Claude v1, which powers its conversational search feature.
c. Reducing the environmental impact and computational cost of LLMs by using more efficient hardware or software architectures. For example, Google

has used specialized hardware, such as TPUs, to speed up the training and inference of its LLMs2. Microsoft has also used parameter-efficient fine-tuning (PEFT) to adapt its LLMs to different tasks and domains using fewer resources.

8.10   LLMs CHALLENGES

LLMs are NLP systems that can generate fluent and diverse text for various tasks and domains. However, they also face many challenges due to the quality of the data they are trained on or use. Some of these challenges are:

a) Data bias: LLMs may learn and reproduce harmful stereotypes, prejudices, or misinformation from the data they are trained on. This can affect their fairness, ethics, and social impacts.

b) Data privacy: LLMs may leak sensitive or personal information from the data they are trained on or use. This can affect their compliance, security, and trustworthiness.

c) Data plagiarism: LLMs may copy or reuse content from the data they are trained on or use without proper attribution or permission. This can affect their originality, creativity, and legality.

To address these challenges, LLMs need to be carefully audited and monitored for data quality issues at different stages of their development and deployment. This can help improve their reliability and safety for various applications and use cases.

8.10.1   LLMs CHALLENGES DUE TO ETHICAL ISSUES AND SOCIAL BIASES

LLMS are LLMs that can process and generate natural language, but they also pose many challenges due to ethical issues and social biases. Some of these challenges are:

a. LLMs can perpetuate harmful stereotypes, unfair discrimination, exclusionary norms, and toxic language that are present in the data they are trained on. For example, LLMs can generate texts that are sexist, racist, homophobic, or xenophobic, or that favor certain groups over others.

b. LLMs can leak private or sensitive information that are contained in the data they are trained on. For example, LLMs can reveal personal names, addresses, phone numbers, email addresses, or credit card numbers that belong to real individuals.

c. LLMs can spread misinformation, false or misleading information that can harm individuals or society. For example, LLMs can generate texts that are inaccurate, incomplete, outdated, or contradictory, or that promote conspiracy theories, propaganda, or extremism.

d. LLMs can plagiarize or misuse copyrighted material that is contained in the data they are trained on. For example, LLMs can generate texts that copy or paraphrase existing works of literature, art, music, or code without proper attribution or permission.

These challenges raise ethical and social questions about the development and deployment of LLMs. For example:

a. How to ensure that LLMs are fair, inclusive, and respectful of human dignity and diversity?
b. How to protect the privacy and security of individuals and groups whose data are used to train LLMs?
c. How to verify the quality and reliability of the information generated by LLMs?
d. How to prevent the misuse or abuse of LLMs by malicious actors or unintended consequences?
e. How to balance the benefits and risks of using LLMs for various applications and use cases?

These questions require careful consideration and action from researchers, developers, users, and policymakers. By understanding and mitigating the ethical issues and social biases of LLMs, we can ensure that they are not only powerful and versatile but also responsible and trustworthy.

8.10.2   LLM CHALLENGES DUE TO ENVIRONMENTAL COSTS

LLMs can process and generate natural language, but they also have high environmental costs due to their energy consumption and carbon emissions. Training and running LLMs require a lot of computational resources and electricity, which can contribute to global warming and climate change. According to a study by Stanford University, the carbon dioxide emissions produced by some of the most popular LLMs in 2022 ranged from 25 to 502 metric tons, depending on the model size, the training data, and the energy source. For comparison, the average annual carbon footprint of an American citizen is about 16 metric tons. The study also found that the carbon emissions of LLMs are equivalent to various real-life examples, such as cars, air travel, or human life.

Some of the challenges that arise from the environmental costs of LLMs are:

a. How to reduce the energy consumption and environmental impact of LLMs without compromising their performance and quality.
b. How to measure and report the carbon emissions of LLMs in a transparent and standardized way.
c. How to balance the benefits and risks of using LLMs for various applications and use cases.
d. How to ensure that LLMs are aligned with ethical, social, and legal values and norms.

There are also some efforts and initiatives that aim to mitigate the environmental costs of LLMs and promote more sustainable AI practices. Some of these are:

a. Using renewable energy sources or low-carbon electricity grids to power LLMs.

b. Using more efficient hardware or software architectures to optimize LLMs.
c. Using smaller or fewer parameters to train LLMs.
d. Using data or feedback to fine-tune or adapt LLMs.
e. Using AI itself to monitor and control the energy usage and carbon emissions of LLMs.

If you want to learn more about the environmental costs and benefits of LLMs, you can check out these web search results:

a. The environmental impact of LLMs—Analytics India Magazine: this article explains the carbon footprint of different LLMs and how AI can be used to reduce energy consumption.
b. Environmental costs and benefits in life-cycle costing: this paper discusses the methodological issues and challenges in incorporating environmental costs and benefits in life-cycle costing.

8.10.3 LLM Challenges due to Data Quality

LLMs are powerful tools for NLP, but they also face many challenges due to data quality issues. Some of these challenges are:

a. Data bias: LLMs may learn and reproduce harmful stereotypes, prejudices, or misinformation from the data they are trained on. For example, LLMs may generate sexist, racist, or homophobic language that can offend or harm users or groups of people.
b. Data privacy: LLMs may leak sensitive or personal information from the data they are trained on. For example, LLMs may reveal names, addresses, phone numbers, or email addresses of individuals that are protected by privacy regulations.
c. Data plagiarism: LLMs may copy or reuse content from the data they are trained on without proper attribution or permission. For example, LLMs may generate text that is identical or similar to existing books, articles, songs, or other copyrighted material.

To address these challenges, LLMs need to be carefully audited and monitored for data quality issues. One possible approach is to use a three-layered framework that consists of:

a. Pre-training audit: This involves checking the quality and diversity of the data sources used to train the LLM, as well as applying techniques such as data filtering, data augmentation, or data anonymization to reduce bias, privacy, or plagiarism risks.
b. Fine-tuning audit: This involves checking the quality and relevance of the data used to adapt the LLM to a specific task or domain, as well as applying techniques such as data selection, data weighting, or data debiasing to improve performance and robustness.

c. Inference audit: This involves checking the quality and appropriateness of the outputs generated by the LLM for a given input or prompt, as well as applying techniques such as output filtering, output ranking, or output rewriting to enhance accuracy and safety.

By using this framework, LLMs can be more reliable and trustworthy for various applications and use cases.

8.10.4   LLM CHALLENGES DUE MODEL ROBUSTNESS, EXPLAINABILITY, AND GENERALIZATION

LLMs face some challenges due to their size, complexity, and data requirements. Some of the main challenges are:

a. Model robustness: LLMs are often trained on large and diverse corpora, which may contain noisy, inconsistent, or contradictory data. This can affect the quality and reliability of the model's outputs, especially when dealing with out-of-distribution or adversarial inputs. LLMs may also suffer from overfitting or underfitting, which can reduce their generalization ability. To improve the robustness of LLMs, some possible solutions are data cleaning, data augmentation, regularization, adversarial training, and model pruning.

b. Model explainability: LLMs are black-box models, which means that their internal workings are not transparent or interpretable. This can make it hard to understand how they produce their outputs, what kind of knowledge they have, and how they reason. This can also raise ethical and social issues, such as accountability, fairness, and trustworthiness. To make LLMs more explainable, some possible solutions are attention mechanisms, probing methods, attribution methods, and retrieval-augmented methods.

Model generalization: LLMs are designed to perform well on a wide range of natural language tasks, such as text generation, text classification, question answering, and summarization. However, LLMs may not be able to adapt well to specific domains or tasks that require specialized knowledge or skills. LLMs may also struggle with low-resource or zero-shot scenarios, where they have limited or no labeled data for the target task. To improve the generalization of LLMs, some possible solutions are pre-training, fine-tuning, transfer learning, meta-learning, and multi-task learning. These are some of the main challenges that LLMs face due to their model robustness, explainability, and generalization. However, these challenges also provide opportunities for further research and innovation in the field of LLMs.

8.11   EXAMPLE APPLICATION

Have you ever wanted to ask questions to a PDF document and get answers in natural language? For example, suppose you have a PDF file of a textbook or a research paper, and you want to quickly find some information without reading the whole document. How can you do that?

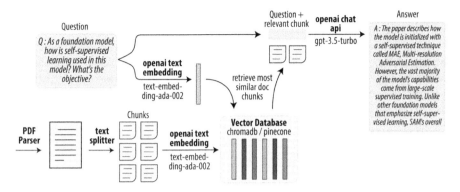

FIGURE 8.3   Overall workflow for a PDF-based LLM application.

(courtesy: harvard.edu.)

One possible solution is to use OpenAI, a powerful platform that provides access to state-of-the-art AI models. In this section, we will show you how to use OpenAI to build a question answering system that can handle PDF documents as input (Figure 8.3).

Here are the main steps for building such an application:

a. First, convert the PDF document into text format, while preserving the layout and structure of the document. You can use tools like PDFMiner or Tika for this task.

b. Next, split the text into smaller chunks, such as sentences or paragraphs, that can be used as input for an embedding model. You can use tools like SpaCy or NLTK for this task.

c. Then, use an embedding model to transform each text chunk into a vector, which is a numerical representation of the meaning and context of the text. You can use models like BERT or USE for this task.

d. After that, store the vectors in a vector database, which is a type of database that allows efficient and fast lookup of nearest neighbors in a high-dimensional space. You can use databases like Faiss or Azure AI Search for this task.

e. Finally, when a question is asked, use a retrieval-augmented generation (RAG) model, which is a type of generative AI model that combines an LLM-like ChatGPT with an information retrieval system. The RAG model will use the question as a prompt and query the vector database to find the most relevant text chunks. Then, it will use the retrieved text chunks as additional input to generate a natural language answer. You can use models like OpenAI RAG or Facebook RAG for this task.

RAG stands for retrieval-augmented generation, which is a method to enhance the output of language models by combining them with an external knowledge retrieval mechanism. RAG models can generate more accurate and informative answers by using relevant data from the vector database as additional input. RAG models can also avoid hallucinations or misleading information that may occur in pure generative models.

A vector database is a type of database that stores and retrieves vectors of data points, such as images, text, or audio. It allows developers to use vector search methods to find similar assets by encoding them into vectors and querying for nearby vectors. Vector databases are important for AI applications that require semantic understanding and similarity search, such as image recognition, NLP, and multimodal search.

8.11.1 Multimodal AI

Multimodal AI is a branch of AI that can process and generate natural language across different modalities, such as text, speech, images, videos, and audio. The high-level architecture of multimodal AI consists of several components, including data acquisition, feature extraction, multimodal fusion, and decision-making (Figure 8.4).

The data acquisition component collects data from various sources, such as cameras, microphones, and sensors. The feature extraction component processes the data to extract relevant features, such as speech patterns, facial expressions, and object recognition. The multimodal fusion component combines the features from different modalities to create a unified representation of the data. Finally, the

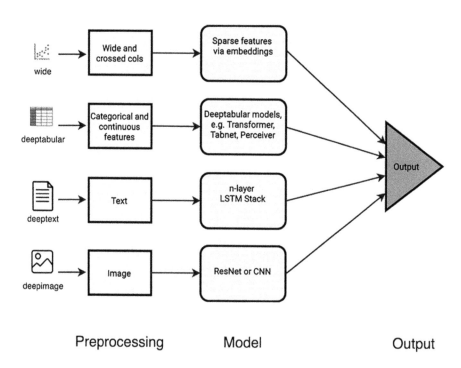

FIGURE 8.4 An example scenario of building multimodal-based LLMs.

(courtesy: Rajiv Shah, Medium.Com.)

decision-making component uses the unified representation to make decisions or generate responses.

Multimodal AI has numerous applications across various industries. For example, in healthcare, multimodal AI can be used to analyze medical images, detect diseases, and monitor patient health. In education, multimodal AI can be used to create personalized learning experiences that adapt to the student's learning style. In finance, multimodal AI can be used to detect fraud and predict market trends. In entertainment, multimodal AI can be used to create immersive experiences that combine audio, video, and haptic feedback.

Research on multimodal AI is ongoing, with many exciting developments on the horizon. Some of the current research topics include multimodal sentiment analysis, multimodal dialogue systems, and multimodal machine translation. Researchers are also exploring the use of multimodal AI in areas such as autonomous vehicles, robotics, and smart cities.

In summary, multimodal AI is a powerful technology that can process and generate natural language across different modalities. Its high-level architecture consists of several components that work together to create a unified representation of the data. Multimodal AI has numerous applications across various industries, and ongoing research is exploring new and exciting ways to leverage this technology.

8.12   IoT AND SENSORS IN AI AND MULTIMODAL SCENARIO

IoT, characterized by a network of interconnected devices embedded with sensors, software, and other technologies, facilitates the collection and exchange of data. These devices, ranging from everyday household items to sophisticated industrial tools, generate a wealth of data that serves as the foundation for AI systems.

LLMs, a subset of AI, are capable of understanding and generating human-like text. When combined with IoT, these models can process and analyze the vast amounts of data generated by IoT devices, leading to more intelligent and context-aware applications.

Sensors, the cornerstone of IoT, provide real-time, high-quality data from the physical world. This data, when processed through LLMs, can lead to valuable insights and actions. For instance, in a smart home scenario, sensors can detect changes in the environment, such as temperature or light, and an LLM can process this data to control the home's heating or lighting system accordingly.

The integration of IoT and sensors with LLMs and AI opens up new avenues for building multimodal applications. These applications can interact with users in multiple ways—through text, speech, images, and even physical actions, providing a more natural and intuitive user experience.

However, building such applications comes with its own set of challenges. These include ensuring the privacy and security of data, dealing with the high dimensionality of sensor data, and managing the computational demands of processing large volumes of data in real-time.

Despite these challenges, the potential benefits of integrating IoT and sensors with LLMs and AI are immense. From smart homes and cities to healthcare and industry, this integration is set to revolutionize the way we interact with the world around us, making our environments more responsive, efficient, and intelligent. This chapter aims to provide a comprehensive understanding of this exciting field, equipping readers with the knowledge to contribute to its development and application.

In a smart city scenario, AI, multimodal systems, and IoT will work together to create a highly interconnected, intelligent urban environment. For example, AI-powered multimodal systems could analyze data from IoT devices to monitor traffic patterns, optimize public transportation routes, and manage energy usage in real-time. These systems could also provide personalized services to city residents, such as recommending optimal travel routes based on real-time traffic data or adjusting home energy usage based on occupancy patterns.

8.12.1 THE INTEGRATION OF MULTIMODAL AI AND IoT

The integration and co-learning of multimodal AI and IoT in smart city applications is a transformative approach that is reshaping various sectors.

Apart from smart cities, there are several other scenarios where multimodal AI and IoT can be integrated. For instance, in healthcare, IoT devices can collect patient data, which can then be analyzed by AI to provide personalized care and treatment. In agriculture, IoT sensors can monitor soil conditions and weather patterns, and AI can analyze this data to optimize crop yields. In manufacturing, IoT devices can track the performance of machinery, and AI can analyze this data to predict maintenance needs and improve efficiency.

In education, this integration facilitates personalized learning experiences, adaptive content delivery, and efficient administrative operations. It enables the development of intelligent tutoring systems that adapt to individual learning styles and paces, enhancing the overall learning experience. In the health sector, the convergence of multimodal AI and IoT is revolutionizing patient care and health management. It enables remote patient monitoring, predictive diagnostics, and personalized treatment plans, improving healthcare outcomes and efficiency. The entertainment industry is also reaping the benefits of this integration. It is being used to create immersive experiences through virtual reality (VR) and augmented reality (AR), personalized content recommendations, and interactive gaming experiences. In the realm of social media, multimodal AI and IoT are being used to analyze user behavior and preferences to deliver personalized content and advertisements. They also facilitate real-time monitoring and analysis of social media trends, aiding in decision-making processes.

8.12.2 ARCHITECTURE OF INTEGRATING MULTIMODAL AI AND IoT

The architecture of integrating multimodal AI and IoT involves several components. At the base level, there are IoT devices, such as sensors and smart meters, that collect data from the physical environment1. This data is then transmitted over a network to a central system for processing1. Here, multimodal AI comes into play, analyzing the data from different modalities to recognize patterns and make predictions. The results of this analysis can then be used to control IoT devices, optimize operations, and make informed decisions. This architecture allows for a seamless integration of multimodal AI and IoT, enabling the creation of intelligent, efficient, and responsive systems.

8.13 FUTURE OF LARGE LANGUAGE MODELS AND MULTIMODAL AI

The future of LLMs is a topic of great interest and debate in the field of AI. LLMs are powerful systems that can process, understand, and generate natural language texts by analyzing vast amounts of data. They have shown remarkable capabilities in various tasks, such as translation, summarization, question answering, text generation, and more.

However, LLMs also face many challenges and limitations, such as bias, inaccuracy, toxicity, data efficiency, and explainability. These issues pose ethical, social, and technical problems that need to be addressed before LLMs can be widely adopted and trusted. Therefore, researchers and practitioners are exploring different approaches and techniques to improve the quality, reliability, and safety of LLMs.

Some of the promising directions for the future of LLMs are:

a. Self-training: This is a technique that allows LLMs to learn from their own generated texts, rather than relying on external data sources. Self-training can help LLMs improve their performance on specific tasks or domains, as well as reduce their data requirements and environmental impact.

b. Fact-checking: This is a technique that enables LLMs to verify the factual correctness of their generated texts, or to flag potentially false or misleading information. Fact-checking can help LLMs avoid generating harmful or inaccurate content, as well as increase their credibility and trustworthiness.

c. Sparse expertise: This is a technique that leverages the knowledge and skills of human experts to guide or correct the outputs of LLMs. Sparse expertise can help LLMs overcome their limitations in certain domains or tasks, as well as enhance their interpretability and accountability.

8.14 CONCLUSION

This chapter discusses the architecture, implementation, and challenges of LLMs. The chapter began with an introduction to LLMs and their working principles. It then delved into the architecture of LLMs, including the role of transformers in generative AI, attention mechanisms, and embedding layers. The chapter also covered the architecture of GPT and T5, two of the most popular LLMs. The chapter then moved on to discuss the promotion of LLMs, including prompt engineering and tuning and fine-tuning. It also covers the open-source culture and HuggingFace, a popular platform for developing LLM applications. The chapter concluded with a deep dive into HuggingFace Model T5, its derivatives, and working with Hugging Face and T5. The chapter also highlighted the latest trends in LLMs in Google, OpenAI, and Microsoft. It concluded with a discussion of the challenges facing LLMs, including ethical issues and social biases, environmental costs, data quality, model robustness, explainability, and generalization. In summary, the chapter provided a comprehensive overview of LLMs, their architecture, implementation, and challenges.

The integration of multimodal AI and IoT can significantly enhance urban living by improving energy efficiency, infrastructure, public services, and overall quality of life. Smart meters, an example of this integration, use IoT sensors to monitor energy usage, providing real-time data that helps optimize consumption and save costs. Similarly, smart poles offer functionalities like lighting, wireless connectivity, and environmental monitoring, and with AI, they can adapt to environmental conditions and usage patterns, thereby improving efficiency. Public services like waste management and public transportation can also be improved with AI and IoT, as AI can analyze data from IoT devices to optimize waste collection routes or predict peak transit times. Furthermore, the efficiency and responsiveness of cities can be enhanced, improving the quality of life of residents. For instance, smart homes equipped with IoT devices can use AI to learn residents' habits and preferences, automating tasks like adjusting the thermostat or turning off lights.

In conclusion, the combination of multimodal AI and IoT holds great promise for creating smart cities that are more efficient, sustainable, and enjoyable to live in. As these technologies continue to evolve, we can expect to see even more innovative applications that enhance urban life.

REFERENCES

1. H. Naveed, A. U. Khan, S. Qiu, M. Saqib, S. Anwar, M. Usman, N. Akhtar, N. Barnes and A. Mian, "A Comprehensive Overview of Large Language Models," arXiv preprint arXiv:2307.06435, 2023

2. M. E. E. Alahi, A. Sukkuea, F. W. Tina, A. Nag, W. Kurdthongmee, K. Suwannarat and S. C. Mukhopadhyay, "Integration of IoT-Enabled Technologies and Artificial Intelligence (AI) for Smart City Scenario: Recent Advancements and Future Trends," *Sensors*, vol. 23, no. 5206, 65, 2023.

3. Y. K. Meena and K. V. Arya, "Multimodal Interaction and IoT Applications," *Multimedia Tools and Applications*, vol. 82, pp. 4781–4785, 2023.

4. I. M. Enholm, E. Papagiannidis, P. Mikalef, and J. Krogstie, "Artificial Intelligence and Business Value: A Literature Review," *Information Systems Frontiers*, vol. 24, pp. 1709–1734, 2022.

5. X. Hou, Y. Zhao, Y. Liu, Z. Yang, K. Wang, L. Li, X. Luo, D. Lo, J. Grundy, and H. Wang, "Machine Learning: Algorithms, Real-World Applications and Research Directions," *SN Computer Science*, vol. 2, article number 160, 2021.

6. G. Giuggioli and M. M. Pellegrini, "Artificial Intelligence as an Enabler for Entrepreneurs: A Systematic Literature Review and an Agenda for Future Research," *International Journal of Entrepreneurial Behavior & Research*, vol. 29, no. 4, pp. 816–837, 2023.

7. B. G. Jayatilleke, G. R. Ranawaka, C. Wijesekera and M. C. B. Kumarasinha, "Development of Mobile Application Through Design-Based Research," *Asian Association of Open Universities Journal*, vol. 13, no. 2, pp. 145–168, 2018.

8. L. Xie, Z. Luo and X. Zhao, "Critical Factors of Construction Workers' Career Promotion: Evidence from Guangzhou City," *Engineering, Construction and Architectural Management*, vol. 30, no. 6, pp. 2334–2359, 2023.

9. Wei, Ammar Rayes, Wei Wang and Yiduo Mei, "Special Section on AI-Empowered Internet of Things for Smart Cities," *ACM Transactions on Internet of Things*, vol. 21, no. 3, pp. 1–3, May 2021.

10. Simon Elias Bibri, Alahi Alexandre, Ayyoob Sharifi and John Krogstie, "Environmentally Sustainable Smart Cities and Their Converging AI, IoT, and Big Data Technologies and Solutions: An Integrated Approach to an Extensive Literature Review," *Energy Informatics*, vol. 6, no. 1, pp. 1–50, April 2023.

11. A. Vaswani, N. Shazeer, N. Parmar, J. Uszkoreit, L. Jones, A. N. Gomez, … and I. Polosukhin, "Attention Is All You Need." In *Advances in Neural Information Processing Systems* (pp. 5998–6008), 2017.

12. T. B. Brown, B. Mann, N. Ryder, M. Subbiah, J. Kaplan, P. Dhariwal, … and D. Amodei, "Language Models Are Few-Shot Learners." In *Advances in Neural Information Processing Systems* (pp. 1877–1901), 2020.

13. A. Radford, J. Wu, R. Child, D. Luan, D. Amodei, and I. Sutskever "Language Models Are Unsupervised Multitask Learners," *OpenAI Blog*, 1(8), 9, 2019.

14. Y. Liu, M. Ott, N. Goyal, J. Du, M. Joshi, D. Chen, … and V. Stoyanov, "RoBERTa: A Robustly Optimized BERT Pretraining Approach." arXiv preprint arXiv:1907.11692, 2019.

15. J. Devlin, M. W. Chang, K. Lee and K. Toutanova, "Bert: Pre-Training of Deep Bidirectional Transformers for Language Understanding." arXiv preprint arXiv:1810.04805, 2018.

16. C. Raffel, N. Shazeer, A. Roberts, K. Lee, S. Narang, M. Matena, … and P. J. Liu, "Exploring the Limits of Transfer Learning with a Unified Text-to-Text Transformer." arXiv preprint arXiv:1910.10683, 2019.

17. M. Lewis, Y. Liu, N. Goyal, M. Ghazvininejad, A. Mohamed, O. Levy, … and L. Zettlemoyer, "BART: Denoising Sequence-to-Sequence Pre-Training for Natural Language Generation, Translation, and Comprehension." arXiv preprint arXiv:1910.13461, 2020.

18. Y. Zhang, Y. Sun, Y. Zhang, P. Qi, and C. D. Manning, "Pre-Training Named Entity Recognition with Large-Scale Noisy Text." arXiv preprint arXiv:2002.07771, 2020.

19. A. Wang, A. Singh, J. Michael, F. Hill, O. Levy, and S. R. Bowman, "GLUE: A Multi-Task Benchmark and Analysis Platform for Natural Language Understanding." arXiv preprint arXiv:1804.07461, 2019.

20. A. Radford, K. Narasimhan, T. Salimans, and I. Sutskever, "Improving Language Understanding by Generative Pre-Training," 2018. https://s3-us-west-2.amazonaws.com/openai-assets/researchcovers/languageunsupervised/language_understanding_paper.pdf

9 Digitally Enabled Labor Market

The Dark Side of Digital Transformation

Kethellen Santana da Silva, Ana Clara Nunes Gomes Cardoso, Selma Regina Martins Oliveira, and José Cláudio Garcia Damaso
Fluminense Federal University, Rio de Janeiro, Brazil

9.1 INTRODUCTION

Although it is widely reported in the literature that digital transformation and emerging digital technologies can improve the performance of organizations [1–4], little is known about how digital transformation is affecting the labor market [1, 4]. This study shows how new technologies are affecting the workforce in a digitally enabled market in an emerging economy. Understanding this relationship is important for several reasons:

- First, pronounced rapid technological change around the world challenges traditional approaches to governing organizations and demands new perspectives [1]. In this context, unprecedented complexity highlights the importance of technologies with a multifaceted and transformative role [4] and announces new ways in which people and organizations solve problems [1].
- Second, the new technologies of Industry 4.0 have the potential to reshape the workforce, the work environment, roles, tasks of the workforce in organizations, and the future of work [4–7], affecting our understanding of work integration-life [1]. These changes have significant implications for how jobs are done and will be done in the future [4, 8–10]. Several questions remain unanswered [4], such as which employee skills gain or lose importance? How do technologies affect workers' mental health? How can technologies increase or harm employees' productivity, motivation, and job satisfaction? How do employees manage the dark side of technological transformation (e.g., deskilling)? [1]. It is critical to advance our understanding of how emerging digital technologies can reshape the workforce

DOI: 10.1201/9781032656830-9

[4, 1]. In this way, new skills gain strength. For example, some existing studies of societal trends [11, 8] to predict changes in labor markets have identified which skills will be most desirable.

- Third, while the digital transformation brings benefits such as increased productivity, innovation, and cost reduction, it also brings concerns such as anxiety, productivity, remuneration, and job security [12, 17]. The unintended negative consequences of technology in the workplace have been referred to as its "dark side" and caused concern among academics [13].
- Fourth, organizations are socially responsible for the workforce. The concepts of sustainability and resilience call on organizations to adopt more holistic, inclusive, and responsible management in relation to the effects of their activities on the well-being of people and the planet [1]. This requires transformation of organizations. Therefore, institutionalizing initiatives to reduce the perverse effects of digital technologies on the workforce is substantive for organizations to reap the benefits of digital transformation.
- Despite continuing efforts to examine the effects of new technologies on the workforce, the relevant body of knowledge remains underdeveloped. Research aimed at the "perverse" effects of digital technologies on the workforce is still in its infancy. Proposals are missing [1.4]. Any discussion on digital transformation that does not consider the "dark side" component for the workforce would be incomplete. The "dark side" of the digital transformation is in the sense that the old approaches are simply not sustainable and require the need for fundamental changes driven by these new technologies [14], such as new skills.

In this context, the technological innovations emerging from Industry 4.0 not only offer great opportunities to improve the performance of organizations but also impose disproportionate challenges to the workforce. These dramatic technological changes, combined with the expansion of social concerns, serve as the backdrop for this study. Given these assumptions, we have the following problem question: How is the workforce being affected in this context of digital transformation? We propose to uncover the dark side of digital transformation through the lens of the workforce.

In this study, digitally enabled workplace is defined as a broad range of Industry 4.0 technologies aimed at unlocking the full potential of the workplace and making organizations economically successful. This chapter aims to broaden understanding of the dark side of digital transformation for a digitally enabled labor market in an emerging economy.

9.1.1 KEY CONTRIBUTIONS OF THE CHAPTER

This study is original, fills a gap in the literature, and makes significant contributions:

i. Helps to better understand how digital transformation affects the workforce in an emerging economy—in this case, Brazil, in terms of workers' capabilities/skills and mental health.

ii. It sheds light on managers of multinational companies regarding decisions to adopt initiatives to mitigate the perverse effects of digital transformation on the workforce, balancing decent and inclusive work with economic growth (SDG 8).
iii. Expands the arguments in the literature about the "dark side" of digital transformation for the workforce.
iv. It draws attention to the human side of change and helps shape the future of work. By reimagining the organization from the inside out and considering the interplay between technology and the workforce, this study seeks to unlock a wave of innovative insights and substantive evidence-based contributions that pave the way to a better future for workers, organizations, managers, and society in general.

We are excited about publishing this chapter and the possibility of exploring the challenges faced by workers in the face of new digital technologies. As empirical locations, we chose multinational corporations in Brazil—a country in South America. We discovered unexpected insights that can challenge existing management in organizations. We hope that the contributions generated by editing this chapter fulfill Taylor & Francis' mission: to publish empirical work that addresses enigmatic, unique, and intriguing phenomena that are not well explained by the existing body of theory. We seek to uncover how the workforce is being affected by digital transformation and produce robust evidence to support claims and insights.

9.1.2 Chapter Organization

The remainder of this chapter is structured as follows. Section 9.2 presents the theoretical foundation. Section 9.3 highlights the research design adopted. Section 9.4 reports the results. Finally, our discussions and conclusions are presented in Section 9.5.

9.2 DIGITAL TRANSFORMATION AND WORKFORCE

The future of work is a topic that has received increasing attention from academics, disciplines, businesses, etc. [9] and concerns changes in work in the coming decades due to advances in technologies (e.g., artificial intelligence and robotics). Inevitably new technologies will bring innovation to the workforce in companies [15, 16]. The dominant literature [15] argues that there are numerous risks and threats, especially for low-skilled workers (routine tasks). Undeniably, new jobs will be created, positions will change, and the most distressing thing is that there will be many job losses in many sectors [17]. The authors point out that digitization needs workers with high levels of specialization to avoid the risk of unemployment, leading to dramatic consequences in the structure of society, economic growth, and personal existence [17]. The substitution potential is related to routine tasks in occupations [15]. Despite the potential benefits [18] of digital transformation, little is known about its fuzzy side for the workforce. Existing studies [11, 8, 15, 16] highlight the concern that new technologies can bring to the workforce.

There are formidable challenges in digitally transitioning through the lens of the workforce that need to be better understood. We argue that one of the main challenges is related to the development of desirable individual capabilities [11, 8] for a dynamic job market [19]. Ref. [20] argues that the evolution of technology imposes new skills on professionals to remain effectively in the labor market. We argue that the impact of digitization on the workforce can scale differently for different skill sets. Despite continued efforts, new skills may be needed in the workforce; but the relevant body of knowledge remains underexplored. Thus, along with the rise of technological innovations in Industry 4.0, the need for more research to deepen the debate on the capabilities of the workforce in this context of digital transformation becomes evident. This study intends to fill this gap. Ref. [21] suggests the capabilities required to assess the technical potential of automation. They are: social and emotional, sensory perception, cognitive, natural language processing, and physical.

- Sensory perception: Sensory perception skills are essential for receiving stimuli from the environment and making sense of what is received is critical to an individual's understanding and response to the environment because they form the basis of each individual's interaction with the world [22, 23]. For example, sensory and perceptual skills are essential to receive stimuli from the environment and make sense of what is received [24].
- Cognitive skills: Cognitive skills involve retrieving information, recognizing known patterns/categories (supervised learning), generating new patterns/categories, logical thinking/problem-solving, optimization and planning, creativity, articulated/displayed output, and multi-agent planning [24]. These skills are widely considered to be the best predictor of job performance [25] and for the development of innovations.
- Natural language processing capabilities: Natural language processing is the study of computer programs that use natural or human language as input and one of its objectives is the simulation of human language skills (reading, writing, listening, and speaking) [26].
- Social and emotional skills: Social skills are different classes of social behavior in the individual's repertoire to deal adequately with the demands of interpersonal situations, for example, communication [27], people interaction behavior, and objectives [28]. Self-concept and self-efficacy are parts of socio-emotional development [29].
- Physical abilities: Physical abilities are motor and navigation skills, motor skills, navigation, and mobility [21].

On the other hand, understanding how digital technologies are affecting the workforce in terms of causes and consequences of stress (the dark side of digital transformation) [30–34] is also a relevant issue and needs to be better understood. In this study, stress is a state characterized by a specific syndrome of biological facts and is the nonspecific response of the body to ex-rigences to which it is being submitted [35]. Existing studies suggest some negative consequences of digital technologies such as unemployment, social and economic inequality, and wage gap [30].

In addition, other consequences such as cognitive overload and distress [36], anxiety, frustration, and stress among employees [37] are also evidenced and cannot be ignored. The cutting-edge literature [38, 39] indicates different symptoms of stress, for example, nervousness, anxiety, irritability, fatigue, anguish, insomnia, pain in neck and shoulder muscles, and stomach upset [40]. Understanding the nuances of the human–computer symbiosis and its consequences is important in this context of companies in emerging economies and needs to be better explored. In this way, we also intend to examine the prevalence of the causes and consequences of stress caused by the adoption of digital technologies in the context of people's work. We consider multinational companies from multiple sectors.

9.3 RESEARCH METHODOLOGY

A questionnaire (Appendix A) was developed for data collection based on the literature and a pilot test was applied with three professionals from the areas of controllership (2) and auditing (1) with the aim of verifying whether the content and structure of the questionnaire were covered. The experts did not present suggestions for improvement. Thus, we perform a refinement for clarity. The value of Cronbach's Alpha obtained in this phase of the test was 0.84, considered almost perfect [41]. The professional skills construct (social and emotional, natural language processing, and cognitive) and causes and consequences of stress were measured using a Likert scale ranging from 0 (strongly disagree) to 5 (strongly agree).

The analysis of the level of stress intensity was performed using the reference scale developed by Ref. [42]: absence of stress < 1.75; mild to moderate stress; 1.75 to < 2.46; intense stress 2.46 to < 3.16; and very intense stress 3.16, on a scale ranging from 1 to 5 points. According to Ref. [42]: *Absence of Stress*—means a state of good balance between the psychic demands arising from the environment and the psychic structure of the individual. *Mild to Moderate Stress*—signals the occurrence of manifestations of stress without generating significant impacts on the individual's various interaction environments. *Intense Stress*—signals the occurrence of manifestations of stress to a high degree and generates important impacts on the individual; organic and psychological conditions may present alterations and, in some cases, individuals may need psychological treatment/monitoring and clinical treatment. *Very Intense Stress*—indicates the occurrence of stress manifestations to a very high degree, which can generate substantive impacts on the various environments in which the individual operates; the organic and psychic conditions present very important alterations, and the cases of this intensity require clinical and/or psychological treatment/monitoring [42].

The professional social network LinkedIn was used to map professionals who work in the accounting/controlling area in multinational companies from different sectors in Brazil. The choice for the accounting and controllership area is justified because it involves different activities and tasks in the organization (strategic planning, economic and financial plans, production costs, economic feasibility studies of new projects, preparation of budgets, forecast, business case, performance indicators, establish and monitor the implementation of policies, procedures and internal controls, economic and financial feasibility studies, prepare proposals and projects with the aim of facilitating strategic decision-making, prepare and monitor plans of business, implement budgetary control policies and guidelines, tax planning, financial accounting management, and, finally, ensure that the accounting information reflects the movements carried out by the company) and needs to integrate reliable data on a large scale and from various sources, in order to generate information and support the decision-making process in the organization [43]. Analysts, auditors, controllers, directors, managers, etc. were identified. 300 questionnaires were submitted through the Google Forms platform. Returned 21 completed questionnaires composing the total sample of this study. Data were analyzed using descriptive statistics techniques. Figure 9.1(a–e) shows the results of the respondents' general information, considering: sectors in which they work, length of experience in the position, educational area, and educational level.

Most respondents work in the food sector (24%)(Figure 9.1a), with a background in accounting sciences (67%) (Figure 9.1c), with an MBA degree (52%)(Figure 9.1d), with experience over 10 years (57%) (Figure 9.1b), and hold the position of accounting and financial analyst (38%).

9.4 RESULTS

We used a descriptive statistics approach to analyze the data. Furthermore, we examined the reliability of the construct using Cronbach's alpha and the value obtained was 0.88, considered an almost perfect result [41]. Figure 9.2 illustrates the results obtained for the measurement of this study. Thus, Figure 9.2 and Table 9.1 show the results for the mean and standard deviation related to the dimensions of social and emotional abilities [21], natural language/technology/data processing abilities [21], and cognitive abilities/learning/knowledge; Figure 9.3(a) highlights the dimensions of the causes of stress caused by the digital transformation; and Figure 9.3(b) illustrates the symptoms caused by the stress of digital transformation.

On average (M = 3.98; SD = 0.97), our results indicate that professionals from the companies comprising the sectors indicated in this sample have social and emotional capacity (M = 4.46) [21] (S1…S4); natural language/technology/data processing (M = 3.68) [21] (S5…S9); cognitive skills/learning and knowledge (M = 4.16) (S10…S16); and sensory perception (M = 3.61) [21] (S17…S22) substantive (Figure 9.2 and Table 9.1), with 72% of responses concentrated at degrees 4 and 5. We also found that the social and emotional capacities (S1… S4) are the most prominent (M = 4.46) (S1…S4) (Figure 9.1 and Table 9.1), with greater intensity (93 %) of responses in degrees 4 and 5. The most outstanding capabilities in this category are: S1—digital communication capabilities and S4—transformative capabilities (M = 4.5; SD = 0.61 and SD = 0.62, respectively) (Figure 9.2 and Table 9.1).

In addition to social and emotional capabilities, cognitive/learning/knowledge capabilities are quite significant (M = 4.16) (S10…S16) (Figure 9.2), with greater intensity (76%) of responses concentrated in grades 4 and 5. In this category, the most prominent capabilities are (Figure 9.2 and Table 9.1): S14—critical and analytical thinking skills for problem-solving (M = 4.57; SD = 0.61); S15—capabilities directed toward optimization and planning (M = 4.5; SD = 0.70). On the other hand, sensory perception (M = 3.61) (S17…S22) and natural language/technology/data processing

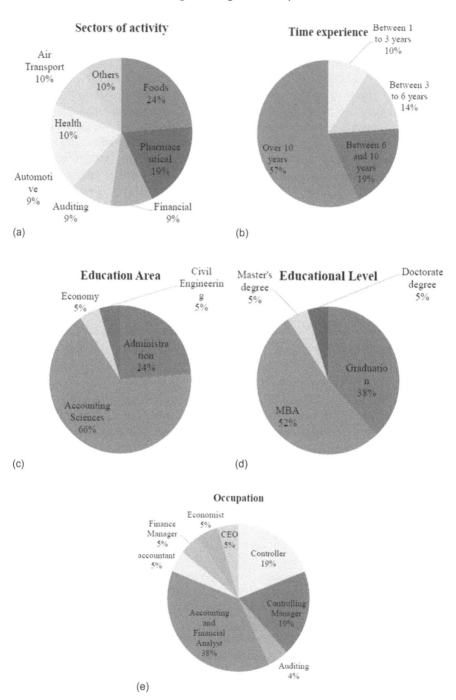

FIGURE 9.1   (a) Sectors of activity, (b) time experience, (c) education area, (d) educational level, and (e) occupation.

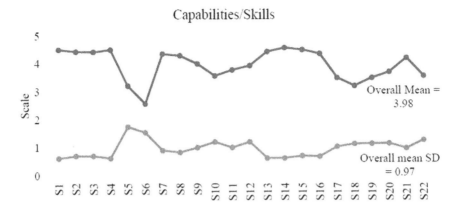

FIGURE 9.2 Results of means (M) and standard deviation (SD) of capabilities: social and emotional [21] (S1...S4); natural language/technology/data processing [21] (S5...S9); cognitive skills/learning and knowledge (S10...S16); and sensorial perception [21] (S17...S22).

TABLE 9.1
Results of Means (M) and Standard Deviation (SD) of Capabilities

Capabilities/Skills		Mean	SD	Capabilities/Skills		Mean	SD
Social and Emotional Capabilities [21] (M = 4.46)				**Cognitive Skills/ Learning/Knowledge [21](M = 4.16)**			
Digital communication	S1	4.5	0.61	Skills to develop creative and innovative technological activities	S10	3.57	1.2
Easily collaborate/ integrate (empathize) with and between other teams	S2	4.43	0.70	Digital/data-driven mindset/culture	S11	3.78	1
Ease of interacting with other multicultural teams oriented to digital technologies	S3	4.42	0.70	Share digital expertise with internal and external sources	S12	3.93	1.2
Transformative, dynamic, proactive, and agile capabilities	S4	4.5	0.62	Ability to absorb digital knowledge	S13	4.43	0.62
Natural Language Processing/ Technology/Data [21] (M = 3.68)				Critical and analytical thinking for problem-solving	S14	4.57	0.62
Logical reasoning to operate algorithms	S5	3.21	1.74	Skills aimed at optimization and planning	S15	4.5	0.70

(Continued)

TABLE 9.1 (CONTINUED)

Results of Means (M) and Standard Deviation (SD) of Capabilities

Capabilities/Skills		Mean	SD	Capabilities/Skills		Mean	SD
Skills to operate with deep learning models/algorithms; natural language processing, Python, SQL, and large-scale data-related technologies (Spark, Pyspark, Impala, Hadoop, etc.).	S6	2.57	1.54	Cognitive flexibility to solve problems in unexpected situations	S16	4.36	0.68
Extract data that is useful and necessary for analysis and interpretation purposes for decision making	S7	4.35	0.90	**Sensory Perception/ Adaptation [21] (M = 3.61)**			
Analytical profile, very focused on information quality (detail-oriented)	S8	4.29	0.83	Skills to detect trends and interpret technological/digital scenarios	S17	3.5	1.03
Prioritize data demands and identify trends and patterns	S9	4	1	Skill to create digital scenario planning	S18	3.21	1.13
				Skills to explore digital opportunities	S19	3.5	1.14
				Skills to adapt to unexpected technological/digital changes in an agile way	S20	3.71	1.15
				Skill for long-term digital vision	S21	4.21	0.98
				Capacity for resilience to technological/digital adversities	S22	3.57	1.27

Overall Mean: 3.98
Overall SD: 0.97

(M = 3.68) (S5…S9) capabilities have more moderate relevance, with prominence for capabilities S21—long-term vision (M = 4.21; SD = 0.98), S7—capabilities to extract data that are useful and necessary for purposes of analysis and interpretation for decision-making (M = 4.35; SD = 0.90) and S8 (M = 4.28; SD = 0.82) (Figure 9.2 and Table 9.1). In short, the research results imply that social and emotional and cognitive/learning/knowledge capabilities are the most relevant. Following the debate on

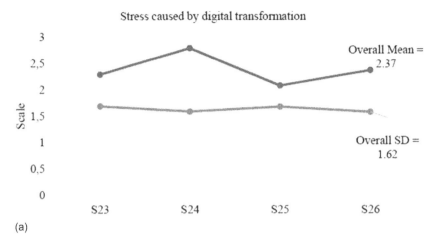

(a)

FIGURE 9.3A Results of means and standard deviation—stress caused by digital transformation in professions (S23...S26): S23 (M = 2.28, SD = 1.68); S24 (M = 2.78, SD = 1.58); S25 (M = 2.07, SD = 1.68); S26 (M = 2.36, SD = 1.57).

Competitive environment caused by new digital technologies.	S23
Pressure from multiple stakeholders (managers, suppliers, investors, shareholders, customers, government, companies, etc.).	S24
Not qualified enough (skills) to work with the new technologies	S25
Nervousness/irritability	S26

the importance of workers' skills, we highlight that communication, transformative, dynamic, proactive, and agile skills are the most substantive for emotional and social skills (M = 4.5). Analytical thinking skills for problem-solving and optimization and planning are the most relevant for cognitive skills, learning, and knowledge (M = 4.57; M = 5.5, respectively). However, the results reveal that other skills are also important, such as natural language processing/technology/data (M = 3.68) and sensory perception/adaptation (M = 3.61).

Figure 9.3a and b highlights the results of the causes [30–34] and symptoms [33, 35, 36, 39] of stress caused by digital transformation, respectively.

According to the characteristics of respondents in this research (Figure 9.1a–e), the causes of stress related to digital transformation (new technologies) indicated in our study were below average (M = 2.37; SD = 1.62) (S23, S24, S25, S26) (Figure 9.3a), as well as stress symptoms (M = 1.18; SP = 1.38) (S27, S28, S29, S30, S31, S32, S33) caused by the digital transformation (Figure 9.3b). In other words, for 80% of the professionals in this sample, digital transformation (new digital technologies) is causing little stress. According to Ref. [42], these results mean absence of stress (<1.75). Interestingly, for the other 20% of the sample of professionals (mainly (auditors and controllers of multinational companies), digital transformation/new technologies are causing stress with a potential drop in productivity (S27), concentration problems in their activities, and a lot of distress (S28) (Figure 9.3b).

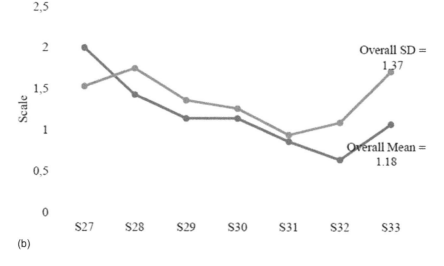

(b)

FIGURE 9.3B Results of means and standard deviation—symptoms of stress caused by digital transformation in professions (S27...S33): S27 (M = 2, SD = 1.53); S28 (M = 1.43, SD = 1.75); S29 (M = 1.14, SD = 1.36); S30 (M = 1.14, SD = 1.26); S31 (M = 0.86, SD = 0.86); S32 (M = 0.64, SD = 1.09); S33 (M = 1.07, SD = 1.71)

Drop in productivity	S27
Concentration/memory deficit problems	S28
Difficulty making a decision	S29
Feeling of loss of control	S30
Make frequent mistakes	S31
Heart problems	S32
Others (fall in immunity, allergies, etc.).	S33

Finally, we asked the professionals in this sample if the companies in which they work implement measures to inhibit the degree of stress of employees with the adoption of new digital technologies of industry 4.0 (AI, big data, blockchain, machine learning, etc.). The results indicated that companies do not implement measures for 44% of respondents; partially implement (some measure) for 28% of respondents and implement measures for 28%. We also asked to what extent the profession they work in is or will be impacted by the new digital technologies of Industry 4.0 (big data, blockchain, machine learning, IoT, etc.). The majority (76%) responded that the impact is or will be substantial (4 and 5). Existing studies address the effects of digital transformation on the labor market on the reality of other countries. These results are in line with the argument of [44], which highlights the disruptive effect of digitization on the workforce, with the creation of new jobs and positions and job losses in many sectors.

9.5   DISCUSSION AND CONCLUSION

Existing literature [45] argues that new digital technologies (artificial intelligence, blockchain, etc.) provoke a lot of uncertainty and concern about the future of the workforce in general [1, 4, 46–51]. The unique contribution of this chapter is to empirically test the evidence of the negative effects of digital transformation on the workforce through the lens of professionals' capabilities/abilities and mental health. We shed light on the scenario of Brazil, an emerging South American economy, which constitutes a group of countries that intend to advance with digital transformation, but with concerns related to the impacts of new technologies on the workforce.

Thus, this work provides expected and unexpected insights that can contribute to the current debate surrounding the dark side of digital transformation for the workforce around the world. Although this investigation confirms the expected results that the workforce is affected by digital transformation, these effects are not substantive for the sample considered in this study. These results can be explained because digital technologies take some time to replace professions (tasks), specifically the more complex ones (e.g., controllership—which is the case of the sample in this study) (according to some authors). Interestingly, we discovered that the professionals in our sample have the capabilities to move forward with digital transformation, especially with regard to social, emotional, and cognitive/learning/knowledge capabilities. We also found that the causes and symptoms of stress caused by new technologies/digital transformation had below average results. There is a reason for this, the first is that the companies in this study's sample may not have reached digital maturity. Second, control activities are considered complex, which would take longer to realize technological effects [52, 21], so the impacts will be medium or long term. A longitudinal study would be interesting to evaluate the behavior of these impacts.

The sample of this research is directed to the area of controllership (tasks), which traditionally deals with large-scale data analysis in an integrated way. This, in turn, means that companies can target their training toward new IT skills, such as skills related to large-scale data; organize and interpret data in order to extract patterns, make them visible and understandable, and indicate paths and solutions to possible problems and decision-making; utilize a broad set of analytical tools and techniques and generate business insights to improve decision-making; build and improve reports and dashboards using visualization and structured programming tools; etc. The results of this study can be useful for professionals, academics, and entrepreneurs who face the challenges of digital transformation and its perverse effects on the job market in emerging economies with characteristics common to Brazil. These findings have implications for theory and practice.

9.5.1   IMPLICATIONS FOR THEORY

This study is one of the first to report digital technologies/transformation and the labor market in multinational companies in emerging markets. The results show how professionals are responding to digital transformation in terms of capabilities, causes, and consequences of stress resulting from the adoption of new technologies.

This study identifies substantive capabilities for the digital transition. Thus, this study fills a theoretical and empirical gap. Our results suggest that companies have no choice when their professionals are pressured by Industry 4.0 technologies but to continually renew and reconfigure their strategies, plans, and policies to reduce the harm caused by emerging digital technologies to the workforce. Furthermore, the capability to create and maintain an environment with a reduced presence of technological stressors is an increasingly growing demand, and managers are asked to help reduce tensions among their collaborators/employees. Therefore, managers must be aware of the need to initiate change for companies that intend to be digitally enabled to reap the benefits of digital transformation. In summary, our findings suggest that companies effectively manage and respond to the demands and demands of their employees to keep their business performance aligned with new technologies. Finally, our results advance the literature's arguments about the "dark side" of digital transformation for the workforce in a digitally enabled labor market in emerging economies.

9.5.2   IMPLICATIONS FOR PRACTICE

The findings of our study have important implications for practice. Drawing an analogy with the literature [17], developing the appropriate capabilities/skills of workers is an essential refinement to match the era of digitalization [19] and turbulent and dynamic environments [51]. We call on managers and leaders to seek strategies to effectively harness the potential capabilities of their employees to generate lasting performance results. Our findings serve as a guide to managers of multinational companies regarding the implementation of initiatives to mitigate the perverse effects of digital transformation on the workforce and ensure dignified and inclusive work, with economic growth (SDG 8), balancing the interests of the organization and of the people who work there.

9.5.3   LIMITATIONS AND SUGGESTIONS FOR FUTURE RESEARCH

Although our study has numerous implications for theory and practice, it is not without limitations. The first is related to the choices of theoretical dimensions made in this work, which although justifiable, other alternatives could be better. We adopt the dimensions of capabilities, causes, and symptoms of stress to examine how the workforce is affected by digital transformation. Future studies may adopt other dimensions from other prestigious literature. The sample was limited in terms of size. Future studies are called to expand the sample. In this study, efforts were made to understand how digital transformation/new technologies affect Brazilian multinational companies in an emerging economy. Therefore, the results of this research should be considered relevant to the sample of companies considered in this study and should not go beyond this limit. We also understand that future studies could focus on other South American countries and other emerging economies to compare results. This research was based on a cross-sectional study in which data were extracted at a specific point in time. It would be interesting for longitudinal studies to be conducted in future research to examine the negative effects of digital transformation on the workforce in digitally enabled markets.

Despite the limitations, the results of this research bring several conclusions relevant to theory, researchers, and organizations about how digital transformation affects the job market through the lenses of social and emotional capabilities, natural language processing, cognitive and sensory capabilities, and causes and effects of stress on some Brazilian multinational companies. We suggest caution in relation to the results obtained and there is certainly a need for further studies in future studies, as the literature relating emerging digital technologies and their impacts on the workforce is still quite limited. Therefore, new studies are suggested, with other designs and variables focused on workers who use new digital technologies as essential tools in their work.

REFERENCES

[1] M. Malgrande. Our Transformation Era: Implications for Management and Organizations - Special Research Forum, *Academy of Management Journal*, 7, 2023.

[2] G. Lanzolla, A. Lorenz, E. Miron-Spektor, M. Schilling, G. Solinas, and C.L. Tucci. Digital Transformation: What Is New if Anything? Emerging Patterns and Management Research, *Academy of Management Discoveries*, 6 (3), 3, 29 Oct 2020.

[3] N. Furr, P. Ozcan, K.M. Eisenhardt. What Is Digital Transformation? Core Tensions Facing Established Companies on the Global Stage. *Global Strategic Journal*, 12 (4), 595–618, 2022.

[4] S. Raisch, R.W. Gregory, K. Leavitt, D. Minbaeva, A. Murray, J.D. Nahrgang, A. Zavyalova. Special Topic Forum – Artificial Intelligence in Management. *Academy of Management Review*, 22, 2024.

[5] A. Colbert, N. Yee, G. George. The Digital Workforce and the Workplace of the Future. *The Academy of Management Journal*, 59 (3), 731–739, Jun 2016.

[6] A. Aloisi, V. de Stefano. Regulation and the Future of Work: The Employment Relationship as an Innovation Facilitator. SPECIAL ISSUE: *Future of Work (Part II): Rethinking Institutions for Social Justice*, 159 (1), 47–69, 2020.

[7] K.F. Pfaffinger, J.A.K. Reif, E. Spieß. Anxiety in a Digitalised Work Environment. *Gr Interakt Org*, 51, 25–35, 2020.

[8] C. Frey, M. Osborne. The Future of Employment: How Susceptible Are Jobs to Computerisation? *Technological Forecasting and Social Change*, 114, 254–280, 2017.

[9] J. Menges, S. Cohen. Special Research Forum: The Human Side of the Future of Work: Understanding the Role People Play in Shaping a Changing World. *Academy of Management Journal*, 7, 4, 2022.

[10] A. Varma, S. Kumar, R. Sureka, W.M. Lim. What Do We Know About Career and Development? Insights from Career Development International at Age 25. *Career Development International*, 27 (1), 113–134, 2022.

[11] D.H. Autor, D. Dorn. The Growth of Low-Skill Service Jobs and the Polarization of the US Labor Market. *American Economic Review*, 103 (5), 1553–1597, 2013.

[12] D. Brougham, J. Haar. Technological Disruption and Employment: The Influence on Job Insecurity and Turnover Intentions: A Multi-Country Study. *Technological Forecasting and Social Change*, 161, 120276, Dec 2020.

[13] E. Marsh, E.P. Vallejos, A. Spence. The Digital Workplace and Its Dark Side: An Integrative Review. *Computers in Human Behavior*, 128, Article 107118, 2022.

[14] G.C.J. Kane. Big Idea: Digital Leadership – The Dark Side of the Digital Revolution, Jan. 29, 2016.

[15] K. Dengler, M. Britta. The Impacts of Digital Transformation on the Labour Market: Substitution Potentials of Occupations in Germany. *Technological Forecasting and Social Change*, 137 (C), 304–316, 2018.

[16] L. Novakova. The Impact of Technology Development on the Future of the Labour Market in the Slovak Republic, *Technology in Society*, 62(C), 5, 2020.

[17] M. Dabić, J.F. Maley, J. Švarc, J. Poček. Future of Digital Work: Challenges for Sustainable Human Resources Management. *Journal of Innovation & Knowledge*, 8 (2), 100353, 2023.

[18] R.S. Hess, R.C. D'Amato. Assessment of Memory, Learning, and Special Aptitudes. In Reynolds, C. R. (Ed.), Vol. 3*: Assessment.* In: Bellack, A. S. and Hersen, M. (edu.), *Comprehensive Clinical Psychology Encyclopedia* (pp. 239–265). Pergamon-Elsevier Science, 1998.

[19] K.S.R. Warner, M. Wäger. Building Dynamic Capabilities for Digital Transformation: An Ongoing Process of Strategic Renewal. *Long Range Planning*, 52 (3), 60, 2018.

[20] J. Luo, Q. Meng, Y. Cai. Analysis of the Impact of Artificial Intelligence Application on the Development of Accounting Industry. *Open Journal of Business and Management*, 6, 850–856, 2018.

[21] J. Manyika, M. Chui, M. Miremadi, J. Bughin, K. George, P. Willmott, M. Dewhurst. *A Future that Works: Automation, Employment and Productivity*. Brief, McKinsey & Company, 2017.

[22] R.C. D'Amato, B.A. Rothlisberg, R.L. Rhodes. Utilizing a Neuropsychological Paradigm for Understanding Common Educational and Psychological Tests. In: C.R. Reynolds, E. Fletcher-Janzen (Eds.), *Handbook of Clinical Child Neuropsychology. Critical Issues in Neuropsychology*. Springer, 1997.

[23] M.D. Lezak. *Neuropsychological Assessment* (3rd ed.). Oxford University Press, 1995.

[24] T. Hess, C. Matt, A. Benlian, F. Wiesböck. Options for Formulating a Digital Transformation Strategy. *MIS Quarterly Executive*, 15 (2), 123–139, 2016.

[25] F.L. Schmidt, J.E. Hunter. The Validity and Utility of Selection Methods in Personnel Psychology: Practical and Theoretical Implications of 85 Years of Research Findings. *Psychological Bulletin*, 124 (2), 262–274, 1998.

[26] J. Cohen. *School Climate Policy and Practice Trends: A Paradox. A Commentary.* Teachers College Record, Feb. 21, 2014.

[27] Z.A.P. Del Prette, A. Del Prette. *Psicologia das habilidades sociais na infância*. Vozes, 2006.

[28] J.I. Pozo. *Aprendizes e Mestres: A Nova Cultura da Aprendizagem*. Porto Alegre. Artmed, 2002.

[29] K.A. Rymanowicz, K.J. Moyses, K.S. Zoromski. School Readiness, Encyclopedia of Infant and Early Childhood Development (2nd ed.). *Reference Module in Neuroscience and Biobehavioral Psychology*, Neuroscience and Biobehavioral Psychology 55–64, 2020.

[30] U. Bamel, S. Kumar, M.L. Weng, N. Bamel, N. Natanya Meyer. Managing the Dark Side of Digitalization in the Future of Work: A Fuzzy TISM Approach. *Journal of Innovation & Knowledge*, 7 (4), 100275, Oct–Dec. 2022.

[31] M. Arntz, T. Gregory, U. Zierahn. OECD Social, Employment and Migration Working Papers. The Risk of Automation for Jobs in OECD Countries: A Comparative Analysis, 2016.

[32] D. Acemoglu, P. Restrepo. Automation and New Tasks: How Technology Displaces and Reinstates Labor. *Journal of Economic Perspectives*, 33 (2), 3–30, 2019a.

[33] A. Aghaz, A. Sheikh. Cyberloafing and Job Burnout: An Investigation in the Knowledge-Intensive Sector. *Computers in Human Behavior*, 62, 51–60, Sep. 2016.

[34] A. Verbeke, T. Hutzschenreuter. Imposing versus Enacting Commitments for the Long-Term Energy Transition: Perspectives from the Firm. *British Journal of Management*, 32 (3), 569–578, Jul. 2021.

[35] H. Selye. *The Stress of Life*. McGraw-Hill, 1956.

[36] R.S. Mano, G.S. Mesch. E-mail Characteristics, Work Performance and Distress. *Computers in Human Behavior*, 26 (1), 61–69, 2010.

[37] A. Jain, S. Ranjan. Implications of Emerging Technologies on the Future of Work. *IIMB Management Review*, 32 (4), 448–454, Dec. 2020.

[38] C. Goldberg. The Interpersonal Aim of Creative Endeavor. *Journal of Creative Behavior*, 20 (1), 35–48, 1986.

[39] C.L. Cooper, S. Crown. Stress and Health: An Introduction, *British Journal of Medical Psychology*, 61 (1), 1–2, 1988.

[40] H.A. Couto. *Stress e qualidade de vida dos executivos*. COP, 1987.

[41] J.R. Landis, G.G. Koch. The Measurement of Observer Agreement for Categorical Data. *Biometrics*, 33 (1), 159–174, 1977.

[42] L.P. Zille. Novas perspectivas para abordagem do estresse ocupacional em gerente: estudos em organizações brasileiras de diversos setores. Belo Horizonte: CEPEAD/FACE/UFMG, (Tese de Doutorado), 2005.

[43] Catho. Vagas abertas de emprego de Controller. https://www.catho.com.br/vagas/controller/. Access in: Febr., 2023

[44] ILO. An Inclusive Digital Economy for People with Disabilities. Publication. International Labour Organization, 10 February 2021. http://www.ilo.org/global/topics/disability-and-work/WCMS_769852/lang--en/index.htm. Access: 3 March, 2023.

[45] F.J. Petani, C. Ramirez, and Y. Gendron. Special issue on Digitalization, work, and professions. *Critical Perspectives on Accounting*, 79 (C), 30, 2021.

[46] E. Dahlin. Are Robots Stealing Our Jobs? *Socius*, 5, 44, 2019.

[47] A. Bhimani, L. Willcocks. Digitisation, 'Big Data' and the Transformation of Accounting Information. *Accounting and Business Research*, 44 (4), 469–490, 2014.

[48] P. Quattrone. Management Accounting Goes Digital: Will the Move Make It Wiser?, *Management Accounting Research*, 31, 118–122, Jun. 2016.

[49] D. Gulin, M. Hladika, I. Valenta. Digitalization and the Challenges for the Accounting Profession. *Proceedings of the ENTRENOVA - ENTerprise REsearch InNOVAtion Conference (2019), Rovinj, Croatia, in: Proceedings of the ENTRENOVA - ENTerprise REsearch InNOVAtion Conference*, Rovinj, Croatia, 12–14 September 2019, 502–511, IRENET - Society for Advancing Innovation and Research in Economy, Zagreb., 2019.

[50] D. R. Knudsen. Elusive Boundaries, Power Relations, and Knowledge Production: A Systematic Review of the Literature on Digitalization in Accounting. *International Journal of Accounting Information Systems*, 36 (1), 100441, 2020.

[51] D. Manciniy. Accounting Information Systems in an Open Society. *Emerging Trends and Issues, Management Control*, 5, 78, May 2016.

[52] J.F. Maley. Preserving Employee Capabilities in Economic Turbulence. *Human Resource Management Journal*, 29 (2), 147–161, 2019.

10 Artificial Intelligence Capability for Auditing

Rafael Pires de Almeida and
Selma Regina Martins Oliveira
Fluminense Federal University, Rio de Janeiro, Brazil

10.1 INTRODUCTION

Although the benefits of artificial intelligence [1] and auditing have been widely reported in cutting-edge literature [2, 3], little is known about the contributions of artificial intelligence to auditing. We examine the current state of companies' artificial intelligence capabilities to achieve audit objectives. We intend to shed light on the context of Brazilian companies that intend to advance with digitally enabled auditing through artificial intelligence. Understanding the current state of artificial intelligence capabilities for auditing is an important priority for several reasons.

First, companies are increasingly being pressured by stakeholders (investors, clients, suppliers, communities, social media, etc.) to incorporate sustainable initiatives into their agenda. The implementation of environmental, social, and governance practices serves as a guide for companies to maximize profits and achieve green and sustainable development [4]. Integrating environmental, social, and governance (ESG) issues into a company's investment decisions will help investors make decisions based on overall performance and not just financial performance [5]. Lack of ESG disclosure by companies can result in unsuccessful investments in high-risk sectors that can pollute the environment or discriminate against workers [5]. In this way, companies are called upon to disclose reliable and transparent information on ESG issues to guide stakeholder decisions [5]. Thus, ESG has received a lot of attention from researchers and the capital market.

Second, the wave of corporate scandals that swept the world severely damaged the reputation of companies, leading to a loss of credibility in the information disclosed, making stakeholders more sensitive regarding the credibility of the information disclosed in company reports [6].

At the same time, the crises of a series of audit failures were noticed around the world [7]. Auditing plays a fundamental role in ensuring that the information provided has reliable credentials for stakeholder decision-making. Modern reporting is data-driven [8]. Academic literature (e.g. Ref. [9] recognizes the growing importance of technology and the need to reimagine the audit work process [10]. Existing studies highlight the role of artificial intelligence in auditing [2, 3].

DOI: 10.1201/9781032656830-10

Artificial intelligence has the potential to reshape audit activities and contribute to achieving your objectives. The emerging generative capabilities of artificial intelligence represent a leap forward in innovative solutions [1] and an opportunity to improve the efficiency, quality, and effectiveness of audits [11]. Existing studies highlight that artificial intelligence can contribute to the flourishing of intelligent auditing [12] in companies. It is worth integrating artificial intelligence into auditing because [13]: (i) artificial intelligence technologies allow the convergence of auditors' audit objectives with the objectives of information demanders; (ii) artificial intelligence technologies create more access to a broader spectrum of data, along with fast and convenient measures for processing; and (iii) auditors can define judgment rules, among other attributes. With the advent of artificial intelligence technologies, audit objectives must be positioned to guarantee the reliability of information [13]. However, existing studies [14, 5] highlight that one of the main reasons why artificial intelligence has not yet produced the expected results is related to delays in implementation and restructuring [15, 10]. Some authors [16] also highlight that the challenges are related to difficulties in data management [16]. Other authors point to the lack of qualification and lack of data-oriented culture in the company, etc. Therefore, organizations need to invest in complementary capabilities to harness the full potential of artificial intelligence and enable auditing to achieve its objectives. In other words, we need to know the current state of companies' artificial intelligence capabilities to achieve audit objectives. There is a lack of empirical studies related to artificial intelligence capabilities and audit objectives. Therefore, this study seeks to answer the following question: What is the current state of artificial intelligence capabilities to achieve audit objectives (transparent, reliable information in compliance with legal and ethical standards and meet the interests of stakeholders)? Primary data was collected from auditors of Brazilian companies.

10.1.1 KEY CONTRIBUTIONS OF THE CHAPTER

This study is original, fills a gap in the literature, and makes substantive contributions:

i. It serves as a guide to managers in allocating and reallocating resources to boost artificial intelligence capabilities aimed at audit objectives in companies that intend to advance with artificial intelligence audit-oriented.
ii. Sheds light on the prominent capabilities of artificial intelligence to achieve intended audit objectives.
iii. Allows improving the quality of information (reliability and transparency) addressed to stakeholders.
iv. Expands the arguments in the literature about the contributions of artificial intelligence to auditing.

10.1.2 CHAPTER ORGANIZATION

The next section presents the theoretical foundation of the research; Section 3 details the methodology; Section 4 presents, analyzes, and discusses the results; Section 5 highlights the conclusions; finally, Section 6 reports the final considerations, implications, limitations, and recommendations for future research.

10.2   THEORETICAL FOUNDATION OF THE RESEARCH

In recent times, the investment world has witnessed a substantive regime change in the values that drive investments through the facets of ESG focused on reinforcing sustainable, socially responsible investment policy and transparent corporate governance, with lasting returns for investors [17]. Therefore, the value of accounting information to create value for stakeholders is widely reported in the literature [10, 8]. The role of the audit is to improve and attest the reliability of the information [18]. The audit is considered a set of intensive information activities involving the collection, organization, processing, evaluation, and presentation of data with the objective of generating a reliable audit opinion (decision). The fulfillment of audit activities is dedicated to obtaining evidence that allows an impartial judgment in relation to the audited organization. Recurring scandals (Enron, Banco PanAmericano, Parmalat, etc.) have broken the confidence of stakeholders about the content of the audited information [19]. The auditor's responsibility is to ensure that the financial statements do not present material distortions caused by fraud or error [20].

Existing studies suggest that auditors are under pressure in the course of their work, compromising audit objectives. At the same time, information technology-based decision support is putting pressure on auditors to play a more effective role in corporate governance and control. Emerging technologies can increase the auditability and transparency of information [21]. Ref. [22] points out that constant advances in computational technologies have driven most large accounting firms to introduce the use of artificial intelligence to make audit judgments as part of their integrated audit automation systems, with the aim of providing opinion on the veracity and fairness of the financial information presented by management and the compliance of this information with applicable accounting standards and relevant legislation. Ref. [13] argues that artificial intelligence has had profound implications for audit objectives and the ways to achieve them, positioning them in ensuring the reliability of accounting information, and not in the compliance of accounting reports with the fundamental principles of report preparation standards.

The use of systems based on artificial intelligence to support auditors in their judgments has been increasing [23, 11, 13, 22]. Artificial intelligence can influence the entire financial statement audit procedure, that is, from establishing audit objectives to ways to achieve them [13]. The purpose of these systems is to support auditors in making better decisions, taking care of possible biases, omissions, and other anomalies that could normally occur in traditional decision-making processes [22]. The prestigious literature [24] suggests that artificial intelligence technologies can identify hidden patterns in data and make predictions or classifications [25, 26], etc. Artificial intelligence can help identify unusual transactions, use data from past transaction behaviors, identify data trends, etc. It is highlighted that the application of artificial intelligence models directed to accounting and auditing tends to improve predictive performance [24], [27]. We advocate a synergy between artificial intelligence and auditing activities to ensure the reliability and impartiality of accounting information.

Given the relevance of artificial intelligence to auditing, challenges prevent orga-nizations from realizing substantive benefits. This needs to be resolved. Ref. [14] points out that implementation and restructuring are among the main challenges to be solved in order to reap the benefits of artificial intelligence. The authors argue that organizations need to invest in their capabilities to boost artificial intelligence. Thus, understanding the current state of artificial intelligence capabilities in order to achieve audit objectives is imperative in the quest to achieve performance gains with artificial intelligence.

This study examines the current state of companies' artificial intelligence capa-bilities to achieve audit objectives. "An artificial intelligence capability is the ability of a company to select, orchestrate and leverage its specific artificial intelligence capabilities" [14]. Ref. [14] highlights the following artificial intelligence capabili-ties: technologies, data, resources, human skills, business skills, interdepartmental coordination, capacity for organizational change, and propensity for risk. That is, artificial intelligence capacity includes tangible resources, which comprise data, technology, and basic resources; human resources, which are business skills and technical skills; intangible resources, which include interdepartmental coordination, capacity for organizational change, and propensity for risk [14]. Additionally, Ref. [28] argues that digital transformation depends on technologies (artificial intelli-gence, big data, data analytics, etc.), basic resources (financial, human, and knowl-edge), dynamic capabilities (e.g., sense, seizing, and reconfiguration), and digital capabilities (e.g., identifying and implementing technologies). Offering keen creativ-ity and oriented imagination to clients and other stakeholders [29, 30] and proper conditions (healthy and growing mind) can be helpful in generating creative ideas [31] aimed at problem-solving [30]. These theoretical underpinnings provide sub-stance to the artificial intelligence capabilities of this study. Based on the theoretical framework presented, we hypothesize that:

H1: Artificial intelligence capabilities can enhance auditing objectives—ensuring the reliability of accounting information.

10.3 METHODOLOGY

A questionnaire for data collection was prepared based on the literature and later tested by two IT and accounting professionals, with the aim of verifying face and content validity. The clarity and understanding of the instrument were also examined. Few adjustment recommendations were suggested by professionals. A pilot test was carried out with three auditors from auditing firms in Brazil. Instrument consistency was tested using Cronbach's Alpha. The Cronbach's Alpha value obtained in this test phase was 0.82. The artificial intelligence capability construct was measured using a five-point Likert scale in relation to capability relevance level (1 = no relevance and 5 = very relevant). Ref. [14] adopted the following variables to measure artifi-cial intelligence capabilities: tangible (data, technology, and basic resources); human skills (technical, business, and interdepartmental coordination); intangibles (capacity for organizational change and propensity to risk). Ref. [28] argues that digital trans-formation depends on technologies, human and financial resources and knowledge;

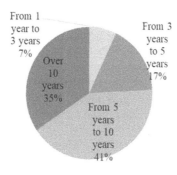

FIGURE 10.1   Experience.

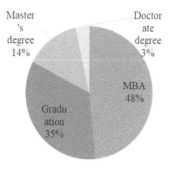

FIGURE 10.2   Educational level.

culture and learning; and dynamic capabilities and digital capabilities, with the aim of improving the value proposition to stakeholders.

We argue that dynamic capabilities can enhance artificial intelligence capabilities (tested in this study). In this study, we adopted the following capacities: tangible (data, technologies, and basic resources); technical, managerial human skills, etc.; intangibles (learning, culture, creativity, propensity for risk, etc.), and dynamic capabilities sense, seizing, and reconfiguration. This study adopts the information reliability variable to measure the audit objective [13]. The professional social network LinkedIn was used to map the auditors. Using the Google Forms platform, 100 questionnaires were submitted and 21 have been answered so far. Descriptive statistics techniques will be used to calculate the mean and standard deviation (Phase 1). Neurofuzzy technology will be used to verify the overall performance of artificial intelligence capabilities to achieve the audit objectives (Phase 2). Figures 10.1 and 10.2 show the characteristics of the sample (experience and educational level).

Most auditors have more than five years of experience (41%) and an MBA degree (48%).

10.4   RESULTS

The results are presented in two phases: phase 1: determining the prominence of artificial intelligence capabilities; and phase 2: assessment of the global performance of artificial intelligence capabilities for audit purposes.

10.4.1 PHASE 1: DETERMINING THE PROMINENCE OF ARTIFICIAL INTELLIGENCE CAPABILITIES

From the existing sample (so far), results tend to indicate moderate relevance (M = 3.85) of artificial intelligence capabilities to achieve audit objectives (Figure 10.3).

Figure 10.4 (a–k) highlights the intensity of concentration of responses.

In the opinion of most auditors (Figure 10.3), the current state of artificial intelligence to achieve the audit objectives is moderate (70%) (Figure 10.4a) of responses concentrated in levels 4 and 5 (data—AIC1–AIC3: 68% (Figure 10.4b), basic resources—AIC4: 48% (Figure 10.4c), technical skills—AIC5–AIC10: 60% (Figure 10.4d), business skills—AIC11–AIC15: 68% (Figure 10.4e), data-driven culture—AIC16: 80% (Figure 10.4f), creativity—AIC17: 67% (Figure 10.4g), propensity to risk—AIC18: 57% (10.4h), and learning—AIC19–AIC22: 75% (10.4i)). The skills with the greatest protagonism are (Figure 10.3): data-driven culture (M = 4.09), learning (M = 4.15), dynamic skills 81% (4j)—AIC23–AIC25: (M = 4.09) and digital (80%) (Figure 10.4k)—AIC26–AIC28: (M = 4.15). However, we highlight that there are still challenges to be overcome, such as the development of capabilities related to data and technology (M = 3.88) (Figure 10.3), the integration of external data with internal data to facilitate high-value analysis with focus on audit objectives, and implementation of artificial intelligence infrastructure to ensure data is protected end-to-end with high-end technology aimed at achieving audit objectives. More and more investments in network infrastructure, processing, and in advanced cloud

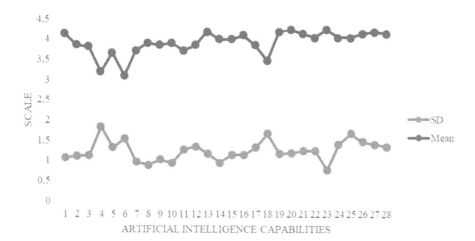

FIGURE 10.3 Prominence of artificial intelligence capabilities to achieve audit objectives—mean (M) and standard deviation (SD)—tangible: AIC1–AIC4; human skills: AIC5–AIC15; intangibles: AIC16–AIC22; dynamic capabilities: AIC23–AIC25; digital capabilities: AIC26–AIC28.

Tangibles: AIC1–AIC4; data/technology: AIC1–AIC3; basic features: AIC4; human skills: AIC5–AIC15; technical skills: AIC5–AIC10; business skills: AIC11–AIC15.

Intangibles: AIC16–AIC22; data-driven culture: AIC16; creativity: AIC17; risk propensity: AIC18; apprenticeship: AIC19–AIC22; dynamic capabilities: AIC23–AIC25; digital capabilities: AIC26–AIC28.

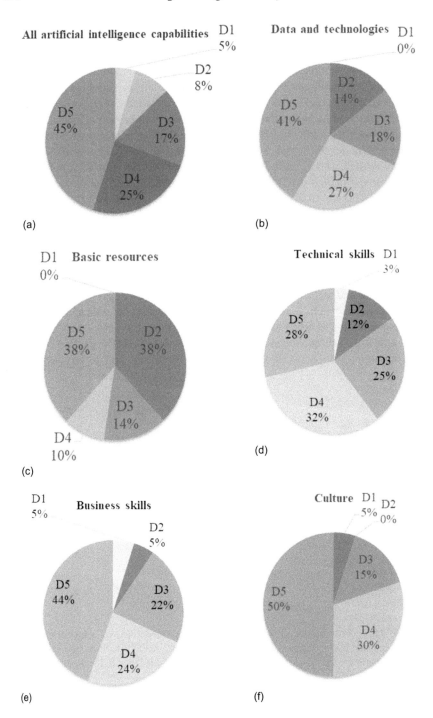

FIGURE 10.4 Intensity of concentration of responses: (a) all artificial intelligence capabilities, (b) data and technologies, (c) basic resources, (d) technical skills, (e) business skills, (f) culture,. *(Continued)*

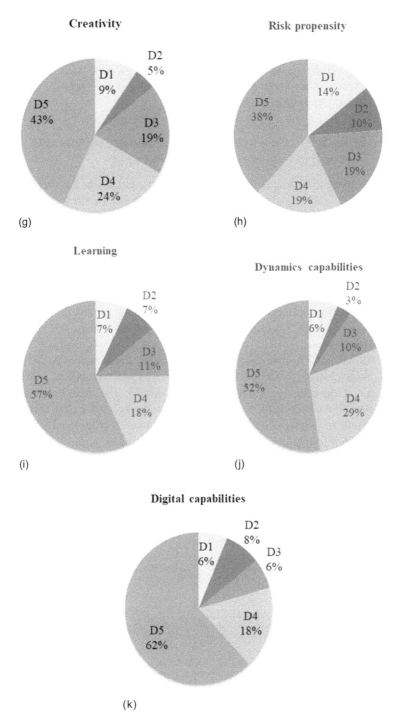

FIGURE 10.4 (CONTINUED) Intensity of concentration of responses: (g) creativity, (h) risk propensity, (i) learning, (j) dynamics capabilities, and (k) digital capabilities.

services to enable complex artificial intelligence skills (e.g., Microsoft Cognitive Services and Google Cloud Vision) [14] aim to achieve audit objectives, etc. In addition, more financial resources (M = 3.19) and technical skills (M = 3.68) (e.g., training and hiring professionals) are nouns to achieve the intended goals. These results are inconsistent with the literature [14], which highlights the relevance of these capabilities to improve the performance of results.

10.4.2   PHASE 2: ASSESSMENT THE GLOBAL PERFORMANCE OF ARTIFICIAL INTELLIGENCE CAPABILITIES FOR AUDIT PURPOSES

Using neurofuzzy modeling (Figure 10.5) was adopted to assess the overall performance of artificial intelligence capabilities (DEAIC) to achieve the audit objectives. This modeling is suitable for studies involving decision-making with a high degree of subjectivity, as is the case of this study [32].

The modeling is structured in three layers: input, intermediate, and output, as described below:

Step 1: *First layer—Definition of the input variables (IV)*: the results found in the previous section will be values of the input variables of the neurofuzzy model. Thus, input variables are data/technology: AIC1–AIC3, basic resources: AIC4, technical skills: AIC5–AIC10, business skills: AIC11–AIC15, data-driven culture: AIC16, creativity: AIC17, risk-taking: AIC18, learning: AIC19–AIC22, dynamic capabilities: AIC23–AIC25, and digital capabilities: AIC26–AIC28. The input variables go through the fuzzification process and the inference block (IB), then producing an output variable (OV), called the intermediate variable, which is added to another variable intermediate, forming a set of new input

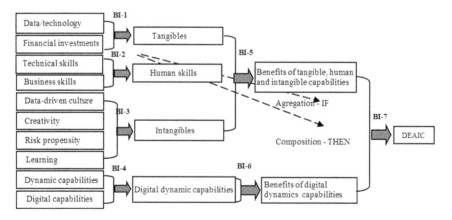

FIGURE 10.5   Neurofuzzy modeling.

variables, in a sequence until the last layer of the network. In the last layer, also composed of intermediate variables, it produces the definitive output variable of the neurofuzzy network. This output variable goes through the defuzzification process so that the final result is obtained: DEAIC. In each network node, other elements are aggregated into a single element, originating a new node, which is added to other nodes, produced in parallel, and give rise to a new node until resulting in a final node.

Step 2: *Second layer—Definition of intermediate variables*: the input variables (artificial intelligence capabilities) (step 1) go through the fuzzy inference process (rule base: IF–THEN), which results in linguistic terms of intermediate variables: low, medium, and high. The intermediate variables determined are: tangibles, human skills, intangibles, and digital dynamic capabilities. These variables are transformed into linguistic variables with their respective Degrees of Conviction or Certainty (DoC). This process is carried out through the judgment of experts (21) giving their opinion in the process.

Step 3: *Third layer—Determination of the output variable*: the modeling output variable was called the degree of assessment of artificial intelligence capabilities—DEAIC. To make comparisons possible, the linguistic vector needs to go through the defuzzification process (inverse fuzzification process where the conversion of fuzzy results into clear results is carried out) to converge in a real number between 0 and 1. Ref. [33] adopted the Center of Maximums (CM) technique in the treatment of the output variable to represent the numerical result [0;1]. The DEAIC value in the range of [0;1] represents a measure of the experts' preference intensity in relation to artificial intelligence capabilities in relation to the audit objectives. Thus, for a DEAIC equal to 1, the preference for capacity is maximum and DEAIC equal to 0, it means that experts have no preference. Summarizing, the last step of the fuzzy logic system (defuzzification) is the linguistic result of the "fuzzy" inference process in a numeric value [33] for comparison purposes. After the "fuzzy" inference, the linguistic values are transformed into numerical values, based on the membership functions [33, 32]. The CM is a method that has been used to determine the exact value for the linguistic vector of the output variable. Using hypothetical degrees to achieve the audit objectives, with the following linguistic vector of the output variable DEAIC, also hypothetical: LOW = 0.32; AVERAGE = 0.53; HIGH = 0.15. The numerical DEAIC value on a scale of 0 to 1 corresponds to 0.672438, resulting from the arithmetic mean of the values resulting from the defuzzification of each of the simulated specialists. This value corresponds to the degree of assessment of the analytical capabilities of artificial intelligence to achieve the audit objectives. Thus, we confirm the study hypothesis: H1: artificial intelligence capabilities can enhance the audit objectives—to ensure the reliability of accounting information.

10.5 DISCUSSION

The contribution of this chapter consists of empirically testing the current state of artificial intelligence capabilities to achieve audit objectives. We shed light on auditing companies in Brazil. The results of this study provide insights that can contribute to the current debate surrounding AI-enabled digital auditing. The preliminary conclusions of this study suggest that artificial intelligence capabilities have moderate relevance for achieving audit objectives (reliable and transparent information). Thus, the single hypothesis is confirmed: H1: artificial intelligence capabilities can enhance auditing objectives - ensuring the reliability of accounting information. The results suggest the most important intangible resources (data-driven culture and learning) and dynamic digital capabilities for achieving audit objectives (in this sample). Our conclusion is in line with the literature by showing the potential of artificial intelligence capabilities. Although the basic resources and risk propensity dimensions have achieved relatively lower results in relation to other capabilities, it is worth investing in artificial intelligence capabilities for audit purposes.

10.6 FINAL CONSIDERATIONS

The conclusions of this study extend the arguments of the existing literature on artificial intelligence capabilities and audit objectives and make interesting contributions: (i) they provide critical insights into where to invest resources for the development of artificial intelligence capabilities for audit objectives; (ii) highlights the current state of artificial intelligence capabilities to achieve audit objectives; (iii) sheds light on which capabilities should be encouraged in decisions to managers to ensure reliability in the information produced for stakeholder decision-making; and (iv) shows managers which paths can still be taken to amplify audit purposes enabled by artificial intelligence, balancing good corporate governance practices, social and environmental responsibility, and stakeholder interests. The findings of this study can be useful to academics, managers, investors, auditors, and other stakeholders who face the challenges of seeking reliable and transparent information for decision-making.

10.6.1 IMPLICATIONS

Artificial intelligence capabilities are crucial for companies in adopting artificial intelligence for auditing to achieve its purposes, which is to ensure the credentials of transparency and reliability of information for decision-making for its stakeholders. Thus, managers are called upon to establish plans to progress toward resources that can amplify artificial intelligence capabilities to reap the benefits that artificial intelligence can offer audit and stakeholders. Interestingly, managers must take advantage of the potential of their dynamic digital capabilities to detect opportunities and trends in audit-oriented (predictive) technologies; develop digital scenario planning/formulate digital strategies focusing on audit objectives; adapt to unexpected environmental changes/ capabilities to adapt to new digital technologies (big data, IoT, etc.) keeping in mind the audit objectives; offer services according to the needs of multiple stakeholders.

10.6.2 Limitations and Suggestions for Future Research

This work is not free from limitations. We adopt intelligence capabilities represented by dimensions related to data, technologies, basic resources, human skills, culture, creativity, risk propensity, learning, and dynamic and digital capabilities. We recognize that the choice of dimensions is justified, other alternatives in the literature could present better results. This study is a first effort to understand the current state of artificial intelligence capabilities to achieve audit objectives (in Brazil). Thus, the findings of this research must be relevant to the sample considered in this study and must not go beyond these boundaries. We highlight that this study is based on a cross-sectional sample in which data were collected at a single point in time. The sample of this study was also limited in terms of size. We call for future studies to increase the sample size. We suggest research opportunities in other developed and emerging economies to compare results. Finally, for illustrative purposes, we use neurofuzzy technology to demonstrate (hypothetical) the degree of global assessment of artificial intelligence capabilities to achieve audit objectives through neurofuzzy technology. We suggest testing the degree of global evaluation in a real situation.

REFERENCES

[1] S. Raisch, R.W. Gregory, K. Leavitt, D. Minbaeva, A. Murray, J.D. Nahrgang and A. Zavyalova. Artificial Intelligence in Management – Special Topic Forum. *Academy of Management Review*, 8, 45, (2024).

[2] A. Fedyk, J. Hodson, N. Khimich and T. Fedyk. Is Artificial Intelligence Improving the Audit Process? *Review of Accounting Studies*, 27(3), 938–985 (September 2022).

[3] J. Kokina and T. H. Davenport. The Emergence of Artificial Intelligence: How Automation is Changing Auditing. *Journal of Emerging Technologies in Accounting*, 14(1), 115–122 (2017).

[4] J. Li, G. Lian and A. Xu. How Do ESG Affect the Spillover of Green Innovation Among Peer Firms? Mechanism Discussion and Performance Study. *Journal of Business Research*, 158, 113648 (March 2023).

[5] W.M.W. Mohammad and S. Wasiuzzaman. Environmental, Social and Governance (ESG) Disclosure, Competitive Advantage and Performance of Firms in Malaysia. *Cleaner Environmental Systems*, 2, 100015 (June 2021).

[6] D.J. Cooper, T. Dacin and D. Palmer. Fraud in Accounting, Organizations and Society: Extending the Boundaries of Research. *Accounting, Organizations and Society*, 38 (6–7), 440–457 (August–October 2013).

[7] M. Harber, W. Maroun, A. and D. Ricquebourg. Audit Firm Executives Under Pressure: A Discursive Analysis of Legitimisation and Resistance to Reform. *Critical Perspectives on Accounting*, 9, 102580 (10 February 2023).

[8] H.K. Duan, M.A. Vasarhelyi, M. Codesso and Z. Alzamil. Enhancing the Government Accounting Information Systems Using Social Media Information: An Application of Text Mining and Machine Learning. *International Journal of Accounting Information Systems*, 48, 100600 (March 2023).

[9] M.G. Alles and G.L. Gray. Will the Medium Become the Message? A Framework for Understanding the Coming Automation of the Audit Process. *Journal of Information Systems*, 34(2), 109–130 (2020).

[10] S. Black and G. Gerard. Advanced Technologies and Decision Support for Audit. *International Journal of Accounting Information Systems*, Editorial, Special Issue. Available in: https://www.sciencedirect.com/journal/international-journal-of-accounting-information-systems/about/call-for-papers [Access in: 15 December 2023].

[11] M. Abdolmohammadi and C. Usoff. A Longitudinal Study of Applicable Decision Aids for Detailed Tasks in a Financial Audit. *Intelligent Systems in Accounting, Finance and Management and International Journal*, 10(3), 139–154 (September 2001).

[12] J. Brazel. How Is Artificial Intelligence Shaping The Audits of Financial Statements? *Forbes*. Available in: https://www.forbes.com/sites/josephbrazel/2022/12/19/how-is-artificial-intelligence-shaping-the-audits-of-financial-statements/?sh=11d19e1e6c28 [Access in: June, 2023].

[13] Y. Gao and L. Han. Implications of Artificial Intelligence on the Objectives of Auditing Financial Statements and Ways to Achieve Them. *Microprocessors and Microsystems*, 104036. Available online 19 January. In Press, Journal Pre-proof What's this? (2021). https://doi.org/10.1016/j.micpro.2021.104036

[14] P. Mikalef, M. Gupta. Artificial Intelligence Capability: Conceptualization, Measurement Calibration, and Empirical Study on Its Impact on Organizational Creativity and Firm Performance. *Information & Management*, 58 (3), 103434 (April 2021).

[15] E. Brynjolfsson, A. McAfee, S. Michael and Z. Feng. Scale Without Mass: Business Process Replication and Industry Dynamics. *Harvard Business School, Technology & Operations Mgt.* Unit Research Paper 07-016 (2008).

[16] W.G. No, K. Lee, F. Huang and Q. Li. Multidimensional Audit Data Selection(MADS): A Framework for Using Data Analytics in the Audit Data Selection Process. *Accounting Horizons*, 33(3), 127–140 (2019).

[17] G. Campbell and K. Kanjilal. Call for Papers on a Special Issue on Sustainable Environmental, Social, and Governance (ESG) Investment, Investors' Sentiment, and Behavioral Finance. *Global Business Review*, Special Issues. (October 2021). Available in: https://journals.sagepub.com/pb-assets/cmscontent/GBR/GBR_CFP_OCT21.pdf [Access in: 15 December 2023].

[18] B. G. Huh, S. Lee and W. Wonsin Kim. The Impact of the Input Level of Information System Audit on the Audit Quality: Korean Evidence. *International Journal of Accounting Information Systems*, 43, 100533 (December 2021).

[19] C. Holm and M. Zaman. Regulating Audit Quality: Restoring Trust and Legitimacy. *Accounting Forum*, 36(1), 51–61 (March 2012).

[20] P. Fusiger and L. Silva. Auditoria Independente: principais infrações que acarretam em processo administrativo sancionador pela Comissão de Valores Mobiliários. XIV Congresso USP de Controladoria e Contabilidade (pp. 1–16). São Paulo (2014).

[21] S. Ølnes, J. Ubacht and M. Janssen. Blockchain in Government: Benefits and Implications of Distributed Ledger Technology for Information Sharing. *Government Information Quarterly*, 34(3), 355–364 (September 2017).

[22] K. Omoteso. The Application of Artificial Intelligence in Auditing: Looking Back to the Future. *Expert Systems with Applications*, 39(9), 8490–8495 (July 2012).

[23] C.E. Brown and M. E. Phillips. Expert Systems for Management Accountants. *Management Accounting* 71, 18–23 (January 1990).

[24] C. Zhang, S. Cho and M. Vasarhelyi. Explainable Artificial Intelligence (XAI) in Auditing. *International Journal of Accounting Information Systems*, 46, 100572 (September 2022).

[25] T. Hastie, R. Tibshirani, J.H. Friedman and J. Franklin. The Elements of Statistical Learning: Data Mining, Inference, and Prediction. *The Mathematical Intelligencer*, 27(2), 83–85 (2004).

[26] E. Alpaydin. *Introduction to Machine Learning*. 4th Edition, MIT Press Academic (2020).

[27] J. Perols, R.M. Bowen, C. Zimmermann and B. Samba. Finding Needles in a Haystack: Using Data Analytics to Improve Fraud Prediction. *The Accounting Review*, 92(2), 34 (2016).

[28] C. Gong and V. Ribiere. Developing a Unified Definition of Digital Transformation. *Technovation*, 102, 102217 (April 2021).

[29] T. Amabile. The Social Psychology of Creativity: A Componential Conceptualization. *Journal of Personality and Social Psychology*, 45(2), 357–376 (1983).

[30] R. W. Woodman, J.E. Sawyer and R.W. Griffin. Toward a Theory of Organizational Creativity. *The Academy of Management Review*, 18(2), 293–321 (1993).

[31] D.K. Simonton and R.I. Damian. Creativity. In D. Reisberg (Ed.), *The Oxford Handbook of Cognitive Psychology* (pp. 795–807). Oxford University Press (2013).

[32] R.L. Oliveira and Q. Cury. Modelo neuro-fuzzy para escolha modal no transporte de cargas, Tese of Doctoral Department Engineering of Transportation of Institute Militar of Engineering (2004).

[33] C. Von Altrock. *Fuzzy Logic and Neurofuzzy Applications in Business and Finance*. Prentice Hall (1997).

11 Smart Green Cities Using IoT-Based Deep Reinforcement Learning Energy Management

Niva Tripathy
DRIEMS University College, Cuttack, India

Subhranshu Sekhar Tripathy
KIIT Deemed to be University, Bhubaneswar, India

11.1 INTRODUCTION

Directives 2012/27/EU and 2010/31/EU of the European Parliament and Council state that the building sector accounts for 40% of the total consumption of energy in the EU and is still expanding, highlighting the need for energy efficiency in buildings [1]. Large cities only make up around 6% of the planet's land area, yet they use more than 75% of its fuel, according to the World Population Index Survey. The ratio is constantly rising, which is causing serious concerns about the increased use of fossil fuels [2]. Furthermore, estimates suggest that roughly 70% of the world's population will reside in big cities by the year 2050. The increasing energy demands will place a significant strain on conventional power plants, forcing them to generate more energy. So, there will be a rise in greenhouse gas emissions [3]. In order to meet the energy needs of buildings, around 40% of fossil fuels are utilized in the production of electricity, especially in large cities. Approximately 27% of the total energy is consumed by residential structures. Development and integration of new, effective energy generating, and distribution technologies will therefore be advantageous. In particular, residential buildings could reduce the enormous global emissions of greenhouse gases and fossil fuels [4].

The development of intelligent energy management systems for smart grids that can predict potential loads has been the primary focus of Internet of Things (IoT) research. In smart cities, smart buildings, smart manufacturing [5], and smart transportation systems [6] all depend on smart grids to supply electricity to their numerous components. Utilization by residential customers [7] or by commercial and manufacturing organizations [8], distribution through smart grids, and energy plant production comprise the retail electrical power network. Users' consumption patterns, many

DOI: 10.1201/9781032656830-11

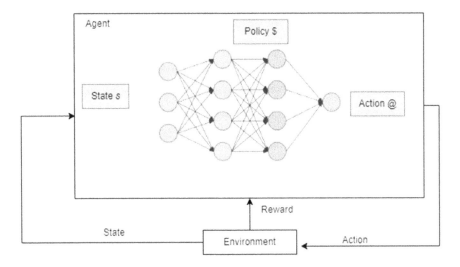

FIGURE 11.1   Structure of deep reinforcement learning.

of whom are unaware of the impact of their energy demands, have a significant influence on the amount of energy produced by energy plants and distributed via smart networks. This leads to inefficient and expensive energy use. Simultaneously, certain users, including factories, aim to reduce costs by optimizing their energy usage through effective management systems (Figure 11.1).

Due to the widespread usage of IoT-based devices in smart cities, which generate and transfer massive volumes of data, it might be challenging to accurately realize data from a complex environment and deliver dependable management actions in response. With its ability to produce almost optimum control actions for time-variant dynamic systems while taking a long-term objective into consideration, deep reinforcement learning (DRL) is a new technology that holds great promise [9].

To accomplish DRL performance, data transit between devices and the DRL agent (cloud server, e.g.) is necessary. With most devices connected wirelessly to the DRL agent, a significant amount of data transfer might rapidly surpass the transmission capacity limit. Because resource-constrained IoT devices are unable to transmit the DRL demand directly to the DRL agent, this poses a major barrier to the deployment of energy management. Edge computing (EC) is an important technology that IoT-based energy management systems can use to solve this problem [10]. Since EC may provide computing services at the network "edge" near IoT devices, it is suitable for DRL utilization in IoT-based systems due to two major advantages. Through preprocessing, EC can dramatically reduce the volume of data sent from devices to the DRL agent. Second, energy conservation technology (ECT) can fill the gap between low-power devices' restricted capabilities and energy management's computing demands [11].

11.1.1   KEY CONTRIBUTIONS OF THE CHAPTER

In this chapter, we proposed an IoT-based deep reinforcement energy management system to improve energy management effectiveness and reduce execution times.

In light of the EC processing capacity constraints, we have here suggested an on-demand energy supply. The outcomes of the trial demonstrate that our approach performs better than previous IoT-based energy scheduling techniques.

11.1.2 CHAPTER ORGANIZATION

The rest of the chapter is organized as follows. Section 11.2 gives the literature survey, Section 11.3 gives the proposed framework, Section 11.4 gives the result analysis, and Section 11.5 gives the conclusion and future scope.

11.2 LITERATURE SURVEY

In this section, a complete literature study is given which will produce a detail idea about the existing work done by researcher so far. In order to stay under the specified tolerable ambient temperature and other constraints, Angelis et al. [12] presented an HEM solution that attempted to lower energy expenditures related to task implementation, energy saving, energy sale, and heat pumps. In an effort to save HVAC and heating costs while accounting for indoor temperature fluctuations, Fan et al. [13] proposed an online HEM technique. In order to lower HVAC and dispatchable load energy expenditures while preserving a tolerable temperature range, Zhang et al. [14] developed an HEM approach. It is challenging to create a building thermal dynamics model for HVAC management that is both dependable and effective.

When temperature ranges are taken into account, Ruelens et al. [15] created a home DR method that uses reinforcement learning to save energy expenses. An automated HVAC system using DRL has been shown to be effective in reducing electricity consumption while maintaining the ideal indoor temperature [16, 17]. For HEMS, Xu et al. [18] suggested a hybrid strategy based on neural networks and Q-learning algorithms.

Building physics is used in engineering calculations and simulation models that are used for benchmarking, falling within the classical modeling category. For classical modeling to produce a trustworthy demand projection, a minimum number of construction parameters must be used [19]. Energy demand prediction in data-driven analysis is carried out by a learning process using historical building energy consumption data together with other relevant elements like climatic data. For these kinds of learning processes, data-driven modeling algorithms can include techniques like regression analysis and artificial neural network (ANN) models. The most popular machine learning (ML) methods, support vector machines (SVM) and artificial neural networks (ANN), are two of the nine time-series prediction strategies for building energy use that Deb et al. [20] discussed.

In comparing ANN and SVM, Ahmad et al. [21] came to the conclusion that it is difficult to determine which algorithm predicts data more accurately. Mocanu et al. [22] employed deep learning auto-regressive estimation techniques to predict building electricity usage, a step toward more sophisticated systems. The forecast, according to the scientists, might be enhanced by including additional model inputs, like time and external temperature data (month and day).

Using machine learning methods and a restricted collection of building-level characteristics, Robinson et al. [23] tried to forecast the annual energy usage of commercial buildings at the city size. The significance of forecasting at the aggregate building scale was underlined in their research in order to prevent individual building prediction errors and to gain insightful knowledge about cities and neighborhoods. Regression trees were employed by Ahmed et al. [24] in their study to anticipate demand at the municipal level. The authors of the study emphasized the significance of short-term day-ahead load prediction for every day of the year.

11.3 PROPOSED FRAMEWORK

This section provides a detailed structure of the proposed framework. The well-known machine learning (ML) idea of reinforcement learning addresses systematic decision-making when faced with uncertainty. The foundation of supervised machine learning is labeled data, which enables the knowledge of the right actions ahead of time. The situation is different in reinforcement learning, though, as the artificial creature updates its effects by learning from experience and making predictions. Unlike unsupervised learning, incentive feedback is given to the agent. Increasing prospective earnings is the primary goal of reinforcement learning. The agent seeks out the appropriate chain of decision-making. The agent takes the possible effects of the current action into account while evaluating a case. The suggested framework is shown in Figure 11.2.

An energy device is any person, thing, or resource inside a network that has the ability to both produce and use energy. These devices are capable of producing or sensing energy information about the connected network. It is the responsibility of the energy edge server to manage device connections or create reliable communication routes for devices. Data querying is among the core elements of the proposed software model that intelligent edge computing services require. Because the energy data acquired in smart cities is heterogeneous, it will be categorized and queued for hierarchical processing by the edge server. The energy edge server can be put at the base station, network gateway, and other locations in a local area network in order to process, cache, and deliver energy data. Data transmission is necessary for both data transfer between energy devices and the energy edge server and task offloading between the edge server and the cloud server. Data transmission can be achieved by a variety of communication technologies, such as power line communications (PLC), Wi-Fi, 5G, and long-term evolution (LTE). The energy edge server can also use the analytical results to determine how a local energy network operates. The energy cloud server and central controller are linked for energy management. The energy cloud server is responsible for both providing energy devices with real-time analysis and computation and satisfying the computing requirements of energy edge servers. Modules for DRL processing are installed in the edge server. The module maintains track of linked users' current status and requirements. The DRL module can then calculate the reward based on the results of the last action that was taken. The DRL module may make decisions with the assistance of a robust DNN that provides accurate estimation and prediction. A DRL cannot function without an optimization function. Energy management is a vital function that allows entities to schedule and control energy from every part of the system without needing to know the state of the underlying layers. Device exiting or entering the network is decided by topology

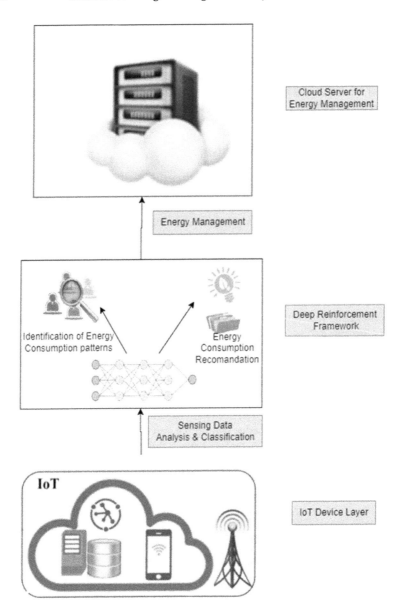

FIGURE 11.2   Proposed IoT-based deep reinforcement energy management framework.

control. The suggested IoT-based energy management system enhances interoperability across various devices and technologies by enabling web-based (e.g., online dashboard for visual management) and mobility applications (e.g., mobile applications on smartphones for facility managers).

11.3.1 Proposed Methodology

Online reinforcement learning is accomplished by the Q-learning algorithm [25]. Quality, represented by Q and initialized to zero at the start of the learning phase, is computed by the algorithm for each state–action pair. Throughout the environment interaction process, the agent keeps an eye on the surroundings and makes decisions about how to modify the system's current state. A reward indicating the worth of the state transition is given to the agent by the new state. A value function $Q^\pi(s(t), a(t))$ is maintained by the agent based on an activity that maximizes the long-term rewards. With discounted reward, the Q-factor update equation looks like this:

$$Q^{t+1}\big(s(t+1),a(t)\big)=Q'\big(s(t),a(t)\big)+\alpha\big(s(t),a(t)\big)\times$$
$$\Big[R(t)+\gamma.MAX\big(Q'\big(s(t+1),a(t)\big)\big)-Q'\big(s(t),a(t)\big)\Big] \quad (11.1)$$

where γ is the discount factor between 0 and 1, and $\alpha(s(t), a(t))$ is the learning rate $(0 < \alpha < 1)$. The agent selects short-term benefits if γ is near zero; otherwise, it explores and aims for long-term rewards. It is demonstrated in [25] that the number of state transitions, k, determines the learning rate α. It fulfills the requirement in that

$$\alpha^k = A\backslash B + k. \quad (11.2)$$

Value of A & B is determined by simulation. Conceptually online learning techniques such as Q-learning work particularly well when used to solve small-scale and discrete state space problems. On the other hand, in more realistic systems, the system becomes stuck in suboptimal policies due to stability concerns, stochastic approximation inefficiencies, and "exploration overhead." The values of $(s(t-1), a(t))$ for every $a \in A$ of a previous state $st-1$ may be affected by updating the Q-value of the state-action pair $(s(t), a(t))$ in time step t.

11.3.2 Online Deep Q-Learning

State space function: The requests $r_u \in [r_{min}, r_{max}]$ of every user u make up the edge DRL agent's state. With a cardinality of U, the state vector is written as $[r1, rU]$.

Action space functions: We allow the edge DRL agent to classify the devices at each decision epoch in order to meet various service needs. The DRL agent specifically chooses which class of devices to service during the present period. The DRL agent can determine an active set of a class of devices after carrying out an action.

Reward function: The incentive is utilized to meet the needs of the devices and attain the lowest possible energy consumption. We define the immediate reward that the cloud DRL agent receives as $dl = e_{max} - e_{real}$, where e_{max} is the maximum possible value of the energy consumption of the devices and e_{real} gives the actual total energy consumption of the devices that can be obtained at each decision epoch.

After the offline DNN training, the deep Q-learning is employed and can be described as follows:

Step 1: The edge DRL agent uses the state–action pair input from devices for each state to determine the predicted Q function value by the DNN at the start of each decision epoch.

Step 2: The e-greedy policy is used to choose the execution action for a certain class of devices. The action in this policy is selected with $(1 - e)$ probability based on the highest assessed Q-value. A random action is chosen from the action set with probability e.

Step 3: Based on the chosen class of devices, the edge DRL agent determines the optimal solution and controls the energy consumption of the devices using the best solution.

Step 4: Following observation of the devices' immediate reward and subsequent state, the experience memory stores the state transition information for the class of devices and samples it.

Step 5: The edge DRL agent uses the sampled state transitions at the conclusion of the decision epoch to produce a loss function that updates the DNN's weights.

11.3.3   ONLINE DEEP Q-LEARNING ON EDGE SERVER

State space function: The state of DRL agent consists of each user $r_u \in [r_{min}, r_{max}]$ of the demands r_i, u from edge server i. The state vector is represented as $[r_{1,1} ... r_{N,U}]$ with a cardinality of $(n + u)$.

Action space function: To satisfy different computation requirements, we let the cloud DRL agent make the priority decision of the edge servers in each decision epoch. Specifically, the DRL agent determines which edge server to be served or not. After executing an action, the DRL agent can derive an active set of edge servers.

Reward function: The incentive is utilized to meet device demands and attain the lowest possible energy usage across the board for the energy management system. We define the immediate reward that the DRL agent receives as $R_e = E_{max} - E_{real}$, where E_{max} is the maximum possible value of total energy consumption and E_{real} gives the actual total energy consumption that can be obtained at each decision epoch.

After the offline DNN tanning, the deep Q-learning is employed at the edge DRL agent for online dynamic control. The process of online deep Q-learning on the edge server can be described as follows:

Step 1: The edge DRL agent uses the state–action pair input for each state to obtain the estimated Q-value by the DNN from the cloud DRL agent at the start of each decision epoch.

Step 2: The e-greedy policy is still applied to choose the execution action.

Step 3: The experience memory of the edge server stores the state transition information after the devices have observed the immediate reward and the subsequent state.

Step 4: Using the provided samples from the experience memory, the edge DRL agent adjusts the DNN's parameters at the conclusion of the decision epoch.

11.4   RESULT AND DISCUSSION

This section provides the comparative analysis of the different methods. We examine five communities, each with 40 dwellings and 10 renewable resources, and each that deploys an edge computing server. The RELOAD database [26] is utilized to simulate various household demands, including HVAC, water heating, lighting, freezing, and drying clothes. It offers hourly load profiles of these realistic demands. The market rates for buying power from the grid are $p(t) = 0.3$ cents during the day, or from 8:00 a.m. to 12:00 a.m., and $p(t) = 0.2$ cents at night, or from 12:00 a.m. to 8:00 a.m. the following day. In the simulation, energy cost and latency are the two metrics taken into account. Furthermore, we contrasted the two suggested DRL-based energy scheduling schemes (referred to as "cooperative DRL scheduling" and "edge DRL scheduling") with the standard approach known as "cloud scheduling," which has the cloud server provide energy scheduling for households without the use of DRL method directly.

The term "energy cost" refers to the daily average overall cost of the homes. Figure 11.3 shows a comparison of the energy costs for the edge DRL scheduling, cloud scheduling, and cooperative DRL scheduling methods. It is demonstrated that the recommended DRL-based systems achieve lower energy costs than the baseline plan. This is so that the optimal energy scheduling strategy can be determined by using the suggested DRL-based method, which can anticipate the profiles of dynamic

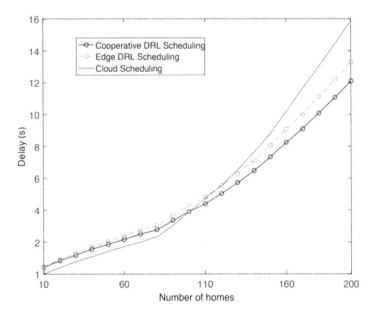

FIGURE 11.3   Delay comparison among different energy scheduling schemes.

demand and renewable energy. The baseline approach may be significantly impacted by the unpredictable nature of user demands and renewable in real-world scenarios. Furthermore, it is observed that the cooperative DRL scheduling system has a lower energy cost than the edge DRL scheduling strategy. That is, the edge server ought to manage the deep Q-learning and offline training for the residences on its own, independent of a cloud server. Some households' scheduling needs will be impeded by the edge server's capacity limit, leading to additional energy expenses.

The delay is the sum of the time it takes for tasks to be finished from each home plus the time it takes for devices to communicate with servers at the higher layer (cloud and edge computing servers). Figure 11.4 shows the comparison of delays for cloud scheduling, edge DRL scheduling, and cooperative DRL scheduling. These three complexes are expected to take longer to finish since more homes are being developed. It is interesting to note that, in comparison to the other two suggested DRL-based schemes, the cloud scheduling technique first has reduced latency and thereafter has larger delay as the number of dwellings increases. Put differently, under the cloud scheduling method, the households will send their computing requirements to the cloud server concurrently, which may cause transmission congestion. Because of this, the transmission latency will rise noticeably as the number of dwellings rises. The edge DRL scheduling technique has a lower latency than cloud scheduling since different classes of houses have different priorities to communicate the requirements. This implies that the residences would not contend for transmission opportunities concurrently. However, because the edge server may not have the computational capability to execute the duties of several devices in a timely way, this technique still causes an execution delay. In the cooperative DRL scheduling scheme,

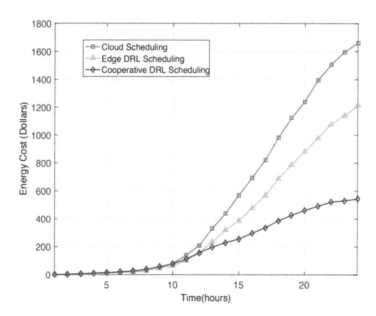

FIGURE 11.4 Energy cost comparison among different energy scheduling schemes.

easy parts (Q-learning) of the scheduling task can be executed at the edge server near the homes, and the complex parts (DNN training) can be sent to the cloud for execution. This scheme can reduce both execution and transmission delay as the number of homes becomes large.

11.5 CONCLUSION

The increasing rate of urbanization makes energy management in smart cities an essential issue to solve. This chapter first provides an overview of energy management in smart houses, with the goal of creating a sustainable and green smart city. It next presents a unified framework for the use of IoT in the construction of green smart cities. We have proposed a novel IoT-based deep reinforcement framework for energy management is proposed. In the developed Q-learning paradigm, the agents for household electric appliances and an energy storage system acquire their behavior separately through interacting with environment until the overall average rewards are maximized. The effectiveness of the energy scheduling plan is examined in the scenarios with and without edge servers, in that order. The illustrative findings show that, in comparison to conventional schemes, the suggested designs can achieve low energy cost and lower latency.

REFERENCES

[1] Ur Rehman, U., et al. (2023). Future of Energy Management Systems in Smart Cities: A Systematic Literature Review. *Sustainable Cities and Society*, 8, 104720.

[2] Amjadipour, F., Ghassemian, H., & Imani, M. (2022). Building Detection Using Very High Resolution SAR Images with Multi-Direction Based on Weighted-Morphological Indexes. *Proceedings of the 2022 International Conference on Machine Vision and Image Processing (MVIP)* (pp. 1–6). https://doi.org/10.1109/MVIP53647.2022.9738776

[3] Shi, Y., Tuan, H. D., Savkin, A. V., Duong, T. Q., & Poor, H. V. (2019). Model Predictive Control for Smart Grids with Multiple Electric-Vehicle Charging Stations. *IEEE Transactions on Smart Grid*, 10(2), 2127–2136. https://doi.org/10.1109/TSG.2017.2789333

[4] Ding, Z., & Lee, W. (2016). A Stochastic Microgrid Operation Scheme to Balance Between System Reliability and Greenhouse Gas Emission. *IEEE Transactions on Industry Applications*, 52(2), 1157–1166. https://doi.org/10.1109/TIA.2015.2490619

[5] Jeschke, S., Brecher, C., Meisen, T., Özdemir, D., & Eschert, T. (2017). Industrial Internet of Things and Cyber Manufacturing Systems. *Industrial Internet of Things: Springer*, 20, 3–19.

[6] Sun, Y., Song, H., Jara, A. J., & Bie, R. J. I. A. (2016). Internet of Things and Big Data Analytics for Smart and Connected Communities. *IEEE Access*, 4, 766–773.

[7] Sajjad, M. et al. (2020). A Novel CNN-GRU-Based Hybrid Approach for Short-Term Residential Load Forecasting. *IEEE Access*, 8, 143759–143768.

[8] Wang, Y. et al. Short-Term Load Forecasting for Industrial Customers Based on TCN-LightGBM. *IEEE Transactions on Power Systems*, 15, 45, 2020.

[9] Li, H. et al. (2017). Deep Reinforcement Learning: Framework, Applications, and Embedded Implementations. *2017 IEEE/ACM Int'l. Conf. Computer-Aided Design* (pp. 847–854).

[10] Ren, J. et al. (Sept./Oct. 2017). Serving at the Edge: A Scalable IoT Architecture Based on Transparent Computing. *IEEE Network*, 31(5), 96–105.

[11] Li, H. et al. (Jan./Feb. 2018). Learning IoT in Edge: Deep Learning for the Internet of Things with Edge Computing. *IEEE Network*, 32(1), 96–101.

[12] Angelis, F, Boaro, M, Fuselli, D, Squartini, S, Piazza, F, & Wei Q. (2013). Optimal Home Energy Management Under Dynamic Electrical and Thermal Constraints. *IEEE Transactions on Industrial Informatics*, 9(3), 1518–1527.

[13] Fan, W., Liu, N, & Zhang J. (2016). An Event-Triggered Online Energy Management Algorithm of Smart Home: Lyapunov Optimization Approach. *Energies*, 9(5), 381–404.

[14] Zhang, D., Li, S, Sun, M, & O'Neill, Z. (2016). An Optimal and Learning-Based Demand Response and HEMS. *IEEE Transactions on Smart Grid*, 7(4), 1790–1801.

[15] Ruelens, F, Claessens, B, Vandael, S, Schutter, B, Babuška, R, & Belmans, R. (2017). Residential Demand Response of Thermostatically Controlled Loads Using Batch Reinforcement Learning. *IEEE Transactions on Smart Grid*, 8(5), 2149–2159.

[16] Wei, T., Wang, Y., & Zhu, Q. (2017). Deep Reinforcement Learning for Building HVAC Control. *The 54th Annual Design Automation Conference*.

[17] Murty, V. V. S. N., & Kumar, A. (2020). Multi-Objective Energy Management in Microgrids with Hybrid Energy Sources and Battery Energy Storage Systems. *Protection and Control of Modern Power Systems*, 5. http://doi.org/10.1186/s41601-019-0147-z

[18] Xu, X., Jia, Y., Xu, Y., Xu, Z., Chai, S., & Lai, C. S. (2020). A Multi-Agent Reinforcement Learning-Based Data-Driven Method for Home Energy Management. *IEEE Transaction on Smart Grid*, 11(4), 3201–3211. http://doi.org/10.1109/TSG.2020.2971427

[19] Seyedzadeh, S., Rahimian, F.P., Glesk, I., & Roper, M. (2018). Machine Learning for Estimation of Building Energy Consumption and Performance: A Review. *Visualization in Engineering*, 6, 5. https://doi.org/10.1186/s40327-018-0064-7

[20] Deb, C., Zhang, F., Yang, J., Lee, S.E., & Shah, K.W. (2017). A Review on Time Series Forecasting Techniques for Building Energy Consumption. *Renewable and Sustainable Energy Reviews*, 74, 902–924. https://doi.org/10.1016/J.RSER.2017.02.085

[21] Ahmad, A. S., Hassan, M. Y., Abdullah, M. P., Rahman, H.A., Hussin, F., Abdullah, H., & Saidur, R. (2014). A Review on Applications of ANN and SVM for Building Electrical Energy Consumption Forecasting. *Renewable and Sustainable Energy Reviews*, 33, 102–109. https://doi.org/10.1016/J.RSER.2014.01.069

[22] Mocanu, E., Nguyen, P. H., Gibescu, M., & Kling, W.L. (2016). Deep Learning for Estimating Building Energy Consumption. *Sustainable Energy, Grids and Networks*, 6, 91–99. https://doi.org/10.1016/J.SEGAN.2016.02.005

[23] Robinson, C., Dilkina, B., Hubbs, J., Zhang, W., Guhathakurta, S., Brown, M. A., & Pendyala, R. M. (2017). Machine Learning Approaches for Estimating Commercial Building Energy Consumption. *Applied Energy*, 208, 889–904. https://doi.org/10.1016/J.APENERGY.2017.09.060

[24] Ahmed, S. S., Thiruvengadam, R., Shashank Karrthikeyaa, A.S., & Vijayaraghavan, V. (2020). *A Two-Fold Machine Learning Approach for Efficient Day-Ahead Load Prediction at Hourly Granularity for NYC*, Springer, pp. 84–97. https://doi.org/10.1007/978-3-030-12385-7_8

[25] Watkins, C. J., & Dayan, P. (1992). Q-learning. *Machine Learning*, 8, 279–292.

[26] N. Hassan et al. (Dec. 2013). Impact of Scheduling Flexibility on Demand Profile Flatness and User Inconvenience in Residential Smart Grid. *Energies*, 6(12), 6608–6635.

12 Technology-Driven Learning System in Higher Education
A Bibliometric Analysis

Risha Thakur, Anita Singh, and Bishwajeet Prakash
Sharda University, Greater Noida, India

12.1 INTRODUCTION

Adoption of LMS in higher education is widely spread [1] and it has proven its worth during pandemic by serving the universities operate smoothly and consistently. There is a shift from traditional educational mode to artificial intelligence (AI)-based education to provide better learning prospects to higher education students [2]. Remote learning has made the education system more adaptable and efficient [3] [4]. Technology should provide strong support in transformation to virtual classrooms. It has been observed in the recent years the significant role AI plays in remote classroom, in virtual education, and also in simplifying different administrative task [5, 6]. AI in higher education institution (HEI) sector is useful in solving problems and complexities related to education in the traditional teaching. It provides smarter educational technology solutions.

Emerging technology and AI integrated with LMS have accelerated the pace of e-learning in HEIs. LMS system helps to manage remote learning as well as support on campus learning. Implementation of the technology has raised several challenges and issues related to the use of technology for the faculty and staff [7]. Several researches have been conducted on different mediums to enhance the effectiveness and quality of using LMS in the universities. LMS plays a significant role in enhancing the teaching and learning experience. It has improved the student learning in collaborative environment and has enabled the instructors to focus on meaningful and effective designing of pedagogical activities [8]. In HEI, different learning management systems (LMSs) have been used such as Moodle, Black board, Canvas, Sakai, Atutor, and Google classroom; generally, they provide similar capabilities. One of them can be content management that allows posting text, sharing links, uploading and downloading files to the users, user account management that enables to control the access to courses and learning content. The activities like discussion forums, chats, private massaging, email, evaluation of assignments, quizzes, and grades are recorded and kept [9]. The methods used for analyzing LMS are productivity tools,

DOI: 10.1201/9781032656830-12

209

learning skills, and communication tools [10]. The devices like smart phones, laptop, desktops, and tablets enhance the flexibility in the learning process. Different terminologies like computer-based instruction (CBI), computer assisted instruction (CAI), and computer-assisted learning (CAL) have been used in reference to LMS to describe computer adoption and have been used in computer application programs, design preparation, teaching, and monitoring distribution of the content [11, 12]. Technology-driven learning system has been used as a platform that enables the learners to engage with each other and other activities like registration for the lectures, grade tracking, and for checking the updates and course announcement [13]. LMS with AI helps the learners in getting access to the content and the information shared by the instructors [12].

The education industry has embraced recent technological advancements and developments. These developments have made the learning process more comprehensive and simplified for students. Nowadays, the focus of information technology (IT) is on remaking colleges and universities through some primary themes such as adaptability, decision-making, and student outcomes [14]. In ICT or modern education, the technologies allow personalization and customization according to students' needs. HEIs have started to use modern technologies to improve efficiency, transformation, completeness, and social experience. Today, HEI face challenges in advancing digital equity [15] and adapting traditional organizational models to advance the future of the workplace. In addition, HEI faces funding, demographics, quality, and competition. Their success requires a change in the form of new teaching and learning that will prepare students for success. HEI continues to be slow to adopt technologies that are more advanced. Improving digital fluency and advancing digital equity are among the most significant challenges of slowing the adoption in higher education. It needs to find ways to adapt technology in order to attract students and teachers, operations, and cut costs.

The basic emphasis of this chapter is to have a look at the literature, extent of studies conducted, identify the growing need, and geographical distribution of technology-driven LMSs in higher education.

12.1.1 Key Contribution of the Chapter

The following are significant contribution of this chapter:

i. This chapter examines the growing pattern of literature on LMS, AI, and technology in higher education over a time period.
ii. The chapter is able to evaluate relevant authors based on the number of production of documents and total citation.
iii. To evaluate most compelling resources with references and citation area, and how countries authorship networking envisages and pertinent positions of developed and developing nations.

12.1.2 Chapter Organization

Section 12.1 presents the work related to technology-driven learning system, and its applicability in HEIs. Section 12.2 elaborates the materials and methods used for extracting the data using PRISMA method. Section 12.3 discusses the analysis and results. For analysis, descriptive bibliometric analysis has been used to have an overview of the research conducted in the past 22 years (2000–2022). Analysis has been done based on authors' contribution in the respective years, keywords used, total citation, affiliation, and co-authorship collaboration network. Section 12.4 indicates the future implication and Section 12.5 concludes the chapter.

12.2 METHODOLOGY

12.2.1 Data Collection

In the current situation Research Gate, Google Scholar, Academia, Web of Science, and Scopus are some of the extensively used sources in field of academia. Research Gate, Google Scholar, and Academia are those individual web-based data platforms which are popular for a wide range of disciplines together with author's network statistics with a forte in recovering literature evidences. Whereas Web of Sciences and Scopus hold a competitive advantage regarding assessing output through analysis of citations and relevancy [16]. Furthermore, Scopus' journal distribution was considerably larger than Web of Sciences' [17]. Thus, for an eminent data collection, Scopus was a superlative choice. The main three keywords in this research are 'higher education', 'learning management system', and 'artificial intelligence'. The preliminary request was entered in the advanced Scopus-based search engine and after that the essential filters were applied as mentioned in Figure 12.1. First keyword input was (learning AND management AND system OR artificial OR intelligence AND higher AND education). The introductory investigations outcome was n = 6,026. The PRISMA procedure was

FIGURE 12.1   Keyword input based investigation.

(Source: Author Computation.)

used for the methodical assessment of research with the aim of attaining the suitable relevance and coalition with the keywords [18]. Few screening filters were applied to eradicate unrelated published documents. The filtration standards are given below:

- Year range: (2000–2022)
- Type of document: (Journal + Book + Book Series + Conference Proceedings)
- Discipline of research: (Business Management Accounting + Social Science + Psychology + Arts & Humanities + Economics, Econometrics and Finance + Decision Science)
- Language: English
- Open access published documents

The entire collection of data was whittled down through filtrations. The titles, abstracts, and full-texts of publications were subsequently scanned to establish relevancy. For the aim of this investigation, the concluding data of (n = 744) was extracted from the Scopus database on 15 August 2023 in CSV format.

12.3 ANALYSIS AND RESULTS

After the extraction, the required dataset was transferred into Microsoft Excel file. The folder contained a total of 774 entries of published documents between the period of 2000 and 2022, it includes detailed archives of data with an annual growth rate of 25.65% related to research authors (2559), authors document keywords (2282), document type (articles—673, books—8, book chapters—11, and conference papers—82), relevant source (350), number of total citations, year of document publication, and published document detailed references (30,054) as stated in Table. 12.1. This data was applied for the completion of a systematic bibliometric analysis. A combination of three software's, R studio 4.3.1 version, (bibliometrics with Biblioshiny package), VOS viewer 1.6.19 version, and Microsoft Excel was used in this research to provide a significant contribution to this chapter mentioned above.

It was published in the journal of vocational education training and absorbed the interplay of information and communications technology systems in vocational education sphere. The paper created a niche for the forthcoming emerging LMS-based research. The embryonic phase was from the year 2000 to 2008 when the number of publications per year was below 10 after then number of publications per year started to rise continuously from the year of 2009 to 2019, the per year publication number of documents was in double digits and then in the thriving phase in the years of 2020, 2021, and 2022, the publications reached to three-digit numbers per year which were 117, 129, and 152 published documents. Thus, the inclusive Scopus indexed open access documents published in English language which are related to LMS, higher education, and AI are 774 in total.

Top 10 highest appropriate authors founded on the quantity of documents published and total citations. Authors total citations and the number of documents published were calculated as in Table 12.2. The most fruitful scholar based on document publications is Bervell. B (6), then both Smith. S (5) and Umar (5) have the same number of publications followed by Kumar Jha (4), subsequently all the succeeding

TABLE 12.1

Elementary Statistics of the Extracted Data

Description	Results
Year range	2000:2022
Document sources	350
Overall documents	774
Annual progress percentage	25.65
Document average phase	4.9
Average citations per document	16.11
Total references	30054
Total author's keywords	2282
Total authors number	2428
Total authors of single-authored documents	131
Single-authored published documents	131
Total co-authors per document	3.3
Total international co-authorships percentage	20.28
Article count	673
Book count	8
Book chapter count	11
Conference paper count	82

Source: Author's computation.

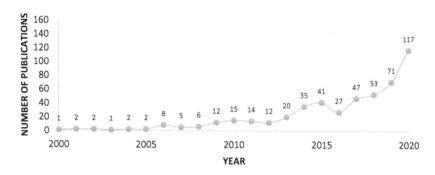

FIGURE 12.2 Growth of literature over the time period.

(Source: Author's computation.)

six scholars from the table have published (3) documents. But when the total citation was projected, the author with highest total citations is Mishra L (785) and then Gupta. T and Shree. A, with a similar total citation of (783) followed by Binu. Vs (268). It is thought provoking to see that the authors with top three total citations have same starting publishing year (2020), this shows that their work has been pragmatically stimulating with comprehensive evidences during the tough COVID-19 period as AI and ICT were offered to all the higher education learning services.

TABLE 12.2

Most Relevant Authors Based on Number of Documents Published Over the Years and Total Citations

Author	Number of Published Documents	Author	Total Citation	Publishing Year start
Bervell B	6	Mishra L	785	2020
Smith S	5	Gupta T	783	2020
Umar In	5	Shree A	783	2020
Kumar Jha	4	Binu Vs	268	2007
Ayub Afm	3	Menezes Rg	268	2007
Dawson S	3	Mukhopadhyay C	268	2007
Kuleto V	3	Ray B	268	2007
Lobos K	3	Shankar Pr	268	2007
Maphalala Mc	3	Sreeram Reddy Ct	268	2007
Mpungose Cb	3	Raelin JA	250	2007

Source: Literatures related to higher education period 2000–2022.

In Table.12.3, the source with the highest number of published documents over the year is IJETL (53) whose scope is 'learning technology', on the second place, it is 'sustainability' (Switzerland) (36) which focuses on the areas of challenges relating to sustainability, socio-economic, scientific, and integrated approaches to sustainable development followed by source IRRODL (19) whose research scope is in distance education. All the relevant top sources have analogous scope of research which is unified to technology, higher education, and societal sustenance[19].

In Table.12.4, the published document in the journal of *IJERO* was authored by Mishra L [20] evinced the highest total citations of 783 with a total citation rate per year of 195.75. This article takes a combination of qualitative and quantitative techniques to study the teachers' and students' perspectives regarding internet-based teaching-learning methods and highlights the steps involved in the execution of technology-aided electronic learning-teaching approaches to deal with the ongoing academic disruption caused by the COVID-19 pandemic.

Reddy CT [21] published document in the journal of *BMC Medical Education* has obtained the second highest total citations of 268 with a total citation rate per year of 15.76. This article attempts to examine the growing incidence of psychological illness, the origins, and seriousness of stress, along with coping techniques in the comprehensive problem-stimulated higher education medicinal education through technical resources. Finally, the third highest total cited is Raelin Ja [22] with a total citation rate per year of 14.71 published in the journal of *Academy of Management Learning and Education*. This study was focused on a novel epistemology grounded in practice that incorporates praxis into classroom instruction to assist students in deconstructing the frameworks and procedures that shape their social surroundings and it also investigates the results and specific qualities that result from hands-on instruction through AI-based means. Shahzada A, article titled, 'Effects of COVID-19 in E-learning on

TABLE 12.3

Top 10 Sources Based on Quantity of Documents Published (2000–2022)

Sources	Articles	Scope
International Journal of Emerging Technologies in Learning	53	Learning technology
Sustainability (Switzerland)	36	Challenges relating to sustainability, Socio-economic, scientific, and integrated approaches to sustainable development
International Review of Research in Open and Distance Learning	19	Distance education
Australasian Journal of Educational Technology	18	Technology in tertiary education
Education Sciences	16	Technology in tertiary education
BMC Medical Education	14	Curriculum development, evaluations of performance, medical education
IFIP Advances in Information and Communication Technology	14	ICT in education
Education and Information Technologies	13	ICT in education
Mediterranean Journal of Social Sciences	11	Social science and educational research
Frontiers in Psychology	10	Psychology and education

Source: Literatures related to higher education period 2000–2022.

higher education institution students: the group comparison between male and female' with 77.33 TC per year has determined the E-learning portal success based on literature [23]. Bond M. published an article titled, 'Digital transformation in German higher education: student and teacher perceptions and usage of digital media' and cited with 36 total citation per year. This study was conducted to understand the teaching and learning in German University. Finding of the study suggests that teachers and the students of the university were making limited use of digital technology for assimilative task and LMS were perceived as most significant and useful tool for HEI [24].

Galway LP in the paper showcased the novel approach on design, implementation, and evaluation. Pre- and post-course surveys were used to assess the changes in the self-perceived knowledge. The study included the comparison between traditional lecture-based model of teaching with flipped classroom. The findings of the study suggested that students rated their flipped classroom teaching higher and their course experience was positive and their self-perceived knowledge enhanced [25].

Martin F. conducted systematic review on online learning research to understand the learners outcome and the tendency to classify research like engagement [26]. Nouri J in his article titled, 'The flipped classroom: for active, effective, and increased learning—especially for low achievers' with 22.38 total citation per year examined the perception of the students toward flipped classroom education in research method course. It was concluded the students perceived the positively toward flipped classroom, Moodle, and use of video. It was observed that student's perception of

TABLE 12.4
Top 10 Highest Citation Documents

Author	Article Title	Total Citations	Total Citations per Year	Publisher
Mishra et al. [20]	Online teaching-learning in higher education during lockdown period of COVID-19 pandemic	783	195.75	Elsevier
Sreerama Reddy et al. [21]	Psychological morbidity, sources of stress, and coping strategies among undergraduate medical students of Nepal	268	15.76	Springer Nature
Raelin [22]	Toward an epistemology of practice	250	14.71	George Washington University
Shahzad et al. [23]	Effects of COVID-19 in E-learning on higher education institution students: the group comparison between male and female	232	77.33	Springer Nature
Bond et al. [24]	Digital transformation in German higher education: student and teacher perceptions and usage of digital media	216	36	Springer Open
Galway et al. [25]	A novel integration of online and flipped classroom instructional models in public health higher education	186	18.6	Springer Nature
Martin et al. [26]	A systematic review of research on online teaching and learning from 2009 to 2018	184	46	Elsevier
Nouri [27]	The flipped classroom: for active, effective, and increased learning—especially for low achievers	179	22.38	Springer Open
Sobaih et al. [28]	Responses to COVID-19 in higher education: social media usage for sustaining formal academic communication in developing countries	145	36.25	MDPI
Barrot et al. [29]	Students' online learning challenges during the pandemic and how they cope with them: the case of the Philippines	133	44.33	Springer Nature

Source: Literatures related to higher education period 2000–2022.

enhanced motivation, engagement effective and enhanced learning [27]. Sobaih AEE conducted an empirical research to examine the degree to which faculty members and the students use social media for academic and official communication. The result indicated that social media usage has been promoted for the effective and sustainable teaching and learning [28]. Barrot JS (2021) has made an attempt to use mix method approach and finding indicates the extent to which online learning challenges were related to learning environment at home, techno literacy, and competency. The finding of the study suggested that there is a strong impact on the quality of learning experience and the mental health of the students [29].

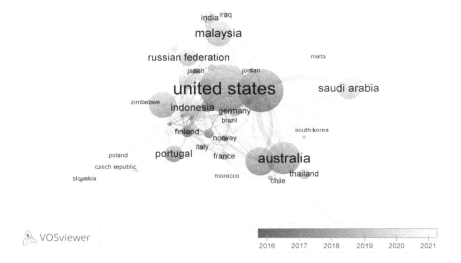

FIGURE 12.3   Country overlay visualization based on co-authorship networking.

Thus, it can be observed from the above studies that technology and LMS play a significant role in enhancing the teaching-learning experience in higher education.

The selection of a minimum of two documents and two citations, which resulted in a threshold of 75 countries (refer Table 12.3). Then a 11-cluster figure is shaped, which includes 69 items. The size density of the color from dark to light portrays the cumulative occurrence of the country, and the number and shape of nodes reveal the dominance of a nation, while the breadth of every link reveals the overlaps among two countries. In the first cluster, there are 11 countries (Canada, China, Hong Kong, Jordan, Kazakhstan, Mexico, the Russian Federation, Taiwan, Ukraine, the United Arab Emirates, and the United States of America), and the United States of America leads the cluster with 17 links, 26 total link strengths and has published 102 documents. In the second cluster, there are 9 countries (Austria, Czech Republic, Hungary, Italy, Poland, Portugal, Romania, Serbia, and Slovakia), and Portugal is governing with 9 links, 10 total link strengths, and has published 28 documents. In the third cluster, there are 9 countries (Brazil, Denmark, Finland, Germany, New Zealand, Norway, Oman, Singapore, and Sweden), and Germany is leading with 28 links, 31 total link strengths, and has published 16 documents. Then in the fourth cluster, there are 8 countries (Croatia, Ethiopia, Greece, Ireland, Latevia, Lithuania, the Netherlands, and Slovenia), and the Netherlands is leading with 19 links, 23 total link strengths, and has published 13 documents. In the fifth cluster, there are 7 countries (Bangladesh, Ghana, India, Iraq, Japan, Malaysia, and Nepal), and Malaysia is foremost with 13 links, 22 total link strengths, and has published 47 documents. In the sixth cluster, there are 6 countries (Belgium, Chile, Ecuador, France, Morrocco, Spain), of which Spain is top with 14 links, 24 total link strengths, and has published 54 documents. Cluster seven comprises six countries (Australia, Colombia, Israel, Palestine, South Korea, and Thailand), of which Australia dominates with 20 links, 29 total link strengths, and has published 63 documents. The eighth cluster has 4 countries (Malta, Nigeria, Turkey, and the United Kingdom), and the United Kingdom has 39 links, 74 total link strengths, and has published 90 documents. The ninth cluster has 4 countries (Bahrain, Egypt, Saudi Arabia, and Tunisia), and Saudi Arabia is leading with 12 links, 21 total link strengths, and has published 37 documents. In the tenth cluster, there are 3 countries (Indonesia, Pakistan, and the Philippines), and Indonesia is leading with 21 links, 24 total link strengths, and has published 31 documents. Then, in the last 11th cluster, there are only Zimbabwe and South Africa, and South Africa is leading with 21 links, 22 total link strengths, and has published 48 documents.

It has been observed that developed nations have secured all the top three ranks in both occurrences and citations category. The 'United States' is a vastly developed nation who has a maximum number of documents published which are 102 and has the second-best number of citations 1517 along with incredible 17 links and 26 total link strength (Table 12.5).

Then the nation with top citation is another developed nation 'United Kingdom' with 1848 citations, 90 occurrences, and 39 links and total link strength of 74. It is stimulating to find 'Australia' on the third rank for both best citation and occurrences of 1262, 63 with total links 20 and 29 total link strength. The flag bearing developing nations with total occurrences and citations of (17, 1135) India, (37, 1002) Saudi Arabia, (47, 606) Malaysia, and (48, 464) South Africa have emerged as significant protagonist who are attempting to engage in the field of LMSs-equipped higher education in order to discover novel approaches through technological advancement for the inclusive economic progress of their respective countries.

The co-occurrence-author keyword scrutiny with a minimum number of 10 was selected in the author's keyword occurrences that exhibit out of 2282 keywords only 25 keywords meet the required threshold.

The dimensions of nodes provide insight into the prevalence of keywords while the width of each link displayed the overlays among two keywords. The authors' keyword with the highest occurrences of 219 is 'higher education' with 23 links and

---

TABLE 12.5
Developed and Developing Countries Overall Document Production and Total Citation Over the Years

Country	Documents	Status of Country	Country	Citations	Status of Country
United States	102	Developed	United Kingdom	1848	Developed
United Kingdom	90	Developed	United States	1517	Developed
Australia	63	Developed	Australia	1262	Developed
Spain	54	Developed	India	1135	Developing
South Africa	48	Developing	Saudi Arabia	1002	Developing
Malaysia	47	Developing	Spain	1000	Developed
Saudi Arabia	37	Developing	Malaysia	606	Developing
Indonesia	31	Developing	Canada	508	Developed
China	30	Developed	South Africa	464	Developing
Russian federation	30	Developing	China	425	Developed

Source: Author computation.

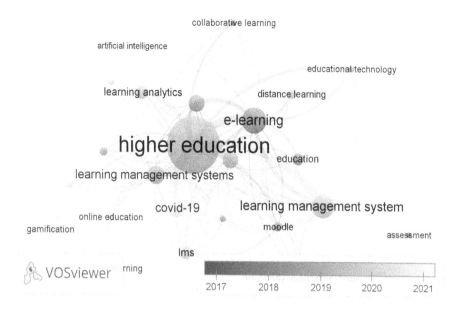

FIGURE 12.4   The global trending research topics.

194 total link strength. Next keyword 'learning' has 75 occurrences with 19 links and 81 total link strength. Closely followed by 'learning management system' which has 67 occurrences with 20 links and 73 total link strength. The progressive frequency of keywords was mirrored through the color encompassing from intense purple to light yellow thus, 'Covid-19' is yellow because it is a recent emerging keyword which has 46 occurrences with 17 links and 66 total link strength along, 'technology acceptance' with 12 occurrences with 10 links and 16 total link strength, 'student engagement' with 11 occurrences with 17 links and 9 total link strength, whereas 'artificial intelligence' has a total of 10 occurrences with 7 links and 6 total link strength.

Thus, 'Artificial Intelligence' became more prominent during COVID-19 times and provided recurrent patronage to 'learning management system' in enabling 'higher education' during tough times. As LMS has proven to be useful in assisting academicians in improving educational and instructional performance.

12.4   FUTURE IMPLICATION

There is continuous demand by society at large, governments, and funding sources for determining the efficiency of research and its significance, as well as the accomplishment of academic endeavors and academic collaborations across the regional and national levels. Through this study, an attempt has been made to project recent additions and performance of developing and developed nations on the mentioned subjects, rising authors, and documents to assist scholars as well as educators in their quest for information regarding LMSs discourse in higher education. Further by referring to the research methodologies of developed nations, the government of developing countries can determine the primary area for advancement and boost the economy's efficiency through the appropriate implementation of technologically oriented AI as well as to enhance the effectiveness by improving higher education policies and schemes. Corporate and managers may project and prepare new improvements with scientific techniques to address gaps and provide novel solutions based on AI. For scholars, this paper can act as a reference to understand the emerging trends and how they are influential in embracing LMSs for the developing nation's higher education along with the recommendation of the most influential sources for the manuscript submission. Apart from this, the chapter provides a thorough outline regarding the diversification of author's keywords in the research which eventually generates an area of future research scope for the scholars. Finally, certain critical themes such as student engagement, AI, technological adoption, and online education in conjunction with LMSs in higher education should be investigated further in order to comprehend the real-world application of these dynamic terms.

12.5   CONCLUSION

Technology is an incipient module in higher education and it will continue to exist for an extended segment. When it comes to technology-enabled higher education research, AI has improved the functioning of LMS drastically which shaped an absolute advanced dimension in the higher education leaning of the students. Major top sources are involved in the fields of technology, learning, and higher education, demonstrating how technical skills can be linked to socially advanced education. The range of publications has increased recently, mainly because of COVID-19, technology acceptance, and the widespread implementation of LMSs in higher education. This study focused on the participation of both developed and developing nations with respect to the field of LMS-based higher education, with the intention of distinguishing the respective nation's status in terms of both document output and the widespread adoption associated with their published work. It was discovered that developed nations outperformed developing nations in the areas of citation and production domains, partly because of an elevated level of comprehension which led to early implementation of AI. Developing nations saw an instant rise in the number of research projects prominently after 2020 and a developing country author Mishra L contributed the most cited document published by Elsevier. This happened because of the problems incurred during COVID-19. During that time period, the governments of several nations launched various approaches that supported the advanced use of LMSs in educational institutions. This study provides an overview of significant writers and documents which influenced the debate on LMSs in higher education since the author's keyword mapping is very dispersed, and there appears to be some strong themes; therefore, the subject matters that are intriguing within the higher education, LMS and technology publications, are diverse.

REFERENCES

[1] Cavus, N., Sani, A.S., Haruna, Y., & Lawan, A.A. (2021), Efficacy of social networking sites for sustainable education in the era of COVID-19: A systematic review. *Sustainability*, 13, 808.

[2] Singh, A., Sharma, R., & Thakur, R. (2022), E-learning: A tool for sustainability in the education industry. *Revolutionizing Business Practices through Artificial Intelligence and Data-Rich Environments* (pp. 73–92). IGI Global.

[3] Meenu, E. (2021), Students can now argue with an AI system for extra marks. *Analytics Insight.* https://www.analyticsinsight.net/students-can-now-argue-with-an-ai-system-for-extra

[4] Claudiu, C., Laurent, G., Luiza, M., & Carmen, S. (2020), Online teaching and learning in higher education during the coronavirus pandemic: Students' perspective. *Sustainability*, 11, 1–24.

[5] Johnson, A. (2019), 5 ways AI is changing the education industry. [online] https://learningindustry.com/ai-is-changing-the

[6] Folorunso, S.O. and All, "Prediction of Student's Academic Performance Using Learning Analytics", Lecture Notes in Networks and Systems, Volume 837 LNNS, Pages 314 – 325, 2024, DOI: 10.1007/978-3-031-48465-0_41

[7] Azlim, M., Husain, K., Hussin, B., & Maksom, Z. (2015), Designing collaborative learning wizard to assist instructors in utilizing tools in LMS. *IEEE Conference on e-Learning, e-Management and e-Services (IC3e)*, Melaka, Malaysia, pp. 81–85, https://doi.org/10.1109/IC3e.2015.7403491

[8] Turnbull, D., Chugh, R., & Luck, J. (2019), Learning management systems: An overview. In A. Tatnall (Ed.), *Encyclopedia of Education and Information Technologies.*

[9] Kaewsaiha, P. (2019), Usability of the learning management system and choices of alternative. *6th International Conference on Education and Psychological Sciences (ICEPS 2019)*. Tokyo, Japan.

[10] Kehrwald, B. A., & Parker, B. (2019), Implementing online learning: Stories from the field. *Journal of University Teaching and Learning Practice*, 16(1), 1.

[11] Bradley, V. M. (2021) Learning Management System (LMS) use with online instruction. *International Journal of Technology in Education (IJTE)*, 4(1), 68–92.

[12] Jung, S., & Huh, J. H. (2019), An efficient LMS platform and its test bed. *Electronics*, 8(2), 154.

[13] Al-Fraihat, D., Joy, M., Masádeh, R., & Sinclair, J., (2020), Evaluating e-learning systems success: An empirical study. *Computers in Human Behavior*, 102(1), 67–86.

[14] Adeniyi, A.E. and All, "Comparative Study for Predicting Melanoma Skin Cancer Using Linear Discriminant Analysis (LDA) and Classification Algorithms", Lecture Notes in Networks and Systems, Volume 837 LNNS, Pages 326 – 338, 2024, DOI: 10.1007/978-3-031-48465-0_42

[15] Elmes, J. (2017), Six significant challenges for technology in higher education in 2017.

[16] Falagas, M. E., Pitsouni, E. I., Malietzis, G. A., & Pappas, G. (2008), Comparison of Pub Med, Scopus, Web of Science, and Google Scholar: Strengths and weaknesses. *The FASEB Journal*, 22(2). https://doi.org/10.1096/fj.07-9492lsf

[17] Hallinger, P., & Nguyen, V. T. (2020), Mapping the landscape and structure of research on education for sustainable development: A bibliometric review. *Sustainability (Switzerland)*, 12 (5). https://doi.org/10.3390/su12051947

[18] Moher, D., Liberati, A., Tetzlaff, J., Altman, D., Galtman, D., Antes, G., Atkins, D., Barbour, V., Barrowman, N., Berlin, J. A., Clark, J., Clarke, M., Cook, D., D'Amico, R., Deeks, J.J., Devereaux, P.J., Dickersin, K., Egger, M., Ernst, E., & Tugwell, P. (2009), Preferred reporting items for systematic reviews and meta-analyses: The PRISMA statement. *PLoS Medicine*, 6(7). https://doi.org/10.1371/journal.pmed.1000097

[19] Oxtoby, R. (2000), Agendas for institutional development: Talking to college principals, England and Australia. *Journal of Vocational Education and Training*, 52(2), 305–328.

[20] Mishra, L., Gupta, T., & Shree, A. (2020), Online teaching-learning in higher education during lockdown period of COVID-19 pandemic. *International Journal of Educational Research*, 1, 100012. https://doi.org/10.1016/j.ijedro.2020.100012

[21] Sreeramareddy, C.T., Shankar, P.R., Binu, V., Mukhopadhyay, C., Ray, B., & Menezes, R.G. (2007), Psychological morbidity, sources of stress, and coping strategies among undergraduate medical students of nepal. *BMC Medical Education*, 7, 26. https://doi.org/10.1186/1472-6920-7-26

[22] Raelin, J.A. (2007), Toward an epistemology of practice. *Academy of Management Learning & Education*, 6(4), 495–519.

[23] Shahzad, A., Hassan, R., Aremu, A. Y., Hussain, A., & Lodhi, R. N. (2021). Effects of COVID-19 in E-learning on higher education institution students: The group comparison between male and female. *Quality & Quantity*, 55, 805–826.

[24] Bond, M., Marín, V. I., Dolch, C., Bedenlier, S., & Zawacki-Richter, O. (2018). Digital transformation in German higher education: Student and teacher perceptions and usage of digital media. *International Journal of Educational Technology in Higher Education*, 15(1), 1–20.

[25] Galway, L.P., Corbett, K.K., Takaro, T.K., Tairyan, K.,& Frank, E. (2014), A novel integration of online and flipped classroom instructional models in public health higher education., *BMC Medical Education*, 14(1), 1–9.

[26] Martin, F., Sun, T., & Westine, C.D. (2020), A systematic review of research on online teaching and learning from 2009 to 2018. *Computers & Education*, 159, 104009.

[27] Nouri, J. (2016), The flipped classroom: For active, effective, and increased learning–especially for low achievers, *International Journal of Educational Technology in Higher Education*, 13(1), 1–10.

[28] Sobaih, A.E.E., Hasanein, A.M., & Abu Elnasr, A.E. (2020), Responses to COVID-19 in higher education: Social media usage for sustaining formal academic communication in developing countries. *Sustainability*, 12(16), 6520.

[29] Barrot, J.S., Llenares, I.I., & Del Rosario, L.S. (2021), Students' online learning challenges during the pandemic and how they cope with them: The case of the Philippines. *Education and Information Technologies*, 26(6), 7321–7338.

13 A Time-Varying JSON Data Model and Its Query Language for the Management of Evolving Green Things in IoGT-Based Systems

Zouhaier Brahmia and Safa Brahmia
University of Sfax, Sfax, Tunisia

Fabio Grandi
University of Bologna, Bologna, Italy

13.1 INTRODUCTION

Nowadays, green computing [1, 2], also called sustainable computing or green information technology (IT), aims at saving the environment, and in particular the environment of computers, servers, and related devices like printers, scanners, monitors, and network and communication equipment. The Internet of Things (IoT) connects (to Internet), within a smart environment, people and things (sensors, devices, etc.) located anywhere, at any time, with anyone and anything, via the use of any link and any service, in order to provide intelligent and advanced services to such people [3]. Hence, since IoT technology requires a lot of energy consumption, green IoT [3–9] was proposed with an initial goal of minimizing power consumption by IoT. Internet of Green Things (IoGT) [10, 11] is a technology that allows exchanging information between people and healthy farm things. It provides information like soil moisture, temperature, humidity, and nutrient level via the use of appropriate sensors. Besides, IoGT aims at being useful in the fight against climate change that is increasing day by day and is a threat to humanity, as shown by several scientific researchers [12–16].

IoGT sensor data represent entities which are evolving over time, and several (smart) farming applications [17, 18] and climate change applications [19, 20] require bookkeeping of the entire history of such data. Furthermore, IoGT data could be considered as big data [21–24] since they satisfy the three Vs of big data (Volume, Velocity, and Variety), and the standard JSON (JavaScript Object Notation) format [25] is considered one of the best data formats to efficiently manage (i.e., model, store, and exchange) big data. Nevertheless, to the best of our knowledge, there is no standard time-varying data model [26] for IoGT data management, on the one hand, and the JSON format does not provide any explicit and native support for handling temporal data, on the other hand. Hence, in order to manage the evolution over time of big data and time-varying IoGT data, both designers and developers of applications that are dedicated to smart farming or climate change should proceed in an ad hoc manner. In order to fill this gap, we propose in this chapter Temporal JSON IoGT (TIoGT) data model, a temporal extension of the JSON data model, for an efficient management of temporal and evolution aspects of data and green things in IoGT-based systems. Moreover, with the aim of making our model useful and practical, we define a user-friendly temporal query language [27] for such a model, named Query Language for TIoGT (QL4TIoGT), to be used for querying time-varying green things and exploiting TIoGT data. QL4TIoGT is built as a temporal extension of the JSONPath language [28, 29], which is a JSON query language that allows to retrieve data from a non-temporal JSON document. The extension consists in augmenting JSONPath with a large set of temporal functions that allow to satisfy any user's requirement concerning TIoGT data querying.

The remainder of this chapter is structured as follows. Section 13.2 studies related work. Section 13.3 proposes TIoGT, our temporal JSON data model for IoGT. Section 13.4 presents QL4TIoGT, our temporal query language for querying TIoGT documents. Section 13.5 summarizes the chapter and provides some remarks on our future work.

13.2 RELATED WORK

In this section, we briefly review the most recent research work related to the main topic of our chapter. Indeed, we deal with green computing (Section 13.2.1), green IoT and IoGT (Section 13.2.2), temporal big data modeling (Section 13.2.3), and temporal big data querying and analytics (Section 13.2.4).

13.2.1 GREEN COMPUTING

Since the literature of green computing is very rich, and due to space limitations, we restrict our study to a subset of the important contributions.

Shree et al. [30] show how cloud computing is dangerous for the environment due to the emission of lots of harmful gases (e.g., carbon dioxide), the lots of wasted energies, and the big number of resources and equipment that are working 24/24 and 7/7 and producing lots of heat. The authors also show how green computing, in cloud computing, can resolve such problems in an eco-friendly way.

In Ref. [31], the authors have defined a dynamic energy-aware cloudlet-based mobile cloud computing model. This latter assigns, manages, and optimizes the cloud framework services and uses, based on cloudlets, to reach green computing.

Recently (in 2023), a comprehensive review of green computing research contributions has been provided in Ref. [32]. Akin [33] has dealt with green computing in hospitality. Josephson et al. [34] have focused on the untapped opportunities for green computing.

13.2.2 Green IoT and IoGT

13.2.2.1 Green IoT

In Ref. [3], the authors have reviewed, for green IoT, (i) its main concepts (e.g., IoT); (ii) the technologies that are necessary to build a green IoT platform, like green RFID (Radio Frequency IDentification) tags, green sensing network, and green cloud computing network; (iii) its applications, like smart cities, smart energy systems, smart grids, smart factories, smart health systems, and smart logistics; and (iv) its open challenges. The life cycle of green IoT, which includes green design, green production, green utilization, and green recycling, has also been investigated.

Sarkar and Misra [4] have modeled fog computing as a green computing framework to support the requirements of IoT applications.

In Ref. [6], the authors have focused on green computing in Social Internet of Vehicles (SIoV) environments.

Tan et al. [8] propose a blockchain-based system for unified access control of (heterogeneous) green smart devices (GSD).

Recently, Shanmugapriya [9] has surveyed energy management evolution and techniques for green IoT environments. Some challenges of green IoT have also been presented.

13.2.2.2 IoGT

In Ref. [10], the authors have proposed, in the IoGT, a system that is based on both the Message Queuing Telemetry Transport (MQTT) protocol and the Flask framework, to access and control low-cost ESP8266 smart robots. Data are gathered by these robots. Some cloud-like database server is used to store the collected data. The authors have also stressed the help of the JSON format for exchanging data between robots and the importance of integrating robotics and IoGT for farming (e.g., for irrigation).

Shetty and Ramaiah [11] deal with the Crop Internet of Things (CIoT), in an Indian context. It aims at monitoring several characteristics related to soil (e.g., temperature, humidity, and pH level), while using appropriate sensors. The authors show how CIoT could improve crop production, at both quantity and quality levels, via the use of big data analytics and machine-learning techniques.

13.2.3 Temporal Big Data Modeling

There are various proposals that concern temporal big data modeling.

Brahmia et al. [35] have proposed the Temporal JSON Schema (τJSchema) infrastructure that allows to create a temporal JSON document and to validate it against some τJSchema. This latter is created from a conventional JSON Schema and a set of temporal logical and physical characteristics specified on this conventional schema. From a temporal database [36] point of view, τJSchema supports both transaction time [37] (i.e., the time when data are current in the database) and valid time [38] (i.e., the time when data are valid in the modeled reality). From a schema evolution [39] point of view, τJSchema supports transaction-time schema versioning [40].

In Ref. [41], the authors have proposed a temporal JSON data model defined in a formal way. This model supports only valid-time data.

Goyal and Dyreson [42] have proposed to model temporal JSON as a virtual document that merges both time metadata and JSON data.

In Ref. [43], the authors have proposed TempoJCM, a temporal JSON conceptual data model. It allows to build, in a graphical way, a (non-temporal) JSON conceptual data model while declaring any concerned model component as being of transaction-time, valid-time, or bitemporal format.

TempoJCM++ [44] extends the TempoJCM model to support conceptual modeling of scheduled schema changes.

Besides, Baumann et al. [45] have dealt (among others) with modeling of spatio-temporal big data in JSON. Mehmood et al. [46] have focused on modeling of temporal sensor data under the MongoDB DataBase Management System (DBMS).

13.2.4 TEMPORAL BIG DATA ANALYTICS AND QUERYING

13.2.4.1 Temporal Big Data Analytics

In Ref. [47], Cuzzocrea has presented temporal big data analytics as a field that focuses on how to model, capture, and analyze temporal aspects of big data during analytics phase. It involves complex tasks like temporal versioning of big data and temporal big data querying. Cuzzocrea [47] has reviewed the different contributions dealing with models, paradigms, and techniques of temporal big data analytics, and published on or before 2020. Some interesting open research challenges for this issue have also been presented like temporal big data representation, time-slice queries on temporal big data repositories, temporal big data analytics that is based on machine learning, uncertain and imprecise (temporal) big data in temporal big data analytics, and preserving privacy of (temporal) big data in temporal big data analytics.

Brahmia et al. [48–50] have dealt with temporal versioning of JSON-based big data, in the τJSchema framework. In Ref. [48], they have studied τJSchema versioning by focusing only on changes to conventional JSON Schema; however, in Ref. [49], they have focused on temporal schema versioning driven by changes to temporal logical and physical characteristics. In Ref. [50], in order to make easy the schema versioning process, they proposed three sets of high-level schema change operations: the first acts on conventional JSON Schema, the second on temporal characteristics, and the third on τJSchema. The authors confirm that these high-level operations are more user-friendly for the designer and preserve the consistency of the changed schema component.

Mach-Król [51] proposed a conceptual framework for the implementation of temporal big data analytics in enterprises. The work was done while applying the Design Science Research in Information Systems (DSRIS) methodology [52].

Besides, it is worth mentioning that several recent works have focused on spatio-temporal big data analytics, like [53], [54], and [55].

13.2.4.2 Temporal Big Data Querying

The literature of big data includes several proposals for querying non-temporal big data like JSONPath [28, 29], JSONiq [56], Jaql [57], JNL [58], and JSON-λ [59].

Moreover, a standard language, called SQL/JSON [60], has been proposed as an extension of SQL to make it supporting JSON data querying.

Besides, each Not Only SQL (NoSQL) DBMS has its own/proprietary non-temporal query language, like MQL of MongoDB, CQL of Cassandra, N1QL of Couchbase, and AQL of AsterixDB.

Notice that Sharma and Gadia [61] have proposed an extension of the ParaSQL language to support querying spatio-temporal big data. Nevertheless, to the best of our knowledge, no temporal query language for (temporal) big data has already been proposed in the literature.

13.3 TIoGT: THE DATA MODEL FOR TIME-VARYING JSON IoGT DATA

In this section, we propose our temporal JSON data model, named TIoGT, which allows to represent temporal aspects of IoGT data. TIoGT is a temporal extension of the JSON data model. The extension consists in enhancing the JSON data model to support the representation of the temporal aspects of JSON model components (i.e., objects, object members, arrays, and array items) that are evolving over time and for which the database administrator or the application developer wants to keep the full history of their evolution in the database.

The temporal dimensions that are supported by TIoGT are transaction time and valid time. Therefore, a temporal JSON model component may have a transaction-time, valid-time, or bitemporal [62] format, according to the temporal axis (or axes) along which its history is kept.

If a JSON model component is not temporal (i.e., its value is not time-varying), it is called a snapshot JSON model component.

In our TIoGT data model, time-varying JSON model components are versioned, and versions are timestamped with transaction- and/or valid-time as follows:

- a transaction-time component version has two timestamps that represent the bounds of its transaction-time interval: TTbegin (transaction time begin) and TTend (transaction time end).
- a valid-time component version has two timestamps that represent the bounds of its valid-time interval: VTstart (validity time start) and VTend (validity time end).
- a bitemporal component version has four timestamps that represent the bounds of its transaction-time interval (TTbegin and TTend) and those of its valid-time interval (VTstart and VTend).

Timestamps are automatically managed by the JSON-based NoSQL DBMS (e.g., MongoDB), in a transparent manner to the user.

A snapshot JSON component has no timestamp.

In particular, in the TIoGT model, an individual JSON model component, named "comp" and with "value" contents, is made temporal by transforming it into an ordered list of temporal (i.e., timestamped) versions, each of which is defined as a JSON object with three (in case of a transaction-time or a valid-time JSON model component) or five (in case of a bitemporal JSON model component) JSON object members, grouped together in a JSON array named "comp" as follows (in this case, "comp" is a valid-time JSON model component):

"comp":[{ "value":…, "VTstart":…, "VTend":… },

{ "value":…, "VTstart":…, "VTend":… },

{ "value":…, "VTstart":…, "VTend":… }]

The object member "value" provides the contents of a version; it can be a simple member (e.g., with integer or string value) or structured one (e.g., a set of object members when the version has several properties like those of a weather observation showed in Figure 13.1: air temperature, relative humidity air pressure, wind speed, and wind direction). The object members "VTstart" and "VTend" represent the valid-time timestamps, "comp" is of valid-time format. If "comp" has transaction-time format, the two timestamp object members "TTbegin" and "TTend" have to be used instead of "VTstart" and "VTend," respectively. If "comp" has bitemporal format, four timestamp object members have to be used: "VTstart," "VTend," "TTbegin," and "TTend."

Notice that the "VTend" object member of a valid-time or a bitemporal JSON model component version can have the special value "now" [63], which means that such a version is currently valid and continues to be valid in the real world until replaced by another version in the future. Besides, the "TTend" object member of a transaction-time or a bitemporal JSON component version can have the special value "UC" [63], which means "Until Changed" and denotes that such a component version is the current one (for such a component) until some change occurs.

13.3.1 ILLUSTRATIVE EXAMPLE

To illustrate the use of our TIoGT temporal JSON data model, let us suppose to deal with an application that stores and manages weather observations received from several stations that have been established for weather monitoring; each station is characterized by its identifier (id) and its geographic coordinates: latitude and longitude. Each weather observation, done at some time (observationTime), provides the air temperature (airTemperature), relative humidity (relativeHumidity), air pressure (air-Pressure), wind speed (windSpeed), and wind direction (windDirection). Each new observation is stored as a new temporal version of observation and timestamped with the transaction time (TTbegin), which represents the recording time of the observation in the database. Figure 13.1 shows an instance of our TIoGT data model, which stores the information of one station whose id is 11223344. Time-varying JSON data and their temporal versions are presented in red bold type.

Notice that in Figure 13.1 the object member "observationTime" could be considered as the valid-time of observation data (i.e., the time on which the weather

```
{ "stations":[
{ "station":{
"id": 11223344,
"latitude": 52.52,
"longitude": 13.41,
"observation":[
{ "observationTime": "2024-01-01T15:00:26",
"airTemperature": 25,
"relativeHumidity": 64,
"airPressure": 970,
"windSpeed": 3,
"windDirection": 95,
"TTbegin": "2024-01-01 15:01:34",
"TTend": "2024-01-01 15:21:33"},
{ "observationTime": "2024-01-01T15:20:26",
"airTemperature": 24,
"relativeHumidity": 69,
"airPressure": 972,
"windSpeed": 2,
"windDirection": 76,
"TTbegin": "2024-01-01 15:21:34",
"TTend": "2024-01-01 15:41:33"},
{ "observationTime": "2024-01-01T15:40:26",
"airTemperature": 26,
"relativeHumidity": 76,
"airPressure": 971,
"windSpeed": 4,
"windDirection": 83,
"TTbegin": "2024-01-01 15:41:34",
"TTend": "UC" }]}}]}
```

FIGURE 13.1   An instance of the TGIoT temporal JSON data model.

observation has really occurred in the reality). In fact, in the temporal database literature, temporal data, like observation data in our case, can be defined either as event
data or as state data.

- Event data are timestamped with a single occurrence time that could be either
  VTstart, that is, the value of the observationTime member, or TTbegin, that is,
  the time of observation data recording in the database, which is the commit/
  execution time of the transaction that writes observation data in the database.
- State data are timestamped with a time interval that could be either a valid-
  time interval [VTstart, VTend], or a transaction-time interval [TTbegin,
  TTend]. Bitemporal data, which are timestamped with two time intervals,
  are state data.

Notice that in our chapter, and for the sake of simplicity, we want to manage observation data as state data, timestamped by transaction-time interval. Hence, in our example, we have kept observationTime not as a valid-time timestamp but as an ordinary object member like the air temperature and the air pressure properties.

13.4   QL4TIoGT: THE TEMPORAL QUERY LANGUAGE FOR TIoGT DATA

In this section, we present QL4TIoGT, the temporal query language that we propose for querying TIoGT documents, and illustrate its use via some examples of temporal queries.

13.4.1   SYNTAX AND SEMANTICS OF QL4TIoGT

We have defined QL4TIoGT as a temporal extension of the JSONPath query language [28, 29]. The temporal extension of JSONPath consists in adding, to the JSONPath syntax, some transaction-time functions, as shown below, in order to allow querying transaction-time JSON data that are stored in TIoGT temporal JSON documents. Notice that the definition of such functions has been inspired by some past approaches dealing with temporal database querying like Refs. [64] and [65].

In general, we have added a set of transaction-time functions in order to make a separation between (i) temporal queries which are supposed to be specified at a high level, via a high-level query language; and (ii) the temporal data model that includes low-level information concerning the representation of time (e.g., if a transaction-time interval is closed at the end or not, the representation of "UC"). Indeed, the implementation aspects should be hidden/transparent to the end user.

QL4TIoGT supports the transaction-time functions as shown in Table 13.1. It is worth mentioning that:

- the *tt-current-asof(t)* function is used to write rollback queries, with TTend set to "t" in the result;
- the difference between *tt-overlaps/tt-includes* and *tt-current-between/tt-current-asof* is that the formers only select data preserving their timestamps in the results, whereas the latter select the same data but restrict their timestamps to the values of the arguments;
- some functions could be expressed via some other functions while using default arguments like the following ones:
  - *tt-history = tt-current-between(0,UC)*;
  - *tt-current-version = tt-includes(UC)*.

13.4.2   EXAMPLES OF QL4TIoGT TEMPORAL QUERIES

The main types of temporal queries are [64]: history selection, rollback (with transaction time only), snapshot (with valid time only), time-point selection, temporal slicing, temporal join, temporal aggregate, and restructuring. In this subsection, we show how to use QL4TIoGT to express, on the temporal JSON document of Figure 13.1, a query example of each one of the three following types: history selection, rollback, and temporal slicing. Notice that, for each query, we provide three things: the QL4TIoGT

TABLE 13.1

Temporal functions supported by QL4TIoGT

Function Signature	Function Semantics
tt-current-version	to return the current version of a component
tt-current-between(a,b)	to return a slice of consecutive versions of a component, which were current consecutively between the temporal value a (the beginning time) and the temporal value b (the ending time)
tt-current-asof(t)	to return the version of a component, which was current at the temporal value t (i.e., the version whose transaction-time contains the temporal value t)
tt-first-version	to return the first transaction-time version of a component
tt-history	to return the full transaction-time history of a component by providing each value of each version of this component, with its transaction-time timestamp
tt-includes(t)	to return the version of a component whose transaction-time interval includes the temporal value t
tt-precedes(a,b)	to return the version of a component whose transaction-time interval precedes the temporal interval [a, b]
tt-meets(a,b)	to return the version of a component whose transaction-time interval meets the temporal interval [a, b]
tt-overlaps(a,b)	to return the version of a component whose transaction-time interval overlaps the temporal interval [a, b]
tt-starts(a,b)	to return the version of a component whose transaction-time interval starts the temporal interval [a, b]
tt-during(a,b)	to return the version of a component whose transaction-time interval is contained in the temporal interval [a, b]
tt-contains(a,b)	to return the version of a component whose transaction-time interval contains the temporal interval [a, b]
tt-finishes(a,b)	to return the version of a component whose transaction-time interval finishes the temporal interval [a, b]
tt-equal-to(a,b)	to return the version of a component whose transaction-time interval is equal to the temporal interval [a, b]

query, its translation in JSONPath, and the result of the execution of the query. Notice also that we have tested and evaluated each JSONPath query using an online tool that supports JSONPath, called JSONPath Online Evaluator [66]. Recall that the result of the execution of a JSONPath query is always a JSON array of values. It is worth mentioning that we have slightly changed the presentation of the query results provided below (e.g., by putting observation versions between brackets, adding the corresponding object member name just before the extracted value) in order that they become more useful and easily interpretable by both application programs and humans.

13.4.2.1 History Selection Query

Query 1: Retrieve the air pressure history of the station 11223344.

This query could be expressed using QL4TIoGT as follows:

tt-history.\$.stations[?(@.station.id == "11223344")].observation[*].airPressure

The translation of this QL4TIoGT query to JSONPath is as follows:

> **$.stations[?(@.station.id == "11223344")].observation[*].airPressure, TTbegin,TTend**

The result of this query is as follows:

> [{airPressure: 970,
> **TTbegin: "2024-01-01 15:01:34", TTend: "2024-01-01 15:21:33"** },
> { airPressure: 972,
> **TTbegin: "2024-01-01 15:21:34", TTend: "2024-01-01 15:41:33"** },
> { airPressure: 971,
> **TTbegin: "2024-01-01 15:41:34", TTend: "UC"** }]

13.4.2.2 Rollback Query

Query 2: Retrieve the weather details collected by station 11223344 and available at "2024-01-01 15:30:00."

This query could be expressed using QL4TIoGT as follows:

> **tt-current-asof("2024-01-01 15:30:00").$.stations[?(@.station.id == "11223344")].observation[*]**

The translation of this QL4TIoGT query to JSONPath is as follows:

> **$.stations[?(@.station.id == "11223344")].observation[?((@.TTbegin <= "2024-01-01 15:30:00") && (@.TTend >= "2024-01-01 15:30:00"))]**

The result of this query is as follows:

> [{ **"observationTime": "2024-01-01T15:20:26",**
> **"airTemperature": 24,**
> **"relativeHumidity": 69,**
> **"airPressure": 972,**
> **"windSpeed": 2,   "windDirection": 76,**
> **"TTbegin": "2024-01-01 15:21:34",**
> **"TTend": "2024-01-01 15:41:33"** }]

13.4.2.3 Temporal Slicing

Query 3: Find the air temperature history provided by station 11223344 between "2024-01-01 15:10:00" and "2024-01-01 16:10:00."

This query could be expressed using QL4TIoGT as follows:

> **tt-current-between("2024-01-01 15:10:00","2024-01-01 16:10:00").**
> **$.stations[?(@.station.id == "11223344")].observation[*].airTemperature**

The translation of this QL4TIoGT query to JSONPath is as follows:

> **$.stations[?(@.station.id == "11223344")].observation[?((@.TTbegin <= "2024-01-01 16:10:00") && (@.TTend >=**

"2024-01-01 15:10:00"))].airTemperature,max(TTbegin," 2024-01-01 15:10:00"),min(TTend,"2024-01-01 16:10:00")

The result of this query is as follows:

```
[ { "airTemperature":25,
  "TTbegin":"2024-01-01 15:01:34", "TTend":"2024-01-01 15:21:33" },
{ "airTemperature":24,
  "TTbegin":"2024-01-01 15:21:34", "TTend":"2024-01-01 15:41:33" },
{ "airTemprature":26,
  "TTbegin":"2024-01-01 15:41:34", "TTend":"UC" } ]
```

13.5   CONCLUSION

In this chapter, we have proposed TIoGT, a temporal JSON data model for representing the evolution of over time of data of internet of green things, and QL4TGIoT, a temporal JSON query language for querying time-varying TIoGT data. The TIoGT model has been defined as a temporal extension of the (non-temporal) standard JSON data model to temporal aspects, and the QL4TIoGT has been designed as a temporal extension of the (non-temporal) JSONPath query language.

In the near future, we plan to develop a prototype tool, on top of some JSON-based document-oriented NoSQL DBMS (e.g., MongoDB and CouchDB), that shows the feasibility of our approach and supports both the TIoGT model and its query language QL4TIoGT. Moreover, we intend to extend QL4TIoGT to also support valid time, in order to allow it to query not only transaction-time JSON data but also valid-time and bitemporal JSON data that are stored in TIoGT documents.

REFERENCES

[1] A. Hooper, "Green computing," *Communication of the ACM*, vol. 51, no.10, pp. 11–13, 2008.

[2] S. Vikram, "Green computing," in *Proceedings of the 2015 International Conference on Green Computing and Internet of Things (ICGCIoT)*. IEEE, 2015, pp. 767–772.

[3] M. A. Albreem, A. A. El-Saleh, M. Isa, W. Salah, M. Jusoh, M. M. Azizan, and A. Ali, "Green internet of things (IoT): An overview," in *Proceedings of the 2017 IEEE 4th International Conference on Smart Instrumentation, Measurement and Application (ICSIMA)*. IEEE, 2017, pp. 1–6.

[4] S. Sarkar, and S. Misra, "Theoretical modelling of fog computing: A green computing paradigm to support IoT applications," *IET Networks*, vol. 5, no. 2, pp. 23–29, 2016.

[5] T. Poongodi, S. R. Ramya, P. Suresh, and B. Balusamy, "Application of IoT in green computing," in *Advances in Greener Energy Technologies*, A. Bhoi, K. Sherpa, A. Kalam, and G. S. Chae, Eds. Springer, 2020, pp. 295–323.

[6] N. Kumar, R. Chaudhry, O. Kaiwartya, N. Kumar, and S. H. Ahmed, "Green computing in software defined social internet of vehicles," *IEEE Transactions on Intelligent Transportation Systems*, vol. 22, no. 6, pp. 3644–3653, 2020.

[7] M. Muniswamaiah, T. Agerwala, and C. C. Tappert, "Green computing for Internet of Things," in *Proceedings of the 2020 7th IEEE International Conference on Cyber Security and Cloud Computing (CSCloud)/2020 6th IEEE International Conference on Edge Computing and Scalable Cloud (EdgeCom)*. IEEE, 2020, pp. 182–185.

[8] L. Tan, N. Shi, K. Yu, M. Aloqaily, and Y. Jararweh, "A blockchain-empowered access control framework for smart devices in green internet of things," *ACM Transactions on Internet Technology (TOIT)*, vol. 21, no. 3, pp. 1–20, 2021.

[9] I. Shanmugapriya, "A survey on energy management evolution and techniques for green IoT environment," in *Proceedings of the 4th International Conference on Communication, Computing and Electronics Systems (ICCCES 2022)*, LNEE vol. 977. Springer, 2023. pp. 155–165.

[10] C. R. S. Ram, S. Ravimaran, R. S. Krishnan, E. G. Julie, Y. H. Robinson, R. Kumar, L. H. Song, P. H. Thong, N. Q. Thanh, and M. Ismail, "Internet of Green Things with autonomous wireless wheel robots against green houses and farms," *International Journal of Distributed Sensor Networks*, vol. 16, no. 6, 2020. doi: 10.1177/1550147720923477

[11] S. Shetty, and N. S. Ramaiah, "CIoT: Internet of Green Things for enhancement of crop data using analytics and machine learning," in *Green Internet of Things and Machine Learning: Towards a Smart Sustainable World*, R. Raut, S. Kautish, Z. Polkowski, A. Kumar, and C.-M. Liu, Eds. John Wiley & Sons, Inc., 2021, pp. 141–162.

[12] M. Kotz, L. Wenz, A. Stechemesser, M. Kalkuhl, and A. Levermann, "Day-to-day temperature variability reduces economic growth," *Nature Climate Change*, vol. 11, no. 4, pp. 319–325, 2021.

[13] S. Sippel, N. Meinshausen, E. M. Fischer, E. Székely, and R. Knutti, "Climate change now detectable from any single day of weather at global scale," *Nature Climate Change*, vol. 10, no. 1, pp. 35–41, 2020.

[14] Reddy, G.V. and et al., "Human Action Recognition Using Difference of Gaussian and Difference of Wavelet," *Big Data Mining and Analytics*, vol. 6, no. 3, pp. 336–346, 1 September 2023, DOI:10.26599/BDMA.2022.9020040

[15] M. M. Majedul Islam, "Threats to humanity from climate change," in *Climate Change: The Social and Scientific Construct*, S. A. Bandh, Ed. Springer, 2022, pp. 21–36.

[16] N. Ahmed, T. I. Khan, and A. Augustine, "Climate change and environmental degradation: A serious threat to global security," *European Journal of Social Sciences Studies*, vol. 3, no. 1, pp. 161–172, 2018.

[17] D. A. Aziz, R. Asgarnezhad, M. S. Mustafa, A. A. Saber, and S. Alani, "A developed IoT platform-based data repository for smart farming applications," *Journal of Communications*, vol. 18, no. 3, pp. 187–197, 2023.

[18] B. Mohamed, E. Abdellatif, M. Mouhajir, M. Zerifi, Y. Rabah, and E. Youssef, "A module placement scheme for fog-based smart farming applications," *International Journal of Electrical & Computer Engineering (2088-8708)*, vol. 13, no. 6, pp. 7089–7098, 2023.

[19] A. E. Hassanien, and A. Darwish, Eds., *The Power of Data: Driving Climate Change with Data Science and Artificial Intelligence Innovations*. Springer Nature, 2023.

[20] D. L. Urban, "Climate change: Adapting for resilience," in *Agents and Implications of Landscape Pattern: Working Models for Landscape Ecology*. Springer, 2023, pp. 287–321.

[21] A. Roukh, F. N. Fote, S. A. Mahmoudi, and S. Mahmoudi, "Big data processing architecture for smart farming," *Procedia Computer Science*, vol. 177, pp. 78–85, 2020.

[22] S. Wolfert, L. Ge, C. Verdouw, and M. J. Bogaardt, "Big data in smart farming—A review," *Agricultural Systems*, vol. 153, pp. 69–80, 2017.

[23] H. Hassani, X. Huang, and E. Silva, "Big data and climate change," *Big Data and Cognitive Computing*, vol. 3, no. 1, Article 12, 2019. doi: 10.3390/bdcc3010012

[24] H. D. Guo, L. Zhang, and L. W. Zhu, "Earth observation big data for climate change research," *Advances in Climate Change Research*, vol. 6, no. 2, pp. 108–117, 2015.

[25] Internet Engineering Task Force, *The JavaScript Object Notation (JSON) Data Interchange Format*, Internet Standards Track document, December 2017. https://tools.ietf.org/html/rfc8259 (accessed: February 19, 2024).

[26] C. S. Jensen, and R. T. Snodgrass, "Temporal data models," in *Encyclopedia of Database Systems* (2nd edition), L. Liu, and M. T. Özsu, Eds. Springer, 2018, pp. 3940–3945.

[27] C. S. Jensen, and R. T. Snodgrass, "Temporal query languages," in *Encyclopedia of Database Systems* (2nd edition), L. Liu, and M. T. Özsu, Eds. Springer, 2018, pp. 4023–4028.

[28] S. Gössner, *JSONPath – XPath for JSON*, 21 February 2007. http://goessner.net/articles/JsonPath/ (accessed: February 19, 2024).

[29] Internet Engineering Task Force, *JSONPath: Query Expressions for JSON*, Internet-Draft, 25 April 2022. https://datatracker.ietf.org/doc/draft-ietf-jsonpath-base/ (accessed: February 19, 2024).

[30] T. Shree, R. Kumar, and N. Kumar, "Green computing in cloud computing," in *Proceedings of the 2020 2nd International Conference on Advances in Computing, Communication Control and Networking (ICACCCN)*. IEEE, 2020, pp. 903–905.

[31] K. Gai, M. Qiu, H. Zhao, L. Tao, and Z. Zong, "Dynamic energy-aware cloudlet-based mobile cloud computing model for green computing," *Journal of network and computer applications*, vol. 59, pp. 46–54, 2016.

[32] S. G. Paul, A. Saha, M. S. Arefin, T. Bhuiyan, A. A. Biswas, A. W. Reza, N. M. Alotaibi, S. A. Alyami, and M. A. Moni, "A comprehensive review of green computing: Past, present, and future research," *IEEE Access*, vol. 11, pp. 87445–87494, 2023.

[33] M. H. Akın, "'Green Computing' in Hospitality," in *Handbook of Research on Sustainable Tourism and Hotel Operations in Global Hypercompetition*, H. Sezerel, and B. Christiansen, Eds. IGI Global, 2023, pp. 118–136.

[34] C. Josephson, N. Peill-Moelter, Z. Pan, B. Pfaff, and V. Firoiu, "The sky is not the limit: Untapped opportunities for Green Computing," *ACM SIGENERGY Energy Informatics Review*, vol. 3, no. 3, pp. 33–39, 2023.

[35] S. Brahmia, Z. Brahmia, F. Grandi, and R. Bouaziz, "τJSchema: A framework for managing temporal JSON-Based NoSQL Databases," in *Proceedings of the 27th International Conference on Database and Expert Systems Applications (DEXA'2016), Part 2, LNCS vol. 9828*. Springer, 2016, pp. 167–181.

[36] C. S. Jensen, and R. T. Snodgrass, "Temporal Database," in *Encyclopedia of Database Systems* (2nd edition), L. Liu, and M. T. Özsu, Eds. Springer, 2018, pp. 3945–3949.

[37] C. S. Jensen, and R. T. Snodgrass, "Transaction Time," in *Encyclopedia of Database Systems* (2nd edition), L. Liu, and M. T. Özsu, Eds. Springer, 2018, pp. 4200–4201.

[38] C. S. Jensen, and R. T. Snodgrass, "Valid Time," in *Encyclopedia of Database Systems* (2nd edition), L. Liu, and M. T. Özsu, Eds. Springer, 2018, pp. 4359–4360.

[39] Z. Brahmia, F. Grandi, B. Oliboni, and R. Bouaziz, "Schema Evolution," in *Encyclopedia of Information Science and Technology* (3rd edition), M. Khosrow-Pour, Ed. IGI Global, 2015, pp. 7641–7650.

[40] Z. Brahmia, F. Grandi, B. Oliboni, and R. Bouaziz, "Schema Versioning," in *Encyclopedia of Information Science and Technology* (3rd edition), M. Khosrow-Pour, Ed. IGI Global, 2015, pp. 7651–7661.

[41] Z. Hu, and L. Yan, "Modeling Temporal Information with JSON," in *Emerging Technologies and Applications in Data Processing and Management*, Z. Ma, and L. Yan, Eds. IGI Global, 2019, pp. 134–153.

[42] A. Goyal, and C. Dyreson, "Temporal JSON," in *Proceedings of the 5th IEEE International Conference on Collaboration and Internet Computing (CIC 2019)*. IEEE, 2019, pp. 135–144.

[43] Z. Brahmia, F. Grandi, S. Brahmia, and R. Bouaziz, "A graphical conceptual model for conventional and time-varying JSON data," *Procedia Computer Science*, vol. 184, pp. 823–828, 2021.

[44] Z. Brahmia, S. Brahmia, F. Grandi, and R. Bouaziz, "TempoJCM++: An extension of TempoJCM to support schema change modeling," in *Proceedings of the 5th International Conference on Artificial Intelligence and Smart Environments (ICAISE'2023), LNNS vol. 837*. Springer, 2024. doi: 10.1007/978-3-031-48465-0_73

[45] P. Baumann, E. Hirschorn, J. Maso, V. Merticariu, and D. Misev, "All in one: Encoding spatio-temporal big data in XML, JSON, and RDF without information loss," in *Proceedings of the 2017 IEEE International Conference on Big Data (Big Data 2017)*. IEEE, 2017, pp. 1–10.

[46] N. Q. Mehmood, R. Culmone, L. Mostarda, "Modeling temporal aspects of sensor data for MongoDB NoSQL database," *Journal of Big Data*, vol. 4, no. 1, Article no. 8, 2017.

[47] A. Cuzzocrea, "Temporal big data analytics: New frontiers for big data analytics research (Panel Description)," in *Proceedings of the 28th International Symposium on Temporal Representation and Reasoning (TIME 2021)*. Leibniz International Proceedings in Informatics (LIPIcs), 2021, pp. 4:1–4:7.

[48] S. Brahmia, Z. Brahmia, F. Grandi, and R. Bouaziz, "Temporal JSON schema versioning in the τJSchema framework," *Journal of Digital Information Management*, vol. 15, no. 4, pp. 179–202, 2017.

[49] S. Brahmia, Z. Brahmia, F. Grandi, and R. Bouaziz, "Versioning temporal characteristics of JSON-based big data via the τJSchema framework," *International Journal of Cloud Computing*, vol. 10, no. 5/6, pp. 406–441, 2021.

[50] Z. Brahmia, S. Brahmia, F. Grandi, and R. Bouaziz, "Versioning schemas of JSON-based conventional and temporal big data through high-level operations in the τJSchema framework," *International Journal of Cloud Computing*, vol. 10, no. 5/6, pp. 442–479, 2021.

[51] M. Mach-Król, "Conceptual framework for implementing temporal big data analytics in companies," *Applied Sciences*, vol. 12, no. 23, Article 12265, 2022.

[52] V. Vaishnavi, and W. Kuechler, *Design Research in Information Systems*, 20 January 2004 (updated in 2017 and 2019 by V. Vaishnavi, and P. Stacey); last updated (by V. Vaishnavi, and S. Duraisamy) on 15 December 2023. http://www.desrist.org/design-research-in-information-systems/ (accessed: February 19, 2024).

[53] F. Baig, P. Nalluri, J. Kong, and F. Wang, "SPEAR-board: cross-platform interactive spatio-temporal big data analytics," in *Proceedings of the 30th International Conference on Advances in Geographic Information Systems (SIGSPATIAL'22)*. ACM, Inc., 2022, Article 105, pp. 1–4.

[54] H. Liang, Z. Zhang, C. Hu, Y. Gong, and D. Cheng, "A survey on spatio-temporal big data analytics ecosystem: resource management, processing platform, and applications," *IEEE Transactions on Big Data*, pp. 1–20, 2023. doi: 10.1109/TBDATA.2023.3342619

[55] V. Uma, and G. Jayanthi, "Scalable spatio-temporal reasoning of sequential events using spark framework," in *Proceedings of the 2018 10th International Conference on Advanced Computing (ICoAC)*. IEEE, 2018, pp. 47–51.

[56] D. Florescu, and G. Fourny, "JSONiq: The history of a query language," *IEEE Internet Computing*, vol. 17, no. 5, pp. 86–90, 2013.

[57] K. S. Beyer, V. Ercegovac, R. Gemulla, A. Balmin, M. Y. Eltabakh, C.-C. Kanne, F. Özcan, and E. J. Shekita, "Jaql: A scripting language for large scale semistructured data analysis," *PVLDB*, vol. 4, no. 12, pp. 1272–1283, 2011.

[58] P. Bourhis, J. L. Reutter, and D. Vrgoč, "JSON: Data model and query languages," *Information Systems*, vol. 89, Article 101478, 2020.

[59] J. Pokorný, "JSON functionally," in *Proceedings of the 24th European Conference on Advances in Databases and Information Systems (ADBIS 2020), LNCS vol. 12245*. Springer, 2020, pp. 139–153.

[60] D. Petković, "SQL/JSON standard: Properties and deficiencies," *Datenbank Spektrum*, vol. 17, no. 3, pp. 277–287, 2017.

[61] S. Sharma, and S. Gadia, "Expanding ParaSQL for spatio-temporal (big) data," *The Journal of Supercomputing*, vol. 75, pp. 587–606, 2019.

[62] C. S. Jensen, and R. T. Snodgrass, "Bitemporal relation," in *Encyclopedia of Database Systems* (2nd edition), L. Liu, and M. T. Özsu, Eds. Springer, 2018, pp. 310.

[63] C. S. Jensen, C. E. Dyreson, M. Böhlen, J. Clifford, R. Elmasri, S. K. Gadia, F. Grandi, P. Hayes, S. Jajodia, W. Käfer, N. Kline, N. Lorentzos, Y. Mitsopoulos, A. Montanari, D. Nonen, E. Peressi, B. Pernici, J. F. Roddick, N. L. Sarda, M. R. Scalas, A. Segev, R. T. Snodgrass, M. D. Soo, A. Tansel, P. Tiberio, and G. Wiederhold, "The consensus glossary of temporal database concepts – February 1998 version," in *Temporal Databases – Research and Practice*, O. Etzion, S. Jajodia, and S. Sripada, Eds. Springer, LNCS vol. 1399, 1998, pp. 367–405.

[64] R.T. Snodgrass, Ed., I. Ahn, G. Ariav, D. S. Batory, J. Clifford, C. E. Dyreson, R. Elmasri, F. Grandi, C. S. Jensen, W. Käfer, N. Kline, K. Kulkarni, T. Y. Cliff Leung, N. Lorentzos, J. F. Roddick, A. Segev, M. D. Soo, and S. M. Sripada, *The TSQL2 Temporal Query Language*. Kluwer Academic Publishers, 1995.

[65] F. Wang, C. Zaniolo, and X. Zhou, "ArchIS: An XML-based approach to transaction-time temporal database systems," *The VLDB Journal*, vol. 17, no. 6, pp. 1445–1463, 2008.

[66] JSONPath online evaluator. https://jsonpath.com/ (accessed: February 19, 2024).

14 Enhancing Financial Transaction Security

A Deep Learning Approach for E-Payment Fraud Detection

Manal Loukili, Fayçal Messaoudi, and Raouya El Youbi
Sidi Mohamed Ben Abdellah University, Fez, Morocco

14.1 INTRODUCTION

Credit card fraud is a pervasive issue that poses significant challenges to financial institutions and customers worldwide [1]. As technology advances and transactions become increasingly digitized, fraudulent activities continue to evolve and become more sophisticated [2]. Detecting and preventing fraudulent credit card transactions is crucial to protect customers, minimize financial losses, and maintain the integrity of the financial system [3].

Traditional rule-based fraud detection methods have limitations in effectively capturing complex patterns and adapting to rapidly changing fraud tactics [4]. As a result, there is a growing need for advanced techniques that can analyze large volumes of data, identify subtle patterns, and make accurate predictions in real-time. Machine learning and deep learning algorithms have proved to be a promising solution and a powerful tool in various tasks in today's e-business world [5], notably customer churn prediction [6], sentiment analysis [7], recommendation systems [8], dynamic pricing, demand prediction [9], and fraudulent payment detection [10].

The objective of this research is to develop a robust deep learning model for the detection of fraudulent credit card transactions. By leveraging the power of deep learning, the model aims to capture intricate patterns and features from transaction data and customer information to accurately classify transactions as fraudulent or legitimate. The study also addresses the challenge of class imbalance, where fraudulent transactions are significantly outnumbered by legitimate transactions, by employing a class weight strategy to ensure balanced learning.

DOI: 10.1201/9781032656830-14

14.1.1 Key Contributions of the Chapter

The chapter makes the following significant contributions:

   i. The chapter provides a comprehensive analysis of the evolving landscape of credit card fraud, highlighting the increasing sophistication of fraudulent activities in the context of digitized financial transactions.
   ii. It addresses the limitations of traditional fraud detection methods and underscores the necessity for advanced techniques like deep learning to analyze large datasets, identify subtle fraud patterns, and make real-time predictions.
   iii. A novel deep learning model is proposed for detecting fraudulent credit card transactions, effectively utilizing transaction data and customer information. The model's design specifically tackles the challenge of class imbalance in fraud detection, employing a class weight strategy to ensure balanced learning and improved accuracy.

14.1.2 Chapter Organization

Section 14.2 delves into the current landscape and trends in mobile payment fraud, offering insights into the challenges and complexities of the domain. Section 14.3 provides an overview of related work in the field of credit card fraud detection, setting the stage for subsequent discussions. Section 14.4 presents the methodology, detailing the data preprocessing steps, the architecture of the proposed deep learning model, the training procedures, and the evaluation metrics. Section 14.5 discusses the experimental results, showcasing the effectiveness of the proposed model in identifying fraudulent transactions. Finally, Section 14.6 concludes the chapter by summarizing the key findings and emphasizing the significance of deep learning in combating credit card fraud, while also suggesting avenues for future research in this rapidly evolving field.

14.2 CURRENT LANDSCAPE AND TRENDS IN MOBILE PAYMENT FRAUD

14.2.1 Growth in Mobile Payment Platforms

The rise of mobile payment systems has been meteoric, driven by the widespread adoption of smartphones and the availability of various digital payment services. This evolution in the financial transaction landscape, while convenient, has opened up new avenues for fraudulent activities, necessitating advanced security measures.

14.2.2 Emerging Trends in Mobile Payment Fraud

With the sophistication of mobile payment systems, fraudsters have developed equally sophisticated methods. Identity theft and account takeovers have become prevalent, with fraudsters using stolen credentials to access mobile payment accounts. SIM swap scams are another emerging trend, where fraudsters deceive mobile operators to gain control over a victim's phone number, thereby manipulating SMS-based verification processes. Phishing attacks have adapted to the mobile environment, with the creation of fake payment apps or deceptive messages to extract sensitive information. Vulnerabilities in mobile payment applications are also being exploited, alongside QR code tampering, which redirects payments to fraudulent accounts.

14.2.3 TECHNIQUES USED BY FRAUDSTERS

Fraudsters employ a range of techniques, including social engineering, where users are manipulated into sharing sensitive information. The use of malware and ransomware to infiltrate mobile devices is another tactic, either to steal credentials or to lock devices for ransom. Exploiting network vulnerabilities to intercept data during transactions also poses a significant risk.

14.2.4 IMPACT OF MOBILE PAYMENT FRAUD

The impact of mobile payment fraud extends beyond financial losses, which affect both individuals and institutions. It leads to an erosion of trust in mobile payment platforms and creates regulatory and compliance challenges, placing additional pressure on providers to adhere to stringent security standards.

14.2.5 RESPONSE TO MOBILE PAYMENT FRAUD

In response, a combination of strategies is being employed. Enhanced security measures, such as biometric authentication and machine learning-based fraud detection systems, are being implemented. Educating users about safe practices is critical, as is the collaboration among financial institutions, payment providers, and regulatory bodies to develop standardized security protocols. Big data analytics is increasingly being leveraged to identify and prevent fraudulent transactions. Strengthening legal and regulatory frameworks also plays a crucial role in deterring fraud and protecting user data.

14.3 RELATED WORK

In modern financial systems, the rise of mobile payment platforms has revolutionized the way consumers conduct transactions, providing convenience and efficiency [11]. However, this advancement has also brought about new challenges, particularly concerning the detection and prevention of fraudulent activities. Mobile payment fraud poses a significant threat to the financial ecosystem, leading to substantial monetary losses for individuals and businesses alike [12]. To combat this menace, efficient and adaptive fraud detection systems are essential.

Detecting fraudulent activities in mobile payment systems is a critical challenge due to the increasing sophistication of fraudsters' tactics [13]. Machine learning, and deep learning in particular, has emerged as a powerful tool in various fields due to its ability to automatically learn intricate patterns and representations from data [14]. In recent years, deep learning has gained significant attention in fraud detection applications, showcasing remarkable success in tackling complex and dynamic fraud schemes. The utilization of deep learning in mobile payment fraud detection offers promising potential for enhancing the security of financial transactions.

Several studies have tackled the problem of fraud detection using machine learning techniques, each presenting different approaches and methods. In a study [15], the authors focused on extending the transaction aggregation strategy and creating a feature set based on analyzing the periodic behavior of transaction time using the von Mises distribution. Their evaluation using real credit card fraud data revealed that incorporating the proposed periodic features resulted in an average improvement in savings of 13%.

In another study [16], addressed the opportunities and challenges associated with electronic payments, emphasizing fraud as a severe risk. They discussed various types of e-payments and their advantages, as well as the prospects for future growth and adoption, particularly in emerging markets.

The authors in Ref. [17] proposed a credit card fraud detection system that utilized a machine learning approach with a genetic algorithm for feature selection. Their system outperformed existing detection systems, as demonstrated by experimental results comparing popular machine learning classifiers.

In Ref. [18], the author proposed a credit card fraud detection system that aimed to improve accuracy and efficiency in imbalanced datasets. The system utilized k-means semantic fusion and the artificial bee colony algorithm to filter features and assess transaction legitimacy. Experimental results showed significant improvements in classification accuracy.

The authors in Ref. [19] suggested fraud prediction models incorporating various machine learning methods such as support vector machines, k-nearest neighbors, and artificial neural networks. These models achieved high accuracy in detecting credit card fraud compared to unsupervised learning methods.

In Ref. [20], the authors introduced a hybrid model combining TabNet and XGBoost algorithms for credit card fraud detection. The TabNet algorithm extracted important features, while XGBoost classified transactions as fraudulent or legitimate. The hybrid model outperformed classical models in terms of accuracy and AUC-ROC score, significantly improving fraud detection in credit card transactions.

The literature emphasizes the significance of utilizing advanced machine learning techniques for credit card fraud detection and prevention. The discussed models and algorithms demonstrate promising results in this field.

14.4 METHODOLOGY

In this paper, we propose a deep learning-based approach for mobile payment fraud detection. The primary objective is to develop a model that can accurately and efficiently identify fraudulent transactions while handling the class imbalance problem. The proposed model considers both transaction data and customer information, allowing it to learn complex patterns and relationships for improved fraud detection performance. Figure 14.1 outlines the methodology adopted in this study.

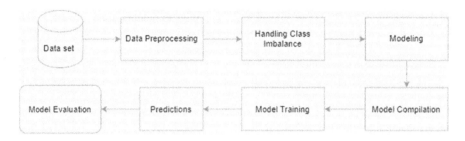

FIGURE 14.1 The methodology adopted.

14.4.1 DATASET

The dataset used in this study contains 50,000 entries with 11 columns. Here's a breakdown of the columns and their descriptions:

- step: The unit of time in the simulated transaction. It is a discrete value representing the elapsed steps during the simulation.
- type: The type of transaction, such as payment, transfer, and cash out.
- amount: The amount of the transaction.
- nameOrig: The customer ID of the originator (sender) of the transaction.
- oldbalanceOrg: The initial balance of the originator's account before the transaction.
- newbalanceOrig: The balance of the originator's account after the transaction.
- nameDest: The customer ID of the recipient (destination) of the transaction.
- oldbalanceDest: The initial balance of the recipient's account before the transaction.
- newbalanceDest: The balance of the recipient's account after the transaction.
- isFraud: A binary indicator (0 or 1) that denotes whether the transaction is fraudulent or not. "1" indicates fraud, and "0" indicates a legitimate transaction.
- isFlaggedFraud: A binary indicator (0 or 1) that identifies transactions flagged as potentially fraudulent by the system. This column may be used as a label for a specific type of fraud.

Table 14.1 shows the first five rows of the dataset used.

The dataset is in a Pandas DataFrame format, with a range index from 0 to 49,999. The columns have different data types, including integers, floats, and objects (strings). The non-null count for all columns is 50,000, indicating that there are no missing values in the dataset (Figure 14.2).

14.4.2 DATA PREPROCESSING

In order to prepare the data for training and evaluating the deep learning model, the following data preprocessing steps were performed:

- Drop unnecessary columns: Remove the "step," "nameOrig," and "isFlagged Fraud" columns from the dataset as they are not required for the analysis.

TABLE 14.1
Dataset Sample

step	amount	Type	oldbalanceOrg	nameOrig	newbalanceOrig	nameDest	oldbalance Dest	newbalance Dest	isFlagged Fraud	isFraud
1	9839.64	PAYMENT	170136.00	C1231006815	160296.36	M1979787155	0.0	0.0	0	0
1	1864.28	PAYMENT	21249.00	C1666544295	19384.72	M2044282225	0.0	0.0	0	0
1	181.00	TRANSFER	181.00	C1305486145	0.00	C553264065	0.0	0.0	0	1
1	181.00	CASH_OUT	181.00	C840083671	0.00	C38997010	21182.0	0.0	0	1
1	11668.14	PAYMENT	41554.00	C2048537720	29885.86	M1230701703	0.0	0.0	0	0

```
<class 'pandas.core.frame.DataFrame'>
RangeIndex: 50000 entries, 0 to 49999
Data columns (total 11 columns):
 #   Column          Non-Null Count  Dtype
---  ------          --------------  -----
 0   step            50000 non-null  int64
 1   type            50000 non-null  object
 2   amount          50000 non-null  float64
 3   nameOrig        50000 non-null  object
 4   oldbalanceOrg   50000 non-null  float64
 5   newbalanceOrig  50000 non-null  float64
 6   nameDest        50000 non-null  object
 7   oldbalanceDest  50000 non-null  float64
 8   newbalanceDest  50000 non-null  float64
 9   isFraud         50000 non-null  int64
 10  isFlaggedFraud  50000 non-null  int64
dtypes: float64(5), int64(3), object(3)
memory usage: 4.2+ MB
```

FIGURE 14.2 Dataset information.

- One-hot encode categorical variables: Perform one-hot encoding on the "type" column to convert it into numerical features. This can be achieved using the onehot_encode() function provided.
- Separate the target variable: Extract the "isFraud" column as the target variable (y) and remove it from the feature set (X).
- Train-test split: Split the dataset into training and testing sets using the train_test_split() function. Set the train size to 0.7 (70% of the data) and use a random state for reproducibility.
- Tokenize and pad sequences: Create a tokenizer object and fit it on the "nameDest" column in the training data. Convert the customer names to sequences using the tokenizer's texts_to_sequences() method. Pad the sequences to a fixed length using the pad_sequences() function.
- Drop the "nameDest" column: Remove the "nameDest" column from the feature sets (X_train and X_test) as it has been tokenized and padded.
- Standardize numerical features: Scale the numerical features in X_train and X_test using a standard scaler. Fit the scaler on X_train and transform both X_train and X_test using the fitted scaler.

The resulting preprocessed data will consist of X_train, X_test, customers_train, customers_test, y_train, and y_test, which can be used for training and evaluating the deep learning model.

14.4.3 THE CLASS IMBALANCE HANDLING

Class imbalance is a common challenge in machine learning, where one class in the target variable has significantly more examples than the other class(es). In this case, the target variable is "isFraud," which has two classes: 0 (non-fraudulent transactions) and 1 (fraudulent transactions). As we can see from the value counts, there are only 71 examples of fraudulent transactions compared to 34,929 non-fraudulent transactions. To address this class imbalance, an oversampling technique was adopted.

Oversampling is a technique where we increase the number of examples in the minority class (isFraud = 1) by generating synthetic samples. This helps in balancing the class distribution and can prevent the model from being biased towards the majority class.

Here are the steps taken to handle the class imbalance:

- Concatenating Data: First, the feature set X_train, the customer names customers_train, and the target variable y_train are combined into a single DataFrame called train_df.
- Counting Class Distribution: The value counts of each class in the target variable isFraud are calculated to understand the class distribution.
- Determining Samples to Oversample: The number of examples to sample from the majority class (class 0) is calculated by subtracting the count of the minority class (class 1) from the count of the majority class (class 0). In this case, we need to generate 34,858 examples of fraudulent transactions to balance the classes.
- Oversampling the Minority Class: A random sample of fraudulent transactions (class 1) is drawn with replacement from the original data using the sample() function. The number of samples to draw is equal to the number of examples we determined in the previous step. The oversampled data is stored in the DataFrame oversampled_data.
- Combining Data and Shuffling: The oversampled data is concatenated with the original data along the row axis using pd.concat(), and the rows are shuffled randomly to avoid any patterns in the data.
- Updating Data Split: The updated DataFrame train_df is used to redefine the feature set X_train, customer names customers_train, and the target variable y_train.

After applying these steps, X_train, customers_train, and y_train are updated with balanced class distribution. This helps ensure that the machine learning model is exposed to a more balanced representation of both classes during training, leading to better generalization and performance in the minority class.

14.4.4 MODEL ARCHITECTURE DESIGN

In this paper, we propose a deep learning-based model architecture that combines transaction data and customer information to enhance the fraud detection capabilities. The model architecture consists of the following components (Figure 14.3).

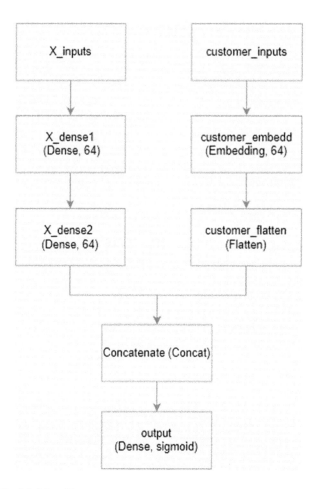

FIGURE 14.3 Model architecture.

Neural network architecture:

- X_inputs: Represents the input layer for the transaction data, with a shape of (None, 10). The input shape (10,) implies that there are 10 features for each transaction.
- customer_inputs: Represents the input layer for customer information, with a shape of (None, 1). The input shape (1,) indicates that there is one feature for each customer.
- X_dense1 and X_dense2: These are dense layers (fully connected layers) with 64 neurons each, and ReLU activation function applied to them. These layers process the transaction data and capture relevant patterns.
- customer_embedding: An embedding layer that converts customer IDs into dense vectors of size 64. This layer is used to learn meaningful representations of customer information.

- customer_flatten: Flattens the embedded customer information into a 1D vector.
- Concatenate: Concatenates the outputs from X_dense2 and customer_flatten into a single vector to combine transaction data and customer information.
- Output: The final output layer with a sigmoid activation function, producing a single probability value representing the likelihood of a transaction being fraudulent.

The adoption of a deep learning approach in mobile payment fraud detection is driven by its ability to autonomously learn and adapt to complex data, making it superior to traditional methods. Deep learning models excel in processing large, unstructured datasets and effectively handle the class imbalance problem, crucial for identifying subtle and evolving fraudulent activities. Their scalability and flexibility in dealing with increasing transaction volumes and diverse data types further enhance their suitability.

The model architecture is designed to leverage the inherent relationships between transaction data and customer information, enabling the model to learn intricate patterns that can help distinguish between fraudulent and non-fraudulent transactions. By incorporating both types of information, the model is able to capture nuanced fraud patterns that may not be evident from transaction data alone.

The proposed model architecture provides a foundation for effectively detecting mobile payment fraud using deep learning techniques. Through training and fine-tuning, the model can learn to accurately classify transactions and identify potentially fraudulent activities, contributing to enhanced security in financial transactions.

14.4.5 TRAINING

The deep learning model was trained and configured with specific parameters to optimize its performance for fraud detection. The model utilized the Adam optimizer, known for its efficiency in handling sparse gradients and adaptive learning rate capabilities, making it well-suited for complex datasets like ours. The loss function chosen was Binary Cross-Entropy, which is particularly appropriate for binary classification tasks such as distinguishing between fraudulent and legitimate transactions. For performance metrics, we focused on accuracy and the area under the receiver operating characteristic curve (AUC), both of which are critical for evaluating the model's ability to correctly classify transactions and differentiate between the two classes.

The training was performed over 10 epochs with a batch size of 32. Early stopping was employed with a patience of 2 to prevent overfitting, ensuring that our deep learning model generalizes well to new, unseen data. This technique involves monitoring the model's performance on a validation set and halting the training process when the performance ceases to improve, thereby avoiding the excessive learning of training-specific patterns. Specifically, we chose to monitor the model's validation loss as our stopping criterion, a common choice given that a decrease in validation loss is indicative of improved generalizability. The training process was configured to stop if the

TABLE 14.2

Training History: Metrics for Each Epoch

Epoch	Loss	Accuracy	AUC	Validation Loss	Validation Accuracy	Validation AUC
1	0.0655	0.9321	0.9955	0.0125	0.9981	0.9990
2	0.0037	0.9981	0.9987	0.0087	0.9984	0.9993
3	0.0032	0.9984	0.9989	0.0095	0.9984	0.9991
4	0.0029	0.9985	0.9991	0.0077	0.9986	0.9995
5	0.0026	0.9987	0.9991	0.0070	0.9989	0.9994
6	0.0024	0.9987	0.9992	0.0088	0.9985	0.9993
7	0.0025	0.9987	0.9993	0.0106	0.9984	0.9992
8	0.0022	0.9988	0.9993	0.0079	0.9987	0.9995
9	0.0021	0.9989	0.9994	0.0068	0.9988	0.9995
10	0.0020	0.9989	0.9994	0.0070	0.9987	0.9996

validation loss did not improve for two consecutive epochs, a parameter known as "patience." This choice strikes a balance between allowing the model enough time to learn complex patterns and preventing it from overfitting to the training data. During training, the model achieved the performance presented in Table 14.2.

14.4.6 EVALUATION

The trained model was evaluated on the test dataset, yielding the following results:

- Test accuracy: 99.720%
- Test AUC: 97.431
- Confusion matrix: Figure 14.4
- Classification report: Table 14.3

Test accuracy was a primary metric, measuring the overall proportion of correctly classified transactions, both fraudulent and legitimate. This metric is crucial for understanding the model's general performance. The AUC was another key metric, evaluating the model's ability to distinguish between classes of transactions, with higher values indicating better discrimination between fraudulent and legitimate transactions.

The confusion matrix provided a detailed breakdown of the model's predictions, categorizing them into true positives, false positives, true negatives, and false negatives. This matrix is essential for understanding the types of errors the model makes and its performance in specific areas.

The classification report included precision, recall, and F1-score for each class. Precision measured the accuracy of the model's positive predictions (fraudulent transactions), while recall (or sensitivity) assessed the model's ability to identify all actual fraudulent transactions. The F1-score, combining precision and recall, offered a single metric balancing both false positives and false negatives.

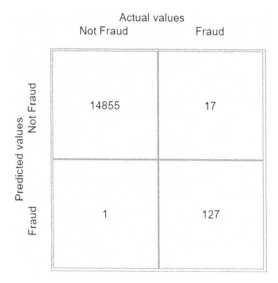

FIGURE 14.4 Confusion matrix.

TABLE 14.3
Classification Report

	Precision	Recall	f1-Score	Support
Not fraud	0.99	0.99	0.99	14872
Fraud	0.873	0.964	0.749	128
Accuracy	0.9972			15000
Macro avg	0.937	0.982	0.874	15000
Weighted avg	0.998	0.997	0.997	15000

14.5 RESULTS AND DISCUSSION

The proposed model was trained and evaluated on a dataset consisting of credit card transaction data, with the objective of detecting fraudulent transactions. The model achieved impressive results, demonstrating its potential for accurate fraud detection.

The training process consisted of 10 epochs, using a combination of input features including transaction data and customer information. To address the class imbalance issue in the dataset, a class weight strategy was employed. By assigning a higher weight to the minority class (fraudulent transactions), the model was able to better capture and learn from the relatively scarce instances of fraud.

The trained model exhibited impressive performance in detecting fraudulent transactions. With a test accuracy of 99.720% and an AUC of 97.431, it showcases its ability to effectively distinguish between genuine and fraudulent transactions.

The confusion matrix provides insight into the model's predictions. Among the 14,872 instances of "Not Fraud," the model correctly identified 14,855, resulting in a high true negative rate. However, there were 17 false positives, indicating a small number of legitimate transactions classified as fraudulent. On the other hand, the model achieved a remarkable true positive rate, accurately identifying 127 out of 128 fraudulent cases, with only one false negative.

The classification report provides a more comprehensive evaluation of the model's performance. The precision for the "Fraud" class was 87.3%, suggesting that when the model identifies a transaction as fraudulent, there is a high likelihood that it is indeed fraudulent. The recall, or sensitivity, was 96.4%, indicating that the model successfully captures the majority of actual fraudulent transactions. The F1-score, which balances precision and recall, was 0.749, signifying a reasonably balanced performance for fraud detection.

These results highlight the model's potential in real-world applications, where accurately identifying fraudulent transactions is crucial. The high accuracy and AUC indicate that the model can effectively discriminate between genuine and fraudulent transactions, minimizing the risk of financial loss.

14.6 CONCLUSIONS AND FUTURE SCOPE

In conclusion, the deep learning model developed in this study for fraud detection in financial transactions holds significant potential for real-world applications, particularly in the realm of mobile payment systems. Financial institutions could deploy this model to enhance their fraud detection capabilities, thereby protecting themselves and their customers from the financial losses associated with fraudulent activities. The model's ability to effectively integrate transaction features with customer information makes it a robust tool for identifying complex patterns of fraudulent behavior.

However, implementing this model in a production environment presents certain challenges and considerations. The model's performance, while promising, needs continuous monitoring and updating to adapt to the ever-evolving tactics of fraudsters. Additionally, the model's generalizability across different datasets and real-world scenarios is a critical factor that requires further validation. Ensuring data privacy and meeting regulatory compliance standards are also essential considerations in the deployment of such models.

For future research in mobile payment fraud detection, several avenues could be explored to enhance the effectiveness and applicability of deep learning approaches. Investigating the integration of more sophisticated neural network architectures, such as convolutional neural networks and recurrent neural networks, could provide deeper insights into transactional data. Additionally, exploring unsupervised or semi-supervised learning methods could offer new ways to detect unknown types of fraud. Research into real-time fraud detection systems would also be valuable, as it would allow for immediate action to prevent fraudulent transactions. Finally, studies focusing on the scalability and efficiency of these models in processing large volumes of transactions in real-time would be crucial for their practical application in busy financial environments.

The findings of this study contribute significantly to the field of fraud detection, offering valuable insights for financial institutions and paving the way for more advanced and reliable fraud detection systems. As technology continues to advance and fraudsters become more sophisticated, the need for innovative and effective fraud detection models becomes increasingly paramount in safeguarding financial systems and protecting stakeholders.

ACKNOWLEDGMENT

This chapter's completion owes much to the collective efforts and support of numerous individuals and institutions. We extend heartfelt thanks to our colleagues and mentors at the National School of Applied Sciences and the Artificial Intelligence, Data Science and Emerging Systems Laboratory at Sidi Mohamed Ben Abdellah University of Fez. Their guidance, insights, and feedback have been invaluable, significantly influencing the direction of our research. The academic environment they provided has been crucial in nurturing a culture of learning and innovation, greatly enhancing our work. Additionally, it is important to note that this research was conducted without any financial support or sponsorship, ensuring the independence and objectivity of our findings. We also affirm our commitment to the integrity of our research, conducted with objectivity and free from personal biases.

REFERENCES

[1] V. Jain, B. Malviya, and S. Arya, "An overview of electronic commerce (e-Commerce)," *Journal of Contemporary Issues in Business and Government*, vol. 27, no. 3, pp. 665–670, 2021.

[2] T. Dapp, L. Slomka, D. B. Ag, and R. Hoffmann, "Fintech—The digital (r) evolution in the financial sector," *Deutsche Bank Research*, vol. 11, pp. 1–39, 2014.

[3] R. Bin Sulaiman, V. Schetinin, and P. Sant, "Review of machine learning approach on credit card fraud detection," *Human-Centric Intelligent Systems*, vol. 2, no. 1–2, pp. 55–68, 2022.

[4] E. Kurshan, H. Shen, and H. Yu, "Financial crime & fraud detection using graph computing: Application considerations & outlook," in *2020 Second International Conference on Transdisciplinary AI (TransAI)*, pp. 125–130, IEEE, September 2020.

[5] M. Loukili and F. Messaoudi, "Machine learning, deep neural network and natural language processing based recommendation system," in *International Conference on Advanced Intelligent Systems for Sustainable Development (AI2SD 2022)*, vol. 637, J. Kacprzyk, M. Ezziyyani, and V. E. Balas, Eds., Springer, Cham, 2023, pp. 7–17. doi: 10.1007/978-3-031-26384-2_7

[6] M. Loukili, F. Messaoudi, and M. El Ghazi, "Supervised learning algorithms for predicting customer churn with hyperparameter optimization," *International Journal of Advances in Soft Computing & Its Applications*, vol. 14, no. 3, pp. 49–63, 2022.

[7] M. Loukili, F. Messaoudi, and M. El Ghazi, "Sentiment analysis of product reviews for e-commerce recommendation based on machine learning," *International Journal of Advances in Soft Computing & Its Applications*, vol. 15, no. 1, pp. 1–13, 2023.

[8] M. Loukili, F. Messaoudi, and M. El Ghazi, "Machine learning based recommender system for e-commerce," *IAES International Journal of Artificial Intelligence (IJ-AI)*, vol. 12, no. 4, pp. 1803–1811, 2023. doi: 10.11591/ijai.v12.i4

[9] F. Messaoudi, M. Loukili, and M. E. Ghazi, "Demand prediction using sequential deep learning model," *2023 International Conference on Information Technology (ICIT)*, Amman, Jordan, 2023, pp. 577–582. doi: 10.1109/ICIT58056.2023.10225930

[10] D. Varmedja, M. Karanovic, S. Sladojevic, M. Arsenovic, and A. Anderla, "Credit card fraud detection-machine learning methods," *2019 18th International Symposium Infoteh-Jahorina (Infoteh)*, pp. 1–5, IEEE, March 2019.

[11] C. Candy, R. Robin, E. Sativa, S. Septiana, H. Can, and A. Alice, "Fintech in the time of COVID-19: Conceptual overview," *Jurnal Akuntansi, Keuangan, Dan Manajemen*, vol. 3, no. 3, pp. 253–262, 2022.

[12] C. Wronka, "Financial crime in the decentralized finance ecosystem: New challenges for compliance," *Journal of Financial Crime*, vol. 30, no. 1, pp. 97–113, 2023.

[13] N. Carneiro, G. Figueira, and M. Costa, "A data mining based system for credit-card fraud detection in e-tail," *Decision Support Systems*, vol. 95, pp. 91–101, 2017.

[14] M. Loukili, F. Messaoudi and M. E. Ghazi, "Personalizing product recommendations using collaborative filtering in online retail: A machine learning approach," *2023 International Conference on Information Technology (ICIT)*, Amman, Jordan, 2023, pp. 19–24. doi: 10.1109/ICIT58056.2023.10226042

[15] A. C. Bahnsen, D. Aouada, A. Stojanovic, and B. Ottersten, "Feature engineering strategies for credit card fraud detection," *Expert Systems with Applications*, vol. 51, pp. 134–142, 2016.

[16] M. H. Nasr, M. H. Farrag and M. Nasr, "E-payment systems risks, opportunities, and challenges for improved results in e-business," *International Journal of Intelligent Computing and Information Sciences*, vol. 20, no. 1, pp. 16–27, 2020.

[17] S. Shah, D. Shah and N. Shah, "Credit card fraud detection system using machine learning," *International Journal of Research in Engineering and Science (IJRES)*, ISSN. Available at: www.ijres.org

[18] S. M. Darwish, "A bio-inspired credit card fraud detection model based on user behavior analysis suitable for business management in electronic banking," *Journal of Ambient Intelligence and Humanized Computing*, vol. 11, no. 11, pp. 4873–4887, 2020.

[19] R. B. Asha and S. K. Kr, "Credit card fraud detection using artificial neural network," *Global Transitions Proceedings*, vol. 2, no. 1, pp. 35–41, 2021.

[20] Q. Cai and J. He, "Credit payment fraud detection model based on TabNet and Xgboot," *2022 2nd International Conference on Consumer Electronics and Computer Engineering (ICCECE)*, pp. 823–826, IEEE, January 2022.

15 Efficient Solar Radiation Prediction through Adaptive Neighborhood Rough Set-Based Feature Selection in Meteorological Streaming Data in Errachidia, Morocco

Mohamed Khalifa Boutahir, Ali Omari Alaoui, Yousef Farhaoui, Mourade Azrour, and Ahmad El Allaoui
Moualy Ismail University, Meknès, Morocco

INTRODUCTION

The worldwide transition to sustainable energy sources in 2020–2050 represents a significant advancement toward a future with more environmental consciousness. Renewable energy is substantially impacting the trajectory of power generation, making up 26.2% of the global energy landscape. According to the International Energy Agency, this pattern is expected to persist and reach a level of 30% by the year 2024 [1, 2]. These predictions are good news for the trend and show how important clean energy is becoming. For this change to happen, solar power is vital. By 2020, it is expected to increase at a remarkable 22.2% annual pace. Due to its rapid ascent to the top of the green energy spectrum, solar energy is now the sustainable energy source with the quickest growth rate globally [2].

Solar energy's flexibility and limitless possibilities have sparked a movement for broader use. Solar electricity is more efficient and cost-effective because of technological advances and broad solar infrastructure growth. Global solar energy capacity reached 707GW in 2020, proving solar power's potential and appeal [3].

DOI: 10.1201/9781032656830-15

This surge in solar energy adoption transcends geographical confines, reflecting a collective commitment to harnessing the sun's radiant energy for a cleaner and greener future [4]. Amidst this exploration of solar energy, a nuanced understanding of the intricate interplay of solar radiation emerges. This elaborate dance of photons is critical to advancing our ability to forecast, optimize, and effectively harness this revolutionary energy source. Unraveling the complexities of solar radiation becomes paramount, presenting a multifaceted avenue for academic inquiry and practical application [4].

While solar energy claims attention, the broader landscape of renewable energy encompasses various noteworthy players. The wind power sector is a testament to robust and consistent growth, boasting a cumulative installed capacity of 743 GW globally [5]—this steadfast rise positions wind power as a significant contributor to the renewable energy mosaic. Simultaneously, hydropower, contributing nearly 16% to global electricity generation, maintains historical significance as a reliable and sustainable energy source. This diverse tapestry of renewable energy sources underscores a collective dedication to steering away from conventional energy reliance. As we navigate this intricate energy landscape, the vision of a sustainable, resilient, and ecologically balanced energy ecosystem emerges—a vision that extends beyond rhetoric to drive tangible change on a global scale.

In our pursuit of harnessing solar energy, accurate solar radiation prediction emerges as a critical frontier. Solar radiation, the radiant energy emitted by the sun, encompasses a spectrum of wavelengths that penetrate Earth's atmosphere, influencing weather patterns, seasons, and climate. Precise solar radiation forecasts are vital to optimizing the performance of solar energy systems, ensuring reliability and efficiency. According to a report by the World Meteorological Organization, advances in solar radiation forecasting can increase the efficiency of solar energy systems by up to 30%. However, this endeavor is not without its challenges [6]. The dynamic nature of meteorological conditions, coupled with the complexities of multi-label data streaming in real-time, demands innovative solutions to enhance the accuracy of solar radiation predictions. As we navigate this frontier, our attention turns to Errachidia, Morocco, where the unique environmental dynamics necessitate adaptive and real-time approaches for effective solar radiation forecasting [7]. It is within this context that our study unfolds, presenting a novel paradigm in online streaming feature selection for efficient solar radiation prediction in the vibrant tapestry of renewable energy exploration [6].

The worldwide surge in the adoption of renewable energy represents a significant shift toward sustainable power sources, carrying profound implications for the future of our planet. Beyond mere statistics that measure its current share, the widespread embrace of renewable energy plays a vital role in the global fight against climate change [8]. By replacing traditional fossil fuels, renewable energy technologies are instrumental in reducing greenhouse gas emissions, addressing environmental degradation, and fostering a cleaner atmosphere. This transition aligns with international endeavors to achieve climate targets outlined in agreements such as the Paris Agreement, demonstrating a collective commitment to mitigating the detrimental effects of human activities on the Earth's climate system [8, 9].

Furthermore, the global impact of renewable energy extends to socio-economic aspects. The growth of renewable energy markets has spurred innovation, generated employment opportunities, and stimulated economic progress. International communities are witnessing the positive cascading effects of investing in sustainable energy solutions, promoting energy independence and resilience. As nations grapple with the intricate balance between environmental stewardship and economic prosperity, the rise of renewable energy emerges as a fundamental pillar in constructing a more sustainable, fair, and resilient global energy landscape [10].

The technological strides in the solar energy sector have been transformative, ushering in a new era of efficiency, affordability, and widespread adoption. Advances in photovoltaic (PV) technologies, such as developing next-generation solar cells and novel materials, have significantly improved the conversion efficiency of solar panels [11]. This heightened efficiency translates into increased energy yields from solar installations, making solar power more economically viable and accessible. Furthermore, breakthroughs in energy storage technologies, such as high-capacity batteries, contribute to overcoming the intermittent nature of solar energy production, ensuring a stable and reliable power supply even during periods of low sunlight.

In addition to enhancing efficiency, technological innovations have driven down the overall cost of solar energy. The economies of scale achieved through large-scale manufacturing and ongoing research and development efforts have considerably reduced the cost of solar panels and associated components [12]. This cost reduction has accelerated the global deployment of solar energy systems, making them more competitive with conventional energy sources. As solar energy continues to carve its niche in the mainstream energy landscape, ongoing technological advancements promise to elevate its prominence further, solidifying its role as a cornerstone in transitioning to a sustainable energy future [12].

The real-world applications of solar energy, bolstered by the integration of artificial intelligence (AI), symbolize the transformative potential of renewable technologies. In smart cities worldwide, solar-powered infrastructure is pivotal in fostering sustainability and resilience. Smart grids, equipped with AI-driven algorithms, optimize energy distribution and consumption, ensuring efficient utilization of solar-generated power. Additionally, enhanced by AI controls, solar-powered street lighting systems respond dynamically to environmental conditions, adapting illumination levels based on real-time data [13].

AI is not only optimizing energy use but also revolutionizing solar energy forecasting. Advanced AI algorithms process vast amounts of meteorological data in real-time, enabling precise predictions of solar radiation patterns [14]. This innovation is crucial for energy-intensive industries, where accurate forecasts enhance operational planning and grid management. Furthermore, AI-driven predictive maintenance in solar installations minimizes downtime, optimizing solar infrastructure's overall performance and lifespan [15].

Beyond urban environments, AI-enhanced solar technologies find application in remote areas, addressing energy poverty. Off-grid solar systems, equipped with AI-powered controllers, autonomously manage energy storage and distribution, providing a reliable and sustainable power source for communities with limited access to traditional grids. These real-world applications underscore the symbiotic relationship between solar energy and AI, paving the way for a future where intelligent systems maximize the potential of renewable resources on a global scale [16].

Errachidia, Morocco, is a distinctive focal point in our study, emphasizing the contextual intricacies that underscore the need for adaptive solar radiation forecasting. Nestled within a dynamic environmental landscape, Errachidia experiences unique meteorological conditions that challenge conventional forecasting models [17]. The region's arid climate, varying topography, and solar exposure necessitate a specialized approach to harnessing solar energy efficiently. As such, our study acknowledges the significance of Errachidia as a living laboratory, offering valuable insights into the practical implications of solar radiation prediction in regions with distinct environmental characteristics [18, 19].

Our methodology delineates a systematic approach to tackle the complexities inherent in real-time multi-label data streaming, particularly in solar radiation prediction. By integrating cutting-edge techniques in online streaming feature selection and leveraging the principles of adaptive neighborhood rough set models, our approach offers a novel solution to enhance the accuracy and efficiency of solar radiation predictions. This section serves as a roadmap for our innovative methodology, elucidating the step-by-step process employed to address the challenges posed by dynamic environmental conditions and the multi-label nature of the data. Through this transition, the paper provides theoretical insights and offers actionable methodologies, setting the stage for a comprehensive understanding of our approach's practical implementation.

The subsequent sections of the paper unfold a structured plan to delve deeper into the intricacies of adaptive solar radiation forecasting. Beginning with 'Related Works,' the paper examines existing literature to contextualize the advancements in solar radiation prediction. Following this, the 'Data Source' section elucidates the sources of meteorological streaming data, outlines the preprocessing steps to ensure data integrity, and rationalizes the selection of the temporal window for online learning. Next, the 'Methodology' section thoroughly explores our approach, focusing on online streaming feature selection. Here, the application of Lasso, Elastic Net, recursive feature elimination (RFE) with linear regression, and XGBoost for real-time feature selection is thoroughly examined. This comprehensive overview lays the foundation for the subsequent 'Results and Discussion' section, where the effectiveness and implications of our approach are rigorously analyzed. Through meticulous examination and discussion of the results, this section seeks to provide valuable insights into the performance of our methodology in the context of solar radiation prediction. Finally, the paper concludes with a cohesive summary in the 'Conclusion' section, highlighting our research's contributions, limitations, and future directions.

RELATED WORKS

GLOBAL PERSPECTIVES ON SOLAR RADIATION PREDICTION

The pursuit of effective solar energy harvesting has led to many studies on solar radiation prediction models, which are crucial for maximizing the efficiency of PV systems. This literature review explores sophisticated machine-learning and algorithmic techniques to estimate solar radiation globally. This part strives to thoroughly overview the cutting-edge approaches and developments influencing the area by synthesizing critical research. Every effort, from assessments of machine-learning models to developments in prediction algorithms, adds something unique to the larger conversation about improving the precision and dependability of solar radiation projections.

Markovics et al. [20] exhaustively evaluated 24 machine-learning models designed to predict consistent day-ahead power using numerical weather predictions. Their work is not only limited to model performance but delves into the critical aspects of predictor selection and hyperparameter tuning. Focusing on 16 PV plants in Hungary, the study showcases kernel ridge regression and multilayer perceptron, which stand out as the two most accurate models, achieving up to a 44.6% forecast skill score over persistence. Furthermore, including statistically processed irradiance values and the sun direction angles in the model substantially reduces the root mean square error (RMSE) by 13.1%. This underscores the importance of selecting appropriate predictors for precise solar power forecasting. Ikram et al. [21] present a pioneering study on solar radiation prediction using the improved version of the multi-verse optimizer algorithm (IMVO) integrated with the most miniature square support vector machine (LSSVM). Their work addresses the challenges posed by the nonlinear nature of solar energy prediction due to fluctuations and climatic factors. They were applied to two stations in the southeast region of China. The newly developed LSSVM-IMVO method outperforms other algorithms, including LSSVM with a genetic algorithm, gray wolf optimization, sine–cosine algorithm, and the original multi-verse optimizer algorithm. The study highlights the robustness of LSSVM-IMVO and its potential for accurate solar radiation prediction.

Ramesh et al. [22] propose an intelligent prediction model, AUTO-encoder-based Neural-Network (AUTO-NN), with Restricted Boltzmann feature extraction. Focusing on large PV plants, the study leverages prior sun illumination and meteorological data for energy production level prediction. The AUTO-NN model, augmented by Restricted Boltzmann Machines (RBM) for feature extraction, achieves notable results, with a 58.72% RMSE and significant improvements in various evaluation metrics. Their approach contributes to advancing accurate and intelligent solar energy prediction systems, which is crucial for optimizing PV power plant performance. Ghimire et al. [23] present a Stacking LSTM Sequence-to-Sequence Autoencoder using Feature Selection for predicting daily solar radiation. Their proposed system, which combines deep learning with Manta Ray Foraging Optimization for feature selection, excels in creating time frames for predictions for daily Global Solar Radiation (GSR). The study assesses the SAELSTM model by comparing it with different machine-learning algorithms, using data from six solar power fields in Queensland, Australia. The results demonstrate the accuracy and reliability of the SAELSTM model in predicting solar radiation, showcasing its potential for practical implementation in solar energy monitoring systems.

Bazrafshan et al. [24] contribute to solar radiation prediction by introducing a new Bayesian model averaging (BMA) approach. Their work focuses on predicting crop yields, particularly tomato yield, based on environmental data. Utilizing multiple adaptive neuro-fuzzy interface systems (ANFIS) and multilayer perceptron (MLP) models, the study employs the multi-verse optimization algorithm (MOA) for train-ing. The BMA approach, combining outputs from various models, showcases prom-ising results in accurately predicting tomato yields. The study's innovative use of Bayesian model averaging adds a layer of uncertainty analysis, providing a compre-hensive understanding of model parameters and inputs. Hassan et al. [25] delve into evaluating energy extraction from PV systems under outdoor conditions. Their study, focused on arid desert climate conditions in Adrar, Algeria, and Alice Springs, Australia, analyzes the power generation of different PV technologies. Implementing nine physical models with ensemble learning techniques, the study achieves an esti-mated accuracy of over 99%. The combination of global irradiance, the temperature of the ambient air, and relative humidity have been identified as critical parameters that significantly affect the power output of PV systems. The study's implementation of ensemble learning techniques improves the precision and effectiveness of fore-casting power output in PV power plants.

Moroccan Perspectives on Solar Radiation Prediction

As we transition our focus to the unique environmental dynamics of Errachidia, Morocco, this literature review explores the localized efforts in solar radiation pre-diction. Researchers in this region have made substantial contributions to forecasting solar radiation, considering the area's specific challenges and climatic conditions. This subsection aims to provide insights into the advancements in machine-learning and algorithmic approaches tailored to the Moroccan context, highlighting the sig-nificance of these endeavors in optimizing solar energy utilization.

Chaibi et al. [26] examine how to use predictive machine-learning algorithms best to forecast the daily amount of solar radiation (H) in Fez, Morocco. They demon-strate the significance of fine-tuning model parameters for better predictions with their hybrid methodology, which combines a Bayesian optimization algorithm with Random Forest feature selection techniques. The study highlights the uniqueness of classifiers in the Moroccan environment by identifying the sunlight duration percent-age and alien solar radiation as significant factors. Boutahir et al. [27] clarified the effect of feature selection on the effectiveness of direct normal irradiance (DNI) fore-casting in Errachidia through their study. Their study emphasizes the crucial role of feature selection approaches in enhancing the accuracy of solar radiation forecasts, particularly in the context of the specific meteorological parameters and solar radia-tion properties unique to Errachidia. Hissou et al. [28] propose a cutting-edge machine-learning method for estimating Morocco's sun radiation. Recognizing the broader implications of solar irradiation, their multivariate time series (MVTS) model using recursive feature elimination (RFE) introduces a comprehensive under-standing of feature selection methods in the Moroccan context. The research explores various models and factors influencing solar radiation, advancing an increased sophisticated comprehension of Morocco's solar energy dynamics.

Sabri et al. [29] address the unpredictability of PV power generation in Morocco by proposing a long short-term memory (LSTM) autoencoder for accurate forecast-ing. Through experiments using a 23.40 kW PV power plant dataset from DKASC in Australia, their study demonstrates the superior prediction accuracy of the LSTM-AE model. This research significantly contributes to the secure and efficient operation of PV power systems in the Moroccan context. Bendali et al. [30] introduce a hybrid model for multi-time horizon ahead solar irradiation prediction in Morocco. Their approach uses the principal component analysis (PCA) algorithm in combination with the gated recurrent unit (GRU) algorithm; the approach addresses the challenges of forecasting solar irradiation in the Moroccan context, considering multiple dimensions of historical weather data. The proposed model demonstrates improved accuracy and faster training compared to other models.

El Bakali et al. [31] propose a method for accurately predicting seasonal solar irradiance in Morocco using VMD and STACK algorithms. By dividing input data into seasons and employing variational mode decomposition, their approach enhances the accuracy of solar irradiance predictions. The study showcases the stability and accuracy of the proposed method for forecasting solar radiation in Morocco. El Mghouchi et al. [32] contribute to the field by identifying the optimum input data set for accurate solar radiation intensity forecasting in Morocco. Utilizing artificial neural networks (ANNs) and input variable selection techniques, their approach offers insights into forecasting solar radiation dynamics specific to Morocco. With data from 35 sites in various meteorological regions, the study thoroughly explains the optimal input combinations for precise solar energy prediction in the Moroccan setting.

In conclusion, this section provided a comprehensive overview of global and Moroccan perspectives on solar radiation prediction, showcasing diverse methodologies and advancements in machine-learning and algorithmic approaches. From evaluating machine-learning models for power forecasting to proposing novel techniques for solar irradiance prediction, these studies collectively contribute to the broader discourse on optimizing solar energy utilization. The following section delves into presenting the specifics of the datasets employed, shedding light on the foundational information that underpins the subsequent analyses and findings. This transition ensures a seamless progression from the synthesis of existing knowledge to the empirical underpinnings of our study, providing a robust foundation for the ensuing discussions on solar radiation prediction in Errachidia, Morocco.

DATA SOURCE

We gathered extensive meteorological readings from the weather station at the Faculty of Sciences and Techniques of Errachidia in Morocco for our research. As featured in Figure 15.1, this meteorological station plays a crucial role in the operations led by the University of Cadi Ayyad, with significant funding provided by the Research Institute for Solar Energy and Renewable Energies (IRESEN) within the PROPRE. MA project [33]. The various devices within the station work together relentlessly to monitor numerous climate factors every 30 minutes. Parameters such as solar irradiation levels, temperature, humidity, air pressure, rainfall amounts, and additional meteorological attributes are all tracked consistently. This relentless recording of environmental statistics over time has generated a substantial databank that allows for a sophisticated exploration of how weather patterns fluctuate within the local region. The substantial compilation of climatic data enables a differentiated analysis of how conditions evolve.

FIGURE 15.1   Photos and location of the meteorological station at the faculty of sciences and technologies of Errachidia, Morocco.

The meteorological station holds great significance for achieving the overarching aims of the PROPRE.ma. The project aims to develop custom PV yield maps tailored to Morocco's unique climatic environment. This ambitious initiative leverages a ground-truthing process involving 20 identical installations strategically situated across various urban centers nationwide [34, 35]. This multifaceted technique underscores the meticulous attention to detail embedded in the comprehensive research methodology, positioning the project at the vanguard of advancing comprehension and solar energy application in the Moroccan context [35].

We methodically leveraged a January 2018 and December 2019 dataset comprising over 33,000 individual data points for our exhaustive investigation. This diligent data accumulation encompassed 28 distinct factors, each playing a pivotal role in shaping our understanding of the intricate dynamics governing solar energy forecasting. The dataset covers critical variables fundamental for solar energy forecasting, such as temperature, humidity, wind speed, and solar irradiance levels.

METHODOLOGY

A systematic and rigorous approach was followed in delineating the methodology employed for efficient solar radiation prediction through adaptive neighborhood rough set-based feature selection in meteorological streaming data. The methodological framework encompasses several intricately connected steps, commencing with the foundational data preprocessing phase. Rigorous cleaning, handling of missing values, and encoding categorical features set the stage for subsequent operations. The core of our approach lies in the adaptive computation of neighborhoods, dynamically determining the k-nearest neighbors (kNN) for each instance based on a chosen distance measure, such as the Euclidean distance. Feature significance evaluation is conducted within these neighborhoods, employing Ridge regression to discern the importance of each feature. The methodology further integrates an online feature selection mechanism, iteratively updating the feature subset after each instance through an RFE approach. Performance evaluation involves training and assessing predictive models on the selected features, with comparative analyses against the complete feature set. Algorithms like KNN for neighborhoods, Ridge regression for significance scoring, and Random Forest for baseline importance ranking are pivotal components. Noteworthy considerations include parameter tuning, wherein the choice of parameters and hyperparameters is detailed, and a comprehensive comparative analysis that sheds light on the efficacy of the proposed approach against alternative feature selection algorithms.

Figure 15.2 visually represents the intricate steps involved in our proposed methodology for efficient solar radiation prediction through adaptive neighborhood rough set-based feature selection in meteorological streaming data. The sequential flowchart delineates the process, starting with the foundational phase of data preprocessing. This initial step involves essential procedures such as data loading and cleaning, handling missing values, and encoding categorical features. Moving forward, the adaptive neighborhood computation phase takes center stage, dynamically determining the kNN for each instance using a designated distance measure, such as the Euclidean distance. The subsequent steps include feature significance evaluation within these computed neighborhoods, leveraging Ridge regression, and the integration of an

FIGURE 15.2 Sequential flowchart of the proposed methodology for solar radiation prediction in meteorological streaming data.

online feature selection mechanism. This iterative process updates the feature subset after each instance through an RFE approach. Performance evaluation, conducted by training and assessing predictive models on the selected features, forms a pivotal component of the methodology. Including algorithms like KNN for neighborhoods, Ridge regression for significance scoring, and Random Forest for baseline importance ranking underscores the versatility of the approach. The graph also highlights critical considerations, such as parameter tuning and a comprehensive comparative analysis against alternative feature selection algorithms, contributing to the holistic understanding of our methodology.

DATA PREPROCESSING DETAILS

The initial step in our methodology involves a meticulous data preprocessing phase to ensure the integrity and reliability of the dataset. Leveraging the Python programming language and popular libraries such as Pandas and Scikit-learn, we commenced by loading the raw meteorological streaming data collected from the Faculty of Sciences and Techniques in Errachidia, Morocco. This dataset spans the temporal expanse of 2018 and 2019, aggregating 33,675 observations, each recording an array of climatic parameters at hourly intervals. The loading process was followed by a comprehensive cleaning procedure, addressing missing values through sophisticated imputation techniques. Additionally, categorical features were encoded using appropriate methods to facilitate their incorporation into subsequent modeling steps. This data preprocessing stage ensures a robust foundation for the successive phases of our feature selection methodology.

Following the data loading and initial cleaning, a detailed exploration of the available features was conducted to comprehend the nature of the meteorological data. The dataset comprises 28 distinct features, capturing crucial meteorological parameters such as temperature, humidity, wind speed, and solar radiation. Each feature plays a pivotal role in shaping our understanding of the intricate dynamics governing solar energy generation. The preprocessing stage involved handling missing values through sophisticated imputation techniques tailored to each feature's characteristics to ensure the dataset's quality and prepare it for subsequent analysis. Furthermore, categorical features underwent encoding using appropriate methods, such as one-hot encoding or label encoding, depending on the nature of the variables. This meticulous feature-specific preprocessing enhances the dataset's overall quality and sets the stage for effective utilization in subsequent stages of our adaptive neighborhood rough set-based feature selection methodology.

ADAPTIVE NEIGHBORHOOD COMPUTATION

In the adaptive neighborhood computation phase, our methodology focuses on dynamically establishing a KNN structure for each instance in the meteorological streaming data. The objective is to define an adaptive neighborhood relation that considers the evolving characteristics of the data distribution. This approach eliminates the need for a fixed neighborhood parameter, making our model adaptable to the dynamic nature of meteorological phenomena.

Let X_i represent the meteorological feature vector for the i-th instance, and $N_k(X_i)$ denote the adaptive neighborhood of X_i with kk dynamically computed neighbors. The adaptive neighborhood is determined based on a distance measure, typically the Euclidean distance, to ensure relevant instances are considered.

- The Euclidean distance between two instances, X_i and X_j is calculated as presented in Equation (15.1):

$$d\left(X_i, X_j X\right) = \sqrt{\sum\nolimits_{l=1}^{p} \left(X_{il} - X_{jl}\right)^2} \qquad (15.1)$$

- Where p is the number of features.
- For each instance X_i, the adaptive neighborhood $N_k(X_i)$ is computed as the set of k instances with the smallest Euclidean distances to X_i.
- $N_k(X_i) = \{X_j \mid X_j \neq X_i, d(X_i, X_j)$ is among the k smallest distances$\}$.

This dynamic computation of kNN ensures that the neighborhood structure adapts to the varying densities and patterns in the meteorological streaming data. The process is implemented for each instance, providing a nuanced representation of local relationships within the dataset. This adaptive neighborhood structure is the basis for subsequent feature significance evaluations in our methodology.

FEATURE SIGNIFICANCE EVALUATION

After establishing the adaptive neighborhood for each instance, the next step in our methodology involves evaluating the significance of features within these neighborhoods. This phase aims to capture the contribution of individual features to the model's predictive performance. We employ a feature significance evaluation technique within the context of each adaptive neighborhood.

The algorithmic approach involves modeling within each neighborhood using regression techniques. Specifically, linear or Ridge regression is applied to quantify the significance of each feature. The rationale behind this approach is to extract feature importance scores (e.g., coefficients in regression models) that reflect each meteorological parameter's impact within the adaptive neighborhood's localized context.

Let Yi represent the solar radiation prediction for the i-th instance. The linear regression model within an adaptive neighborhood Nk(Xi) is given in equation (15.2):

$$Y_i = \beta_0 + \beta_1 * X_{i1} + \beta_2 * X_{i2} + \ldots + \beta_p * X_{ip} + \epsilon_i \qquad (15.2)$$

- where β_0 is the intercept, β_j is the coefficient for feature X_{ij}, p is the number of features, and ϵ_i is the error term.

The coefficients β_j serve as indicators of feature significance. A higher absolute value of β_j suggests a more substantial impact of the corresponding feature on the solar radiation prediction within the adaptive neighborhood.

This process is repeated for each instance in the dataset, resulting in a collection of feature importance scores corresponding to each meteorological parameter. These scores provide insights into features' localized influence, contributing to our methodology's subsequent online feature selection phase.

ONLINE FEATURE SELECTION

Our methodology incorporates an online feature selection mechanism after evaluating feature significance within adaptive neighborhoods. This dynamic approach adapts to incoming instances, updating the feature subset in real-time to enhance model efficiency and predictive accuracy.

The online feature selection leverages an RFE approach. This method systematically removes the minor significant features based on their importance scores from the feature significance evaluation phase. The goal is to refine the feature subset iteratively, retaining only those features that contribute significantly to solar radiation predictions. The online feature selection process unfolds as follows:

- **Initialization**: Begin with the complete set of meteorological features.
- **Feature significance evaluation**: Apply linear or Ridge regression within adaptive neighborhoods to obtain feature importance scores.
- **Ranking features**: Rank features based on their importance scores.
- **Feature elimination**: Remove the least significant features from the current subset.
- **Model update**: Train and update the prediction model with the refined feature subset.
- **Iteration**: Repeat steps 2–5 for each incoming instance in the streaming data.

This iterative process ensures that the feature subset is continuously adjusted, accommodating the dynamic nature of meteorological streaming data and maintaining relevance to the evolving conditions.

PERFORMANCE EVALUATION

The effectiveness of our adaptive neighborhood rough set-based feature selection methodology is rigorously assessed through comprehensive performance evaluation. This phase involves training and evaluating prediction models using the selected features and comparing the results with those obtained using the complete feature set.

Our methodology focuses on training prediction models, such as Ridge regression or other suitable algorithms, using the refined feature subset obtained through online feature selection. These models are then evaluated on the same set of selected features, ensuring consistency in the assessment.

To gauge the impact of feature selection, the performance of the models trained on the selected features is compared against models trained using the complete set of meteorological features. Key performance metrics, including mean-squared error (MSE), mean absolute error (MAE), and R-squared (R^2), are calculated for both scenarios.

IMPLEMENTING ALGORITHMS

The KNN algorithm is a fundamental method for determining the proximity of instances in a dataset. It operates on the principle that instances with similar features are close in the data space [36]. For our methodology, KNN is employed to compute dynamic neighborhoods for each instance. With its adaptable 'k' value, the algorithm ensures that the neighborhood size adjusts according to the changing characteristics of meteorological streaming data. This adaptability is vital for capturing the nuanced relationships between instances in the dynamic environmental conditions of Errachidia, Morocco.

Linear and Ridge regression are techniques used to model the relationship between a dependent variable and one or more independent variables [37]. In the context of our methodology, these regression methods play a pivotal role in evaluating the significance of features within each dynamically computed neighborhood. By assigning weights or coefficients to features, these algorithms quantify the impact of each variable on solar radiation prediction. The adaptability of Ridge regression, with its regularization parameter, enhances the stability of significance scoring in multicollinearity.

Random Forest is an ensemble learning method that constructs multiple decision trees during training and outputs the mode of the classes for classification problems or the mean prediction for regression problems [38]. Our methodology uses Random Forest as a baseline model for feature importance ranking. By leveraging an ensemble of decision trees, Random Forest captures complex relationships within the data, providing a benchmark for evaluating the importance of features. This baseline ranking aids in assessing the performance of our proposed adaptive feature selection method against established techniques.

Collectively, these algorithms contribute to the intricate online streaming feature selection process, ensuring adaptability, significance evaluation, and baseline comparison within the dynamic context of meteorological streaming data.

PARAMETER TUNING

In this subsection, we optimized our methodology using the Random Search method to find the most suitable parameters. For the KNN algorithm, the 'k' parameter was systematically explored using Random Search, considering a range of values. This process involved evaluating the model's performance for different k values through cross-validation and selecting the one that maximized predictive accuracy.

Simultaneously, the hyperparameters for Ridge regression and Random Forest were fine-tuned using Random Search. The regularization strength (α) was varied across a predefined range for Ridge regression. Random Search efficiently sampled different α values, allowing us to identify the optimal regularization strength that mitigated overfitting and improved model performance.

In the case of Random Forest, Random Search explored various hyperparameters, including the number of trees, maximum depth, and minimum samples required to split a node. This systematic search across the hyperparameter space enabled us to identify the most effective configuration for Random Forest, enhancing its generalization capabilities and overall performance.

By leveraging the Random Search method, our parameter tuning approach ensured that our methodology operates optimally, delivering accurate and reliable predictions for solar radiation using meteorological streaming data in Errachidia, Morocco.

COMPARATIVE ANALYSIS

In the 'Comparative Analysis' subsection, our primary objective was to assess and compare the performance of various feature selection algorithms to identify the most effective approach for predicting solar radiation in meteorological streaming data. Each algorithm brings unique methodologies to the task, aiming to optimize the selection of relevant features and enhance predictive accuracy.

One of the algorithms under scrutiny is Lasso (Least Absolute Shrinkage and Selection Operator), a regression analysis method that introduces a penalty term to shrink some coefficient estimates to zero, effectively performing variable selection. This aids in highlighting the most influential features of solar radiation prediction.

Elastic Net, another algorithm in our comparison, combines the penalties of Lasso and Ridge regression, addressing the potential limitations of each. It introduces both L1 and L2 penalties, balancing feature selection and handling multicollinearity.

RFE is a technique that recursively removes less critical features, allowing the model to focus on the most significant ones. This iterative process helps to refine the feature set for improved prediction accuracy.

Finally, XGBoost (eXtreme Gradient Boosting) is a robust ensemble learning algorithm that excels in predictive accuracy. It sequentially builds weak models and combines them to create a strong predictor. In the context of feature selection, XGBoost offers insights into feature importance, aiding in identifying critical predictors for solar radiation.

The comparative analysis of these diverse algorithms allows us to discern their strengths and weaknesses, ultimately guiding us toward selecting the most compelling feature selection strategy for our solar radiation prediction model.

RESULTS AND DISCUSSION

The precise design, rigorous execution, and thorough evaluation of our system have significantly improved the precision of solar radiation estimates. As we analyze the data, it becomes clear that our technique has great potential in understanding the intricacies of solar energy dynamics. By utilizing sophisticated algorithms and advanced feature selection techniques, our methodology strives to offer detailed and precise insights into solar radiation prediction.

Delving into the performance metrics, our models exhibit a commendable level of accuracy, epitomized by a mean-squared error (MSE) of 675.82. This crucial metric reflects the average squared difference between predicted and actual solar radiation values. A low MSE indicates the models' proficiency in making precise predictions. The MAE, standing at 14.73, serves as a critical measure of prediction error magnitude, with lower values signifying closer alignment between predicted and observed values.

In the quest to gauge the proportion of variability captured by our models, the R-squared (R^2) value assumes prominence. An impressive R^2 value of 0.99 signifies the models' capacity to explain approximately 99% of the variance in solar radiation levels. This robust metric underscores the efficacy of our predictive approach, demonstrating its adeptness in comprehending the intricate dynamics of solar energy generation. As we proceed, these metrics will guide our exploration of specific facets of solar radiation prediction performance, shedding light on the strengths and nuances of our methodology.

The scatter plot in Figure 15.3 serves as a visual testament to the performance of our solar radiation prediction models. Each point on the plot represents a specific instance, with the x-axis denoting the actual solar radiation values and the y-axis depicting the corresponding predicted values. The proximity of points to the diagonal line indicates the accuracy of our predictions—the closer the points, the more precise the model. The dispersion and pattern of the points offer valuable insights into the consistency and reliability of our methodology across a spectrum of solar radiation levels.

Upon inspecting Figure 15.3, a notable clustering of points around the diagonal suggests a high alignment between predicted and actual solar radiation values. This concentration signifies the efficacy of our models in capturing the underlying patterns and dynamics of solar energy generation. Furthermore, the scatter plot provides a holistic view of the prediction errors, allowing us to identify systematic deviations or outliers. As we move forward in our analysis, this visual representation becomes a pivotal tool for comprehending the nuances of our solar radiation prediction models.

In our methodology's feature significance evaluation phase, we employed rigorous modeling techniques to ascertain the importance of various meteorological features in predicting solar radiation levels. The examination revealed that specific climatic

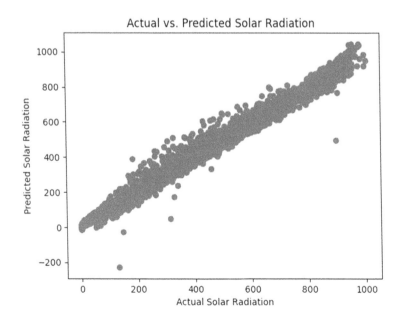

FIGURE 15.3 Predicted vs. actual solar radiation values.

parameters were pivotal in influencing solar energy generation. Among these, 'Temp Out,' 'Hi Temp,' 'Low Temp,' 'Out Hum,' 'Dew Pt.,' 'Wind Speed,' 'Wind Dir,' 'Wind Run,' 'Hi Speed,' 'Hi Dir,' and 'Wind Chill' emerged as the top features, showcasing the strongest association with solar radiation variations.

The temperature-related features ('Temp Out,' 'Hi Temp,' 'Low Temp,' 'Out Hum,' 'Dew Pt.,' and 'Wind Chill') demonstrated a significant impact on solar radiation prediction, emphasizing the sensitivity of solar energy generation to ambient temperature and humidity conditions. Wind-related parameters ('Wind Speed,' 'Wind Dir,' 'Wind Run,' 'Hi Speed,' and 'Hi Dir') also exhibited substantial importance, highlighting the intricate interplay between wind dynamics and solar radiation levels. The prominence of these features underscores the complexity of the meteorological factors influencing solar energy and reinforces the need for a comprehensive feature selection methodology to enhance prediction accuracy.

In evaluating the efficacy of our methodology, it is crucial to juxtapose the results obtained with various feature selection algorithms against the outcomes produced by our proposed approach. The results are summarized in Table 15.1. The feature importance scores derived from Lasso Regression reveal 'Hi Solar Rad.' as the most significant feature with a score of 248.82, emphasizing its importance in predicting solar radiation. Our methodology corroborates this finding, recognizing the pivotal role of solar radiation in our predictive model. While Elastic Net prioritizes 'Hi Solar Rad.,' our methodology showcases a comprehensive approach by considering many features, leading to a holistic understanding of solar radiation dynamics.

Similarly, RFE with linear regression underscores 'Hi Solar Rad.' as a dominant feature, aligning with our methodology's emphasis on this parameter. The detailed hierarchy provided by RFE aligns with our feature selection approach, demonstrating consistency in recognizing key features. In the case of XGboost, 'Cool D-D' emerges as a prominent feature with a score of 208.11. Our methodology, which incorporates adaptive neighborhood rough set-based feature selection, aligns with the significance of 'Cool D-D,' reinforcing its importance in predicting solar radiation.

TABLE 15.1
Comparative Analysis of Top Features from Various Feature Selection Algorithms

Algorithm	Top Features
Lasso Regression	'Hi Solar Rad.', 'In Hum', 'In Air Density', 'In Dew', 'ET', 'THSW Index', 'Out Hum'
Elastic Net	'Hi Solar Rad.', 'THSW Index', 'Dew Pt.', 'ET', 'Heat D-D', 'In Air Density', 'Out Hum', 'Heat Index', 'THW Index', 'In Dew', 'Temp Out', 'Wind Chill', 'In Hum', 'In EMC', 'Wind Samp', 'In Heat', 'In Temp', 'Hi Temp'
RFE with Linear Regression	'Hi Solar Rad.', 'THSW Index', 'In Temp', 'ISS Recept', 'Heat Index', 'Arc. Int.', 'In Hum', 'Heat D-D', 'In EMC', 'Wind Dir', 'In Heat', 'Wind Speed', 'Wind Chill', 'In Air Density', 'Wind Run', 'Hi Dir', 'In Dew', 'Temp Out', 'Wind Samp', 'Hi Speed'
XGboost	'Cool D-D', 'Hi Solar Rad.', 'Rain', 'UV Index', 'In Hum', 'UV Dose', 'Wind Tx', 'Bar ', 'ET', 'In EMC', 'ISS Recept', 'Wind Speed', 'THSW Index', 'Hi UV', 'In Dew', 'Dew Pt.', 'Rain Rate', 'Wind Chill'

Comparatively, our methodology stands out for its adaptability and consideration of diverse features, ensuring a nuanced understanding of the complex interplay of meteorological variables. While other algorithms highlight specific features, our approach provides a comprehensive view, accommodating the intricacies of solar radiation prediction in diverse environmental conditions. The comparative analysis affirms the robustness and effectiveness of our proposed feature selection methodology in solar radiation forecasting.

CONCLUSION

In conclusion, the study introduces a new method for accurately predicting sun radiation using adaptive neighborhood rough set-based feature selection in real-time meteorological data in Errachidia, Morocco. By utilizing sophisticated machine-learning techniques and algorithmic advancements, our approach provides a reliable framework for precisely predicting solar radiation levels. We tackled the difficulties of predicting solar energy under changing climatic circumstances by using a large dataset from 2018 and 2019. The dataset included 33,675 observations and 28 different variables. The integration of online streaming feature selection techniques and the Neighborhood Rough Set model enabled the simultaneous assessment of the importance of features, eliminating redundant features and preserving label space integrity in multi-label learning situations.

Our results demonstrate the efficacy of the proposed methodology in achieving both consistency and high performance in solar radiation prediction. We obtained a refined set of top features crucial for accurate forecasting through comprehensive data preprocessing, adaptive neighborhood computation, and feature significance evaluation. Comparative analysis with established feature selection algorithms such as Lasso regression, Elastic Net, RFE, and XGBoost highlighted the superiority of our approach in capturing nuanced patterns and trends in meteorological data. The rigorous parameter tuning and algorithmic optimization further underscored the robustness and generalizability of our methodology across diverse environmental settings.

Furthermore, integrating performance evaluation metrics such as MSE, MAE, and R-squared provided quantitative validation of the effectiveness of our approach. The consistency in model performance across different evaluation criteria reaffirms the reliability and accuracy of our solar radiation prediction framework. Overall, our study contributes to advancing the state-of-the-art solar energy forecasting by offering a practical and effective methodology that can be applied in real-world settings. As the demand for renewable energy sources continues to grow, our findings hold significant implications for optimizing the efficiency and reliability of solar power systems, ultimately driving the transition toward a sustainable energy future.

In summary, this research lays the groundwork for future endeavors in solar radiation prediction. It underscores the importance of data-driven approaches in enhancing the efficiency and reliability of solar energy generation. By leveraging innovative machine-learning techniques and algorithmic advancements, our methodology offers a promising avenue for optimizing solar power systems and accelerating the adoption of renewable energy technologies globally.

REFERENCES

1. Androniceanu, A., & Sabie, O. M. (2021). Overview of green energy as a real strategic option for sustainable development. *Energies*, 15(22), 8573. https://doi.org/10.3390/en15228573
2. IRENA. (2018). Global energy transformation: A roadmap to 2050, International Renewable Energy Agency, Abu Dhabi.
3. IEA. Renewables 2020 – Analysis and forecast to 2025. https://iea.blob.core.windows.net/assets/1a24f1fe-c971-4c25-964a-57d0f31eb97b/Renewables_2020-PDF.pdf
4. Boulakhbar, M., Lebrouhi, B., Kousksou, T., Smouh, S., Jamil, A., Maaroufi, M., & Zazi, M. (2020). Towards a large-scale integration of renewable energies in Morocco. *Journal of Energy Storage*, 32, 101806. https://doi.org/10.1016/j.est.2020.101806
5. Energy5. Impact of wind energy on job creation and economic growth. https://energy5.com/impact-of-wind-energy-on-job-creation-and-economic-growth
6. Paletta, Q., Terrén-Serrano, G., Nie, Y., Li, B., Bieker, J., Zhang, W., Dubus, L., Dev, S., & Feng, C. (2023). Advances in solar forecasting: Computer vision with deep learning. *Advances in Applied Energy*, 11, 100150. https://doi.org/10.1016/j.adapen.2023.100150
7. Khalifa Boutahir, M., Hessane, A., Lasri, I., Benchikh, S., Farhaoui, Y., & Azrour, M. Dynamic threshold fine-tuning in anomaly severity classification for enhanced solar power optimization. *Data and Metadata*. https://dm.saludcyt.ar/index.php/dm/article/view/94
8. Hoa, P. X., Xuan, V. N., & Thu, N. T. P. (2024). Determinants of renewable energy consumption in the fifth technology revolutions: Evidence from ASEAN countries. *Journal of Open Innovation: Technology, Market, and Complexity*, 10(1), 100190. https://doi.org/10.1016/j.joitmc.2023.100190
9. Chen, X. H., Tee, K., Elnahass, M., & Ahmed, R. (2023). Assessing the environmental impacts of renewable energy sources: A case study on air pollution and carbon emissions in China. *Journal of Environmental Management*, 345, 118525. https://doi.org/10.1016/j.jenvman.2023.118525
10. JCW Resourcing. The growing demand for renewable energy careers. https://www.jcwresourcing.com/de/insights/blog/the-growing-demand-for-renewable-energy-careers--1940895236/
11. Dada, M., & Popoola, P. (2023). Recent advances in solar photovoltaic materials and systems for energy storage applications: A review. *Beni-Suef University Journal of Basic and Applied Sciences*, 12(1), 1–15. https://doi.org/10.1186/s43088-023-00405-5
12. Kavlak, G., McNerney, J., & Trancik, J. E. (2018). Evaluating the causes of cost reduction in photovoltaic modules. *Energy Policy*, 123, 700–710. https://doi.org/10.1016/j.enpol.2018.08.015
13. Moghaddam, S. M. S. H., Dashtdar, M., & Jafari, H. (2022). AI applications in smart cities' energy systems automation. *Repa Proceeding Series. Okinawa, Japan*, 3(1), 1–5. https://doi.org/10.37357/1068/CRGS2022.3.1.01
14. Mohammad, A., & Mahjabeen, F. (2023). Revolutionizing solar energy: The impact of artificial intelligence on photovoltaic systems. *International Journal of Multidisciplinary Sciences and Arts*, 2, 56.
15. Utilities One. The potential of artificial intelligence in solar panel maintenance. https://utilitiesone.com/the-potential-of-artificial-intelligence-in-solar-panel-maintenance
16. Steculła, K., Wolniak, R., & Grebski, W. W. (2022). AI-driven urban energy solutions—From individuals to society: A review. *Energies*, 16(24), 7988. https://doi.org/10.3390/en16247988

17. Boutahir, M. K., Farhaoui, Y., & Azrour, M. (2022). Machine learning and deep learning applications for solar radiation predictions review: Morocco as a case of study. In: S.G. Yaseen (Eds.), *Digital economy, business analytics, and big data analytics applications* (vol. 1010). Studies in Computational Intelligence. Springer, Cham. https://doi.org/10.1007/978-3-031-05258-3_6

18. Mohamed, K. & Abouzid, H., & Teidj, S. (2021). Prédiction de Rayonnement Solaire Global (RSG): Par les Réseaux de Neurones Artificiels Cas d'étude: la ville d'Er-Rachidia, Maroc.

19. Paoli, C., Voyant, C., Muselli, M., & Nivet, M. (2009). Solar radiation forecasting using ad-hoc time series preprocessing and neural networks. ArXiv. /abs/0906.0311

20. Markovics, D., & Mayer, M. J. (2022). Comparison of machine learning methods for photovoltaic power forecasting based on numerical weather prediction. *Renewable and Sustainable Energy Reviews*, 161, 112364. https://doi.org/10.1016/j.rser.2022.112364

21. Ikram, R. M. A., Dai, H., Ewees, A. A., Shiri, J., Kisi, O., & Zounemat-Kermani, M. (2022). Application of improved version of multi verse optimizer algorithm for modeling solar radiation. *Energy Reports*, 8, 12063–12080. https://doi.org/10.1016/j.egyr.2022.09.015

22. Ramesh, G., Logeshwaran, J., Kiruthiga, T., & Lloret, J. (2023). Prediction of energy production level in large PV plants through AUTO-encoder based neural-network (AUTO-NN) with restricted Boltzmann feature extraction. *Future Internet*, 15(2), 46. https://doi.org/10.3390/fi15020046

23. Ghimire, S., Deo, R. C., Wang, H., et al. (2021). Stacked LSTM sequence-to-sequence autoencoder with feature selection for daily solar radiation prediction: A review and new modeling results. *Energies*, 15(3), 1061. https://doi.org/10.3390/en15031061

24. Bazrafshan, O., Ehteram, M., Dashti Latif, S., Feng Huang, Y., Yenn Teo, F., Najah Ahmed, A., & El-Shafie, A. (2022). Predicting crop yields using a new robust Bayesian averaging model based on multiple hybrid ANFIS and MLP models. *Ain Shams Engineering Journal*, 13(5), 101724. https://doi.org/10.1016/j.asej.2022.101724

25. Hassan, M. A., Bailek, N., Bouchouicha, K. et al. (2022). Evaluation of energy extraction of PV systems affected by environmental factors under real outdoor conditions. *Theoretical and Applied Climatology*, 150, 715–729. https://doi.org/10.1007/s00704-022-04166-6

26. Chaibi, M., Benghoulam, E. M., Tarik, L., Berrada, M., & El Hmaidi, A. (2022). Machine learning models based on random forest feature selection and Bayesian optimization for predicting daily global solar radiation. *International Journal of Renewable Energy Development*, 11(1), 309–323. https://doi.org/10.14710/ijred.2022.41451

27. Boutahir, M. K., Farhaoui, Y., Azrour, M., Zeroual, I., & El Allaoui, A. (December 2022). Effect of feature selection on the prediction of direct normal irradiance. *Big Data Mining and Analytics*, 5(4), 309–317. https://doi.org/10.26599/BDMA.2022.9020003

28. Hissou, H., Benkirane, S., Guezzaz, A., & Azrour, M. (2022). A novel machine learning approach for solar radiation estimation. *Sustainability*, 15(13), 10609. https://doi.org/10.3390/su151310609

29. Sabri, M., & El Hassouni, M. (2023). Photovoltaic power forecasting with a long short-term memory autoencoder networks. *Soft Computing*, 27, 10533–10553. https://doi.org/10.1007/s00500-023-08497-y

30. Bendali, W., Saber, I., Bourachdi, B., Amri, O., Boussetta, M., & Mourad, Y. (2022). Multi time horizon ahead solar irradiation prediction using GRU, PCA, and GRID SEARCH based on multivariate datasets. *Journal Européen des Systèmes Automatisés*, 55(1), 11–23. https://doi.org/10.18280/jesa.550102

31. El Bakali, S., Hamid, O., & Gheouany, S. (2023). Day-ahead seasonal solar radiation prediction, combining VMD and STACK algorithms. *Clean Energy*, 7(4), 911–925. https://doi.org/10.1093/ce/zkad025

32. El Mghouchi, Y. (2022). Best combinations of inputs for ANN-based solar radiation forecasting in Morocco. *Technology and Economics of Smart Grids and Sustainable Energy* 7, 27. https://doi.org/10.1007/s40866-022-00152-z

33. Aarich, N., Erraïssi, N., Akhsassi, M., Lhannaoui, A., Raoufi, M., & Bennouna, A. "Propre.Ma" project: Roadmap & preliminary results for grid-connected PV yields maps in Morocco. *2014 International Renewable and Sustainable Energy Conference (IRSEC)*, Ouarzazate, Morocco, 2014, pp. 774–777. https://doi.org/10.1109/IRSEC.2014.7059764

34. Halimi, M., Outana, I., El Amrani, A., Diouri, J., & Messaoudi, C. (2018). Prediction of captured solar energy for different orientations and tracking modes of a PTC system: Technical feasibility study (Case study: South eastern of Morocco). *Energy Conversion and Management*, 167, 21e36.

35. Elabbassi, I. et al. (2023). Adaptive Neural Fuzzy Inference System (ANFIS) in a grid connected-fuel cell-electrolyser-solar PV-battery-super capacitor energy storage system management. In: Y. Farhaoui, A. Rocha, Z. Brahmia, & B. Bhushab (Eds.), *Artificial Intelligence and Smart Environment* (vol. 635). ICAISE 2022. Lecture Notes in Networks and Systems. Springer, Cham. https://doi.org/10.1007/978-3-031-26254-8_21

36. IBM Topics = K-nearest neighbors algorithm. https://www.ibm.com/topics/knn

37. btd. An overview of key concepts and types of regression. https://baotramduong.medium.com/regression-fe5a2d1813ca

38. Chakure, A. Random forest regression in Python explained. https://builtin.com/data-science/random-forest-python

16 Artificial Intelligence-Empowered Date Palm Disease and Pest Management
Current Status, Challenges, and Future Perspectives

Abdelaaziz Hessane, Ahmed El Youssefi, Yousef Farhaoui, and Badraddine Aghoutane
Moualy Ismail University, Meknès, Morocco

16.1 INTRODUCTION

Before integrating new technologies, traditional agriculture relied on methods honed over centuries, characterized by manual labor and empirical knowledge [1]. This labor-intensive approach was subject to the whims of nature, with farmers dependent on natural weather patterns for irrigation and pest control. The productivity and efficiency of these methods varied widely. They needed to be improved to meet the increasing demands of a rapidly growing global population, projected to reach nearly 10 billion by 2050 [2]. Moreover, climate-related challenges, such as extreme weather conditions, strain these systems. Additionally, water scarcity significantly impacted agricultural productivity. Furthermore, the lives of 5 billion people, equivalent to two-thirds of humanity, will be disrupted by at least one month of water shortages, as a new UN report reveals the alarming consequences of climate change on our global water resources [3].

Oasis agricultural systems are not immune to these challenges. Instead, it can be considered one of the most affected systems, given the specific characteristics of this ecosystem. While ancient oasis systems thrived for centuries in harsh environments, modern challenges make them some of the most vulnerable ecosystems. Climate change acts as a central threat, driving factors like knowledge loss, migration, water scarcity, and soil degradation, leading to irrigation inefficiencies and, ultimately, infrastructure collapse [4].

Rising temperatures are not merely a shift in average weather patterns; they represent a multifaceted threat to the delicate ecology of date palm cultivation, upon

DOI: 10.1201/9781032656830-16

which oasis agricultural systems heavily rely for their very existence. Higher temperatures create an environment conducive to the proliferation of numerous pests, including scales, mealybugs, and borers. This exponential growth in pest populations leads to increased damage to date palms and subsequent reductions in yield. Furthermore, warmer temperatures exacerbate the spread of fungal and bacterial diseases, posing a significant threat to palm health and longevity [5]. The challenges, however, extend beyond simple temperature increases. Changes in rainfall patterns also play a crucial role. Extended periods of drought weaken date palms, rendering them more susceptible to pest infestations and disease outbreaks. Additionally, drought can deplete the availability of crucial fungicides and pesticides, hindering effective control measures. Conversely, excessive rainfall or high humidity can favor the development of fungal diseases, adding to the myriad challenges faced by date palm growers.

Extreme weather events further compound the issue. Strong winds and hailstorms inflict physical damage on palms, creating entry points for opportunistic pathogens and pests. Floods, too, wreak havoc by damaging root systems and disrupting nutrient uptake, ultimately diminishing palm health and productivity. Beyond direct physical damage, extreme weather events can disrupt the delicate ecological balance of predators and parasites that naturally control pest populations, leading to unchecked insect growth and increased agricultural losses.

Due to these climatic shifts, farmers may resort to increased pesticide use as pest and disease pressure mounts. However, this reliance on chemical control carries its own set of risks. Frequent pesticide applications can lead to pest resistance, rendering current control measures ineffective and necessitating the development of even more potent chemicals. Furthermore, excessive pesticide use can harm beneficial insects and disrupt the natural equilibrium of the ecosystem, potentially creating unforeseen consequences in the long run.

Considering the various challenges presented, exploring innovative solutions that transcend the constraints inherent in conventional methodologies for managing palm diseases is imperative, which necessitates a multi-faceted approach, illustrated in Figure 16.1, and encompassing the following key aspects:

1. **Comprehensive analysis of climate change impacts**: There is a pressing need for an in-depth understanding of the intricate dynamics between climate change and the proliferation of pests. This knowledge is crucial for devising proactive and productive strategies to mitigate the impact of these changes. By gaining insights into these relationships, it becomes possible to anticipate and counteract the detrimental effects of climate change on palm health.

2. **Rapid response to disease outbreaks**: A swift and effective response mechanism is essential to address the onset of diseases in palm crops by involving continuous monitoring and immediate intervention in sudden disease emergence. The ability to quickly detect and respond to disease outbreaks can significantly reduce the spread and severity of infections, thereby safeguarding the health and productivity of palm plantations.

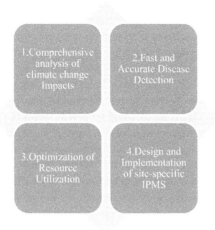

FIGURE 16.1 Multi-faceted approach for date palm disease and pest management.

3. **Optimization of resource utilization**: Efficient management of resources, notably water and pesticides, is critical to this strategy. Optimal resource allocation contributes to the sustainability of palm cultivation and minimizes the ecological footprint of these practices. By optimizing the use of these resources, it is possible to achieve a balance between effective disease control and environmental stewardship.

4. **Site-specific integrated pest management strategies**: Replacing the broader pest management system (PMS) approach with a more focused, site-specific strategy enhances the effectiveness of pest management. This approach involves tailoring pest control methods to the unique conditions and challenges of each palm plantation. Site-specific integrated PMS (SSIPMS) can offer more precise and effective solutions for managing pests and diseases by considering local environmental factors, pest populations, and specific cultivation practices.

Integrating advanced technologies is a pivotal solution for realizing these objectives. Incorporating state-of-the-art tools such as remote sensing (RS), unmanned aerial vehicles (AUVs), the Internet of Things (IoT), and computer vision (CV) marks a significant leap forward in this domain. These technologies, particularly when combined with artificial intelligence (AI) advancements, have demonstrated considerable efficacy [6].

AI models, with their capacity to analyze and uncover intricate relationships within vast datasets, often outperform traditional data analysis techniques. The efficiency and precision with which AI-powered algorithms discover these patterns render them an essential tool for any system aimed at leveraging and maximizing the potential of big data [7]. Its applicability extends beyond agriculture, encompassing critical sectors like medicine, engineering, and education. AI is a game-changer in agricultural evolution, powered by intelligent farming advancements. Its knack for crunching data from sensors and satellites empowers intelligent decision-making. AI's sophisticated analysis drives precise crop management, from predicting bountiful harvests to nipping diseases and pests in the bud. By weaving robotics and drones into the fabric of agriculture, AI automation streamlines harvesting, upping efficiency and productivity. This potent force bolsters agriculture against environmental challenges and paves the way for a sustainable future, feeding future generations.

This study explores the rising trend of AI in date palm farming, mainly focusing on disease and pest management, a crucial topic in oasis farming development. It provides a detailed review of AI's current state and future potential in this domain, recognizing the importance of date palms and the threats they face. This research appeals to a broad audience invested in sustainable agriculture and AI development. Addressing challenges, outlining future directions, and promoting effective AI solutions for date palm protection demonstrate a direct impact in advancing sustainable practices.

The remainder of this paper is structured as follows: Section 16.2 establishes the foundational concepts of AI, setting the stage for further discussion. This is followed by Section 16.3, which examines the overarching role of AI in the cultivation of date palms, providing a broad perspective on its applications. Section 16.4 offers an in-depth analysis of AI's specific applications in managing diseases and pests affecting date palms, highlighting how AI contributes to addressing these challenges. Section 16.5 discusses the limitations and challenges of integrating AI into date palm cultivation practices while proposing future research directions. The paper concludes with Section 16.6, summarizing the essential findings and underscoring the significance of AI in enhancing the sustainability of date palm cultivation practices.

16.2 ARTIFICIAL INTELLIGENCE

The concept of "Artificial intelligence" hinges on two key aspects: "intelligent," reflecting its ability to mimic human reasoning capabilities, and "artificial," emphasizing that machines, typically computers or robots, exhibit this intelligence [8]. In simpler terms, AI can be understood as the ability of machines to process information and solve problems in a way that resembles human intelligence. A more technical definition describes AI as "a collection of theories, techniques, and scientific disciplines, such as mathematical logic, statistics, probability, and computer science, that enable the design of complex computer programs capable of simulating aspects of human intelligence, like learning and reasoning.

The emergence of "Artificial Intelligence" (AI) can be traced back to the 1950s, spearheaded by mathematician Alan Turing [9]. In his influential work, "Computing Machinery and Intelligence," Turing (1950) explored the possibility of imbuing machines with a form of intelligence. He introduced the now-famous "Turing test," where a human converses with another human and a machine, both hidden from view. If the human cannot reliably distinguish between the two based on their responses, the machine is deemed "intelligent" [10]. However, it was in the 2010s that AI truly experienced a resurgence, primarily driven by exponential increases in computing power and access to vast datasets.

Beyond replicating human-like intelligence, AI is a powerful tool for expanding and refining existing approaches in various fields. Fueled by research and development, it fuels simulations that allow for testing and exploration of new possibilities, ultimately leading to advancements in theories, methods, techniques, and even entire application systems related to human intelligence. Notably, AI plays a pivotal role in driving technological progress within the IoT era, where its core principles of reasoning, knowledge processing, and learning find numerous applications [11]. From representing and managing knowledge to automating reasoning and search processes, machine learning (ML) and knowledge acquisition, and natural language processing (NLP), AI's influence extends to diverse areas like computer vision, intelligent robots, automated programming, and cognitive automation.

16.2.1 AI Approaches

Robust and adaptable AI-based tools are required to perform several tasks, mainly classification and regression, various data types, diverse problems, and complex situations. Consequently, AI offers a diverse landscape of approaches, each with unique strengths and limitations. This section delves into four prominent AI approaches applicable to this domain: supervised learning, unsupervised learning, semi-supervised learning, and reinforcement learning.

16.2.1.1 Supervised Learning

Supervised learning, a cornerstone of AI, involves training a model on a labeled dataset, where the correct output is provided for each input. This approach enables the model to learn the mapping from inputs to outputs, making it ideal for tasks such as classification and regression. The effectiveness of supervised learning has been demonstrated in various applications, from image recognition to natural language processing. Recent advancements have focused on deep learning (DL) models, significantly improving the accuracy and efficiency of supervised learning tasks [12, 13].

16.2.1.2 Unsupervised Learning

Unsupervised learning explores how AI can learn patterns and structures from data without explicit labels. This approach is vital for discovering hidden correlations, clustering similar data points, and reducing dimensionality. Unsupervised learning algorithms have been instrumental in anomaly detection, customer segmentation, and feature learning [14].

16.2.1.3 Semi-supervised Learning

Semi-supervised learning occupies the middle ground between supervised and unsupervised learning, leveraging labeled and unlabeled data for training. This approach is particularly beneficial when acquiring a fully labeled dataset is impractical due to cost or time constraints. Semi-supervised learning has shown promise in improving learning efficiency and model performance, especially in domains with limited labeled data [15].

16.2.1.4 Reinforcement Learning

Reinforcement learning (RL) is a pivotal AI approach distinct from supervised, unsupervised, and semi-supervised paradigms. It is characterized by an agent learning to make decisions by interacting with an environment to achieve a goal. The agent learns from the consequences of its actions, rather than from pre-provided data, through a system of rewards and penalties. This learning process enables the agent to develop a strategy for selecting actions that maximize the cumulative reward over time. RL has been instrumental in solving complex decision-making problems, including robotics control, game playing, and autonomous vehicle navigation [16].

The diversity of AI approaches is crucial for developing AI-based tools that are both robust and adaptable. These methodologies cater to varied data types, problem complexities, and specific tasks, ensuring that AI applications can effectively address the broad spectrum of challenges encountered in different domains.

16.2.2 AI CONCEPTS

AI, ML, and DL represent a nested hierarchy of concepts and technologies driving the current wave of innovation in computational problem-solving. At the broadest level, AI encompasses the computing technology field simulating human intelligence. These systems are designed to perform complex tasks that typically require human cognitive functions such as learning, reasoning, problem-solving, perception, and understanding natural language. AI applications range from simple, rule-based systems like chatbots and virtual assistants to complex decision-making algorithms in autonomous vehicles and strategic game systems [17–19].

16.2.2.1 Machine Learning

ML is a subset of AI that includes algorithms and statistical models that enable computers to improve at tasks with experience. ML systems learn from and make predictions or decisions based on data. As illustrated in Figure 16.2, unlike traditional computing systems, which operate based on explicit rules programmed by humans, ML algorithms generate rules by identifying patterns in the data. This paradigm

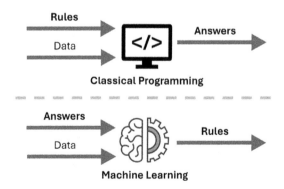

FIGURE 16.2 Traditional programming vs. machine learning.

allows ML systems to adapt to new scenarios and perform tasks without explicitly being programmed for each [20].

16.2.2.2 Deep Learning

DL, a subset of ML, involves a particular set of algorithms inspired by the structure and function of the brain called artificial neural networks. DL models, or deep neural networks (DNNs), can learn from unstructured or unlabeled data using a layered architecture of algorithms called a neural network. These networks can know and make intelligent decisions by analyzing data with complexity and abstraction similar to the human brain [21]. They comprise layers of interconnected nodes, or "neurons," each layer designed to perform specific transformations on input data. Starting with an input layer that receives the raw data, the network processes the information through multiple hidden layers where neurons weigh the input, apply a transformation function (often a non-linear one like rectified linear unit (ReLU) or sigmoid), and pass the result to the next layer. The final layer, the output layer, produces the predictions or classifications [21]. DNNs learn to perform tasks by adjusting the weights of the connections through a process called backpropagation [22], which uses gradient descent [23] to minimize the difference between the network's predictions and the actual data during training, allowing the network to make increasingly accurate predictions over time. DL has been responsible for many breakthroughs, such as computer vision, natural language processing, and audio recognition (Figure 16.3).

As illustrated in Figure 16.4, AI, ML, and DL are interconnected fields that form a hierarchy of concepts and technologies. This hierarchical structure encompasses a rich diversity of algorithms, methods, and approaches, each suited to different types of problems and data. Consequently, AI is highly applicable across numerous domains due to its versatility and adaptability [25, 26]. This flexibility is primarily attributed to AI's ability to learn from multiple data types, enabling it to function effectively in diverse domains. AI's prowess in processing and interpreting complex,

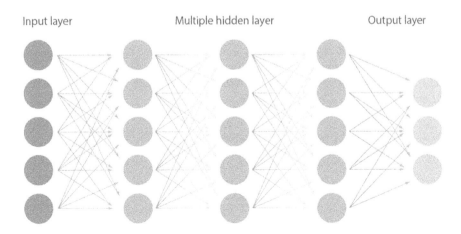

FIGURE 16.3 Overall architecture of DNNs [24].

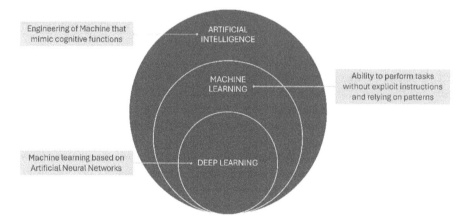

FIGURE 16.4   Conceptual hierarchy of AI, ML, and DL in computational systems.

large data volumes far exceeds human capabilities, making it essential in data-intensive fields. Moreover, AI's capacity to automate repetitive and mundane tasks enhances efficiency.

Additionally, AI plays a significant role in decision-making and predictive analysis, leveraging past data to forecast future trends. Furthermore, AI's capability for customization and personalization is evident. AI augments human capabilities, offering tools for more accurate diagnoses, aiding in complex research, and inspiring innovative designs. Scalable to various problem sizes and complexities, AI is equally effective in personal assistants and large-scale applications [25]. Its continuous learning ability ensures continuous improvement, which is vital in dynamic environments. Finally, AI enhances accessibility and user experiences, and although initially costly, it promises long-term cost efficiencies through automation, improved performance, and error reduction. These attributes underscore AI's transformative potential across different sectors, addressing challenges and creating novel opportunities [12].

The date palm tree (Phoenix Dactylifera L) production and valorization sectors are not exempt from the "Rush to AI" phenomenon. Over the last decade, significant development has been observed through hundreds of research papers discussing integrating AI techniques to valorize palm cultivation and develop fruit productivity [26]. The following section will explore how AI is integrated into date palm production and cultivation.

16.3   ROLE OF AI IN THE DATE PALM FARMING: A GLOBAL VISION

The date palm is a globally significant crop, valued for its nutritional, medicinal, and socio-economic importance, particularly in North Africa and the Middle East [27–29]. By 2020, global date production reached approximately 9.45 million tons, marking a 2.3% increase since 2010 [30]. Producers are challenged to meet the growing global demand by ensuring high-quality yields for self-sufficiency and enhanced export capabilities. Furthermore, the date palm industry is adapting to Agriculture

TABLE 16.1
AI-Powered Date Palm Farming Management-Related Research Topics

References	Research Topic	Main Objectives
[32–50]	Date fruit classification	Classify date fruit according to its variety, maturity stages, and hardness
[51–60]	Date palm tree detection	Inventory and mapping
[61–67]	Date fruit quality assessment	Classify date fruit according to its moisture, thinning, and grading
[60], [68–83]	Date pests and diseases management and control	Detection and recognition of diseases; pest forecasting
[84–86]	Yield estimation	Prediction of the outcome of yield
[87]	Resources management	Predict optimum water and energy requirements for sensor-based micro irrigation systems powered by solar PV

4.0, which involves process automation, climate change adaptation, and efficient resource allocation [31].

Over the past few years, precision agriculture has significantly benefited from the advancements in AI, creating a robust area for research. In response, the date palm industry has begun integrating modern technologies to align with global sustainability standards. Table 16.1 summarizes the main topics where AI was used in date palm farming management over the last decade.

Statistically, the examined research papers indicate that the dominant research topic in applying AI to date palm farming in recent years is "Date fruit classification," accounting for 35% of the studies. However, among the topics presented, "Resources management" and "Yield estimation" are the least researched areas in integrating AI into date palm farming, comprising 2% and 5% of the papers, respectively. Furthermore, classification techniques are the most utilized field, comprising approximately 94.55% of the research. Regarding regression techniques, they are primarily employed in yield estimation and some insect forecasting studies (Figure 16.5).

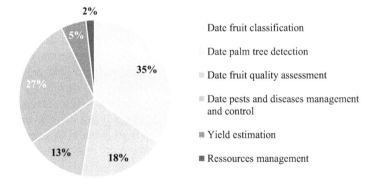

FIGURE 16.5 AI applications in date palm farming management, categorized by research topic.

Despite the critical importance of managing and controlling diseases in date palms, our comprehensive survey reveals that research targeting disease management and control constitutes only 27.3% of the examined studies. This proportion underscores a significant gap in the literature, highlighting the need for a more focused investigation into this crucial area. In the following section, we delve deeper into this discrepancy, providing a thorough analysis to identify the specific research gaps in date palm disease management and control.

16.4   ROLE OF AI IN DATE PALM DISEASE AND PEST MANAGEMENT

The adoption of AI in the date palm industry has seen remarkable development over the past decade, evidenced by the increasing number of research studies annually. However, its application in date palm disease and pest control and management still needs to be more widespread. A rigorous analysis of select papers is essential to understand and gather insightful evidence accurately. The following section presents a comprehensive overview of recent advances in the rate of date palm disease and pest management and control.

The analysis of the investigated research papers on the "Date pests and diseases management and control" research topic reveals a dynamic landscape in date palm agriculture, marked by an emphasis on pest detection and identification. Notably, the Red Palm Weevil (RPW), known for its detrimental impact on date palms, has taken center stage in multiple studies. Researchers have leveraged diverse datasets, from palm leaf images to acoustic recordings, to develop robust models. These models, including various ML and DL techniques, have consistently demonstrated high accuracy levels, with some achieving exceptional results. Additionally, the timeline of this research reflects the ongoing interest and commitment to harnessing AI technologies to improve date palm cultivation.

The preliminary findings of this survey underscore the predominant role of DL methodologies within the realm of date palm disease and pest management research. Among these techniques, convolutional neural networks (CNNs) stand out, being employed in approximately 73% of the analyzed studies. This substantial percentage highlights the widespread adoption of CNNs and underscores their effectiveness and efficiency in addressing complex problems within this domain.

CNNs, with their deep, multi-layer architectures, are specifically designed to process multi-dimensional data such as images. This capability is particularly relevant for detecting and diagnosing diseases and pest infestations in date palms, where visual symptoms are critical indicators of health issues. The superiority of CNNs over traditional ML algorithms—such as Random Forests (RFs), Decision Trees (DTs), and Support Vector Machines (SVMs)—is primarily attributed to their ability to extract features automatically and hierarchically from images. This leads to more accurate and reliable identification of diseases and pests than methods requiring manual feature extraction and selection.

Furthermore, the survey reveals a significant shift toward using transfer learning (TL) approaches in recent research. TL involves adapting models trained on one task to perform on another related task. This technique has gained traction for several reasons:

1. It allows researchers to leverage pre-trained models, which have already learned general features from large datasets, thereby reducing the need for extensive data collection in the specific context of date palm management.
2. It significantly decreases training time and computational resources, making DL more accessible.
3. TL has proven effective in enhancing model performance, especially in scenarios where available data is limited or highly specific.

However, despite the promising advantages of CNNs and the growing interest in TL, this survey also identifies several concerns and challenges associated with their application. The following section will delve into these concerns in detail, offering insights into the limitations of current approaches and suggesting future research directions.

16.5 CHALLENGES, LIMITATIONS, AND FUTURE PERSPECTIVES

Statistical analysis of the examined research shows some critical limitations. First, within the "Date pests and diseases management and control" category, 53% of papers focus on disease detection and classification, while 47% concentrate on insect identification. Specifically, the Red Palm Weevil Insect is the most studied, with about 71,43% of the insect-related papers addressing this insect. Furthermore, most of the datasets used are not open-access. Among the used datasets, open-access datasets represent only 18%. Moreover, DL methods are favored over traditional ML methods.

Given the diversity of diseases affecting date palm trees, it is imperative to expand the scope of research to include a broader range of diseases. Investigating the feasibility of automatic diagnostics using ML and DL techniques can lead to more comprehensive solutions. Researchers should develop models to identify and classify various diseases, ensuring a holistic approach to date palm disease management.

To facilitate further research, there is a pressing need for more open-access datasets related to date palm diseases and pests. These datasets should encompass various conditions and scenarios encountered in date palm cultivation. The research community can benefit immensely from the availability of such datasets, which can serve as valuable resources for training and testing models.

While DL has demonstrated the potential to enhance accuracy for date palm disease and pest management, its computational demands are substantial. To navigate this, researchers are encouraged to pioneer solutions that achieve high accuracy and maintain computational efficiency. Exploring avenues such as developing optimized, lightweight AI models and their integration within IoT ecosystems could offer real-time, efficient monitoring solutions. When implemented in embedded systems, such strategies promise a harmonious balance between precision and resource utilization, charting a sustainable path forward in agricultural technology applications.

16.6   CONCLUSION

In this paper, we have explored the current landscape and future potential of AI in date palm farming, focusing on the critical areas of disease and pest management. Our findings reveal that while AI offers significant promise in enhancing the efficiency and sustainability of date palm cultivation, notable challenges must be addressed. Among these challenges, the need for well-annotated, open-source datasets is a significant barrier to advancing AI applications in agriculture. Additionally, greater collaboration among diverse stakeholders, including researchers, technologists, farmers, and government entities, is paramount to driving innovation and adoption of AI technologies.

This study also highlights the paramount importance of government and institutional support in fostering the integration of AI into agriculture. Such support is crucial for overcoming the current obstacles and maximizing AI's benefits, including optimizing resources like water and pesticides, which are vital for the sustainability of date palm farming. The insights provided in this paper aim to shed light on the potential of AI in agriculture and encourage a collaborative effort to address the challenges faced. By emphasizing the need for open-access data, cross-disciplinary partnerships, and policy support, this research advocates for a strategic approach to leverage AI to improve date palm cultivation, ultimately contributing to the broader goals of sustainable agriculture and food security.

REFERENCES

[1] "Origins of agriculture I History, Types, Techniques, & Facts I Britannica." Accessed: Feb. 01, 2024. [Online]. Available: https://www.britannica.com/topic/agriculture

[2] "World Population Prospects - Population Division - United Nations." Accessed: Feb. 01, 2024. [Online]. Available: https://population.un.org/wpp/

[3] "UN World Water Development Report 2022 I UN-Water." Accessed: Feb. 01, 2024. [Online]. Available: https://www.unwater.org/publications/un-world-water-development-report-2022

[4] A. Santoro, "Traditional oases in Northern Africa as multifunctional agroforestry systems: A systematic literature review of the provided Ecosystem Services and of the main vulnerabilities," *Agroforestry Systems*, vol. 97, no. 1, pp. 81–96, Jan. 2023. doi: 10.1007/s10457-022-00789-w

[5] M. El Bouhssini and J. R. H. Faleiro De Socorro, *Date Palm Pests and Diseases: Integrated Management Guide*. International Center for Agricultural Research in the Dry Areas (ICARDA), 2018.

[6] A. Sharma, A. Jain, P. Gupta, and V. Chowdary, "Machine Learning Applications for Precision Agriculture: A Comprehensive Review," *IEEE Access*, vol. 9, pp. 4843–4873, 2021. doi: 10.1109/ACCESS.2020.3048415

[7] Y. Farhaoui, Ed., *Big Data and Networks Technologies*, vol. 81. Springer International Publishing, 2020. doi: 10.1007/978-3-030-23672-4

[8] J. McCarthy, M. L. Minsky, N. Rochester, and C. E. Shannon, "A Proposal for the Dartmouth Summer Research Project on Artificial Intelligence," 2006.

[9] A. M. Turing, "I.—Computing machinery and intelligence," *Mind*, vol. LIX, no. 236, pp. 433–460, Oct. 1950. doi: 10.1093/mind/LIX.236.433

[10] A. M. Turing, "Computing Machinery and Intelligence (1950)," in *The Essential Turing*, Oxford University Press, 2004. doi: 10.1093/oso/9780198250791.003.0017

[11] M. Merenda, C. Porcaro, and D. Iero, "Edge Machine Learning for AI-Enabled IoT Devices: A Review," *Sensors*, vol. 20, no. 9, p. 2533, Apr. 2020. doi: 10.3390/s20092533

[12] Y. Farhaoui, A. Hussain, T. Saba, H. Taherdoost, and A. Verma, Eds., *Artificial Intelligence, Data Science and Applications*, vol. 838. Springer Nature Switzerland, 2024. doi: 10.1007/978-3-031-48573-2

[13] Y. Farhaoui, A. Rocha, Z. Brahmia, and B. Bhushab, Eds., *Artificial Intelligence and Smart Environment*, vol. 635. Springer International Publishing, 2023. doi: 10.1007/978-3-031-26254-8

[14] H. U. Dike, Y. Zhou, K. K. Deveerasetty, and Q. Wu, "Unsupervised learning based on artificial neural network: A review," in *2018 IEEE International Conference on Cyborg and Bionic Systems (CBS)*, IEEE, Oct. 2018, pp. 322–327. doi: 10.1109/CBS.2018.8612259

[15] R. Jiao et al., "Learning with limited annotations: A survey on deep semi-supervised learning for medical image segmentation," *Computers in Biology and Medicine*, vol. 169, p. 107840, Feb. 2024. doi: 10.1016/j.compbiomed.2023.107840

[16] I. G. T. Isa, A. Effendi, and Suhartono, "Deep reinforcement learning in agricultural IoT-based: A review," *E3S Web of Conferences*, vol. 479, p. 07004, Jan. 2024. doi: 10.1051/e3sconf/202447907004

[17] M. Javaid, A. Haleem, I. H. Khan, and R. Suman, "understanding the potential applications of artificial intelligence in agriculture sector," *Advanced Agrochem*, vol. 2, no. 1, pp. 15–30, Mar. 2023. doi: 10.1016/j.aac.2022.10.001

[18] M. Soori, B. Arezoo, and R. Dastres, "Artificial intelligence, machine learning and deep learning in advanced robotics, a review," *Cognitive Robotics*, vol. 3, pp. 54–70, 2023. doi: 10.1016/j.cogr.2023.04.001

[19] B. Rawat, A. S. Bist, D. Supriyanti, V. Elmanda, and S. N. Sari, "AI and nanotechnology for healthcare: A survey," *APTISI Transactions on Management (ATM)*, vol. 7, no. 1, pp. 86–91, Jan. 2022. doi: 10.33050/atm.v7i1.1819

[20] J. Kufel et al. "What is machine learning, artificial neural networks and deep learning?—Examples of practical applications in medicine," *Diagnostics*, vol. 13, no. 15, p. 2582, Aug. 2023. doi: 10.3390/diagnostics13152582

[21] W. Liu, Z. Wang, X. Liu, N. Zeng, Y. Liu, and F. E. Alsaadi, "A survey of deep neural network architectures and their applications," *Neurocomputing*, vol. 234, pp. 11–26, Apr. 2017. doi: 10.1016/j.neucom.2016.12.038

[22] D. E. Rumelhart, G. E. Hinton, and R. J. Williams, "Learning representations by back-propagating errors," *Nature*, vol. 323, no. 6088, pp. 533–536, Oct. 1986. doi: 10.1038/323533a0

[23] S. Ruder, "An overview of gradient descent optimization algorithms," Sep. 2016.

[24] "What are Neural Networks? | IBM." Accessed: Feb. 07, 2024. [Online]. Available: https://www.ibm.com/topics/neural-networks

[25] I. H. Sarker, "AI-based modeling: Techniques, applications and research issues towards automation, intelligent and smart systems," *SN Computer Science*, vol. 3, no. 2, p. 158, Mar. 2022. doi: 10.1007/s42979-022-01043-x

[26] A. Hessane, A. EL Youssefi, Y. Farhaoui, B. Aghoutane, and Y. Qaraai, "Artificial intelligence applications in date palm cultivation and production: A scoping review," 2023, pp. 230–239. doi: 10.1007/978-3-031-26254-8_32

[27] A. Temitope Idowu, O. Osarumwense Igiehon, A. Ezekiel Adekoya, and S. Idowu, "Dates palm fruits: A review of their nutritional components, bioactivities and functional food applications," *AIMS Agriculture and Food*, vol. 5, no. 4, pp. 734–755, 2020. doi: 10.3934/agrfood.2020.4.734

[28] M. Q. Al-Mssallem, "The role of date palm fruit in improving human health," *Journal of Clinical and Diagnostic Research*, 2020. doi: 10.7860/JCDR/2020/43026.13442

[29] A. El Hadrami and J. M. Al-Khayri, "Socioeconomic and traditional importance of date palm," *Emirates Journal of Food and Agriculture*, vol. 24, no. 5, pp. 371–385, 2012.

[30] "FAOSTAT.[Production/Yield quantities of Dates in World + (Total)]." Accessed: Jun. 16, 2022. [Online]. Available: https://www.fao.org/faostat/

[31] Farhaoui, Y., Lecture Notes in Networks and Systems Volume 837 LNNS, Pages v – vi, *2024 5th International Conference on Artificial Intelligence and Smart Environment, ICAISE 2023, Errachidia 23 November 2023 through 25 November 2023*, Code 309309, ISSN 23673370, ISBN 978-303148464-3

[32] W. S. N. Alhamdan and J. M. Howe, *Classification of Date Fruits in a Controlled Environment Using Convolutional Neural Networks*, vol. 1339. 2021. doi: 10.1007/978-3-030-69717-4_16

[33] M. Koklu, R. Kursun, Y. S. Taspinar, and I. Cinar, "Classification of date fruits into genetic varieties using image analysis," *Mathematical Problems in Engineering*, vol. 2021, 2021. doi: 10.1155/2021/4793293

[34] M. A. Khayer, M. S. Hasan, and A. Sattar, "Arabian date classification using CNN algorithm with various pre-trained models," in *Proceedings of the 3rd International Conference on Intelligent Communication Technologies and Virtual Mobile Networks, ICICV 2021*, 2021, pp. 1431–1436. doi: 10.1109/ICICV50876.2021.9388413

[35] B. D. Pérez-Pérez, J. P. García Vázquez, and R. Salomón-Torres, "Evaluation of convolutional neural networks' hyperparameters with transfer learning to determine sorting of ripe Medjool dates," *Agriculture*, vol. 11, no. 2, p. 115, Feb. 2021. doi: 10.3390/agriculture11020115

[36] L. Khriji, A. C. Ammari, and M. Awadalla, "Artificial intelligent techniques for palm date varieties classification," *International Journal of Advanced Computer Science and Applications*, vol. 11, no. 9, pp. 489–495, 2020. doi: 10.14569/IJACSA.2020.0110958

[37] A. A. Abi Sen, N. M. Bahbouh, A. B. Alkhodre, A. M. Aldhawi, F. A. Aldham, and M. I. Aljabri, "A classification algorithm for date fruits," in *Proceedings of the 7th International Conference on Computing for Sustainable Global Development, INDIACom 2020*, 2020, pp. 235–239. doi: 10.23919/INDIACom49435.2020.9083706

[38] L. Khriji, A. C. Ammari, and M. Awadalla, "Hardware/software co-design of a vision system for automatic classification of date fruits," *International Journal of Embedded and Real-Time Communication Systems*, vol. 11, no. 4, pp. 21–40, 2020. doi: 10.4018/IJERTCS.2020100102

[39] M. Faisal, M. Alsulaiman, M. Arafah, and M. A. Mekhtiche, "IHDS: Intelligent harvesting decision system for date fruit based on maturity stage using deep learning and computer vision," *IEEE Access*, vol. 8, pp. 167985–167997, 2020. doi: 10.1109/ACCESS.2020.3023894

[40] O. Aiadi, M. L. Kherfi, and B. Khaldi, "Automatic date fruit recognition using outlier detection techniques and Gaussian mixture models," *ELCVIA Electronic Letters on Computer Vision and Image Analysis*, vol. 18, no. 1, p. 52, Jun. 2019. doi: 10.5565/rev/elcvia.1041

[41] H. Altaheri, M. Alsulaiman, and G. Muhammad, "Date fruit classification for robotic harvesting in a natural environment using deep learning," *IEEE Access*, vol. 7, pp. 117115–117133, 2019. doi: 10.1109/ACCESS.2019.2936536

[42] M. S. Hossain, G. Muhammad, and S. U. Amin, "Improving consumer satisfaction in smart cities using edge computing and caching: A case study of date fruits classification," *Future Generation Computer Systems*, vol. 88, pp. 333–341, 2018. doi: 10.1016/j.future.2018.05.050

[43] O. Aiadi and M. L. Kherfi, "A new method for automatic date fruit classification," *International Journal of Computational Vision and Robotics*, vol. 7, no. 6, p. 692, 2017. doi: 10.1504/IJCVR.2017.087751

[44] A. Manickavasagan, N. H. Al-Shekaili, N. K. Al-Mezeini, M. S. Rahman, and N. Guizani, "Computer vision technique to classify dates based on hardness," *Journal of Agricultural and Marine Sciences [JAMS]*, vol. 22, no. 1, p. 36, Jan. 2018. doi: 10.24200/jams.vol22iss1pp36-41

[45] G. Muhammad, "Date fruits classification using texture descriptors and shape-size features," *Engineering Applications of Artificial Intelligence*, vol. 37, pp. 361–367, Jan. 2015. doi: 10.1016/j.engappai.2014.10.001

[46] A. Manickavasagan, H. N. Al-Shekaili, G. Thomas, M. S. Rahman, N. Guizani, and D. S. Jayas, "Edge detection features to evaluate hardness of dates using monochrome images," *Food Bioproc Tech*, vol. 7, no. 8, pp. 2251–2258, Aug. 2014. doi: 10.1007/s11947-013-1219-0

[47] V. Nozari and M. Mazlomzadeh, "Date fruits grading based on some physical properties," *Information Processing in Agriculture*, vol. 9, no. 7, pp. 1703–1713, 2013.

[48] M. P. M. R. Benam and S. M. Mazloumzadeh, "Using adaptive neuro-fuzzy inference system for classify date fruits," *Information Processing in Agriculture*, vol. 9, no. 5, pp. 1309–1318, 2013.

[49] A. Haidar, Haiwei Dong, and N. Mavridis, "Image-based date fruit classification," in *2012 IV International Congress on Ultra Modern Telecommunications and Control Systems*, IEEE, Oct. 2012, pp. 357–363. doi: 10.1109/ICUMT.2012.6459693

[50] Y. Noutfia and E. Ropelewska, "Innovative models built based on image textures using traditional machine learning algorithms for distinguishing different varieties of Moroccan Date Palm fruit (*Phoenix dactylifera L.*)," *Agriculture*, vol. 13, no. 1, p. 26, Dec. 2022. doi: 10.3390/agriculture13010026

[51] E. A. Alburshaid and M. A. Mangoud, "Developing date palm tree inventory from satellite remote sensed imagery using deep learning," in *2021 3rd IEEE Middle East and North Africa COMMunications Conference, MENACOMM 2021*, 2021, pp. 54–59. doi: 10.1109/MENACOMM50742.2021.9678262

[52] M. B. A. Gibril, H. Z. M. Shafri, A. Shanableh, R. Al-Ruzouq, A. Wayayok, and S. J. Hashim, "Deep convolutional neural network for large-scale date palm tree mapping from uav-based images," *Remote Sensors (Basel)*, vol. 13, no. 14, 2021. doi: 10.3390/rs13142787

[53] L. El Hoummaidi, A. Larabi, and K. Alam, "Using unmanned aerial systems and deep learning for agriculture mapping in Dubai," *Heliyon*, vol. 7, no. 10, 2021. doi: 10.1016/j.heliyon.2021.e08154

[54] H. Rhinane, A. Bannari, M. Maanan, and N. Aderdour, "Palm trees crown detection and delineation from very high spatial resolution images using deep neural network (U-Net)," in *2021 IEEE International Geoscience and Remote Sensing Symposium IGARSS*, IEEE, Jul. 2021, pp. 6516–6519. doi: 10.1109/IGARSS47720.2021.9554470

[55] M. Culman, S. Delalieux, and K. V. Tricht, "Palm tree inventory from aerial images using retinanet," in *2020 Mediterranean and Middle-East Geoscience and Remote Sensing Symposium, M2GARSS 2020 - Proceedings*, 2020, pp. 314–317. doi: 10.1109/M2GARSS47143.2020.9105246

[56] R. Cousin and M. Ferry, "Automatic localization of phoenix by satellite image analysis," *Arab Journal for Plant Protection*, vol. 37, no. 2, pp. 83–88, Jun. 2019. doi: 10.22268/AJPP-037.2.083088

[57] R. Al-Ruzouq, A. Shanableh, M. A. Gibril, and S. Al-Mansoori, "Image segmentation parameter selection and ant colony optimization for date palm tree detection and mapping from very-high-spatial-resolution aerial imagery," *Remote Sensors (Basel)*, vol. 10, no. 9, 2018. doi: 10.3390/rs10091413

[58] K. Djerriri, M. Ghabi, M. S. Karoui, and R. Adjoudj, "Palm trees counting in remote sensing imagery using regression convolutional neural network," in *International Geoscience and Remote Sensing Symposium (IGARSS)*, 2018, pp. 2627–2630. doi: 10.1109/IGARSS. 2018.8519188

[59] A. Almaazmi, "Palm trees detecting and counting from high-resolution WorldView-3 satellite images in United Arab Emirates," in *Proceedings of SPIE - The International Society for Optical Engineering*, 2018. doi: 10.1117/12.2325733

[60] M. Ahmed and A. Ahmed, "Palm tree disease detection and classification using residual network and transfer learning of inception ResNet," *PLoS One*, vol. 18, no. 3, p. e0282250, Mar. 2023. doi: 10.1371/journal.pone.0282250

[61] T. Shoshan, A. Bechar, Y. Cohen, A. Sadowsky, and S. Berman, "Segmentation and motion parameter estimation for robotic Medjoul-date thinning," *Precision Agriculture*, vol. 23, no. 2, pp. 514–537, 2022. doi: 10.1007/s11119-021-09847-2

[62] M. Keramat-Jahromi, S. S. Mohtasebi, H. Mousazadeh, M. Ghasemi-Varnamkhasti, and M. Rahimi-Movassagh, "Real-time moisture ratio study of drying date fruit chips based on on-line image attributes using kNN and random forest regression methods," *Measurement*, vol. 172, p. 108899, Feb. 2021. doi: 10.1016/j.measurement.2020.108899

[63] M. Faisal, F. Albogamy, H. Elgibreen, M. Algabri, and F. A. Alqershi, "Deep learning and computer vision for estimating date fruits type, maturity level, and weight," *IEEE Access*, vol. 8, pp. 206770–206782, 2020. doi: 10.1109/ACCESS.2020.3037948

[64] H. Raissouli, A. Ali, S. Mohammed, F. Haron, and G. Alharbi, "Date grading using machine learning techniques on a novel dataset," *International Journal of Advanced Computer Science and Applications*, vol. 11, no. 8, 2020. doi: 10.14569/IJACSA.2020. 0110893

[65] A. Hakami and M. Arif, "Automatic inspection of the external quality of the date fruit," in *Procedia Computer Science*, 2019, pp. 70–77. doi: 10.1016/j.procs.2019.12.088

[66] A. Nasiri, A. Taheri-Garavand, and Y.-D. Zhang, "Image-based deep learning automated sorting of date fruit," *Postharvest Biology and Technology*, vol. 153, pp. 133–141, 2019. doi: 10.1016/j.postharvbio.2019.04.003

[67] N. Alavi, "Quality determination of Mozafati dates using Mamdani fuzzy inference system," *Journal of the Saudi Society of Agricultural Sciences*, vol. 12, no. 2, pp. 137–142, Jun. 2013. doi: 10.1016/j.jssas.2012.10.001

[68] M. E. Karar, O. Reyad, A.-H. Abdel-Aty, S. Owyed, and M. F. Hassan, "Intelligent IoT-Aided early sound detection of red palmWeevils," *Computers, Materials and Continua*, vol. 69, no. 3, pp. 4095–4111, 2021. doi: 10.32604/cmc.2021.019059

[69] B. Wang et al., "Towards detecting red palm weevil using machine learning and fiber optic distributed acoustic sensing," *Sensors*, vol. 21, no. 5, p. 1592, Feb. 2021. doi: 10.3390/ s21051592

[70] H. Alaa, K. Waleed, M. Samir, M. Tarek, H. Sobeah, and M. A. Salam, "An intelligent approach for detecting palm trees diseases using image processing and machine learning," *International Journal of Advanced Computer Science and Applications*, vol. 11, no. 7, pp. 434–441, 2020. doi: 10.14569/IJACSA.2020.0110757

[71] A. Magsi, J. A. Mahar, M. A. Razzaq, and S. H. Gill, "Date palm disease identification using features extraction and deep learning approach," in *Proceedings - 2020 23rd IEEE International Multi-Topic Conference, INMIC 2020*, Nov. 2020. doi: 10.1109/ INMIC50486.2020.9318158

[72] M. Teena, A. Manickavasagan, A. M. Al-Sadi, R. Al-Yahyai, M. L. Deadman, and A. Al-Ismaili, "Near infrared imaging to detect Aspergillus flavus infection in three varieties of dates," *Engineering in Agriculture, Environment and Food*, vol. 11, no. 4, pp. 169–177, Oct. 2018. doi: 10.1016/j.eaef.2018.04.002

[73] I. Ashry et al., "CNN–Aided optical fiber distributed acoustic sensing for early detection of red palm weevil: A field experiment," *Sensors*, vol. 22, no. 17, p. 6491, Aug. 2022. doi: 10.3390/s22176491

[74] K. M. Al-Kindi, Z. Alabri, and M. Al-Farsi, "Geospatial detection of Ommatissus lybicus de Bergevin using spatial and machine learning techniques," *Remote Sens Appl*, vol. 28, p. 100814, Nov. 2022. doi: 10.1016/j.rsase.2022.100814

[75] M. A. Arasi, L. Almuqren, I. Issaoui, N. S. Almalki, A. Mahmud, and M. Assiri, "Enhancing red palm weevil detection using bird swarm algorithm with deep learning model," *IEEE Access*, vol. 12, pp. 1542–1551, 2024. doi: 10.1109/ACCESS.2023.3348412

[76] M. Mohammed, H. El-Shafie, and M. Munir, "Development and validation of innovative machine learning models for predicting date palm mite infestation on fruits," *Agronomy*, vol. 13, no. 2, p. 494, Feb. 2023. doi: 10.3390/agronomy13020494

[77] W. Boulila, A. Alzahem, A. Koubaa, B. Benjdira, and A. Ammar, "Early detection of red palm weevil infestations using deep learning classification of acoustic signals," *Computers and Electronics in Agriculture*, vol. 212, p. 108154, Sep. 2023. doi: 10.1016/j.compag. 2023.108154

[78] M. Al-Shalout and K. Mansour, "Detecting date palm diseases using convolutional neural networks," *Lecture Notes in Networks and Systems*, 2022, pp. 1–5. doi: 10.1109/acit53391. 2021.9677103

[79] S. Nuruzzaman Nobel et al., "Palm leaf health management: A hybrid approach for automated disease detection and therapy enhancement," *IEEE Access*, pp. 1–1, 2024. doi: 10.1109/ACCESS.2024.3351912

[80] A. Hessane, A. El Youssefi, Y. Farhaoui, B. Aghoutane, and F. Amounas, "A machine learning based framework for a stage-wise classification of date palm white scale disease," *Big Data Mining and Analytics*, 2023. doi: 10.26599/BDMA.2022.9020022

[81] A. Hessane, M. Khalifa Boutahir, A. El Youssefi, Y. Farhaoui, and B. Aghoutane, "Empowering date palm disease management with deep learning: A comparative performance analysis of pretrained models for stage-wise white-scale disease classification," *Data and Metadata*, vol. 2, p. 102, Dec. 2023. doi: 10.56294/dm2023102

[82] A. Hessane, M. K. Boutahir, A. El Youssefi, Y. Farhaoui, and B. Aghoutane, "Deep-PDSC: A deep learning-based model for a stage-wise classification of parlatoria date scale disease," *Lecture Notes in Networks and Systems Artificial Intelligence, Data Science and Applications*, pp. 345–353, 2023. doi: 10.1007/978-3-031-25662-2_17

[83] A. Hessane, A. El Youssefi, Y. Farhaoui, and B. Aghoutane, "Toward a stage-wise classification of date palm white scale disease using features extraction and machine learning techniques," in *2022 International Conference on Intelligent Systems and Computer Vision (ISCV)*, IEEE, May 2022, pp. 1–6. doi: 10.1109/ISCV54655.2022.9806134

[84] K. Heyns, "Estimation methods for date palm yield – A feasibility study," Mar., 2021.

[85] M. Husain and R. A. Khan, "Date palm crop yield estimation a framework," *SSRN Electronic Journal*, 2019. doi: 10.2139/ssrn.3509195

[86] H. Dehghanisanij, N. Salamati, S. Emami, H. Emami, and H. Fujimaki, "An intelligent approach to improve date palm crop yield and water productivity under different irrigation and climate scenarios," *Applied Water Science*, vol. 13, no. 2, p. 56, Feb. 2023. doi: 10.1007/s13201-022-01836-8

[87] M. Mohammed, H. Hamdoun, and A. Sagheer, "Toward sustainable farming: Implementing artificial intelligence to predict optimum water and energy requirements for sensor-based micro irrigation systems powered by solar PV," *Agronomy*, vol. 13, no. 4, p. 1081, Apr. 2023. doi: 10.3390/agronomy1304108

17 Trends in Green Technology for Clean Energy

Aishat Titilola Rufai, Adewole Usman Rufai,
and Agbotiname Lucky Imoize
University of Lagos, Lagos, Nigeria

17.1 INTRODUCTION TO GREEN TECHNOLOGY

The advent of technological development within the 18th and 19th centuries has brought about significant changes across most industries. This massive development bolstered different industries ranging from building heavy machineries for mass production for the textile industry to agriculture, transportation, to energy sector among others [1]. These machines aim to increase production as well as to bring about effective and efficient ways to perform humans' daily tasks. In addition, these heavy machineries require fuel such as coal, fossil fuels, and other energy sources to operate [1]. Due to rising population and development of new technologies, there has been an increase in energy demand. However, the overuse and overexploitation of these non-renewable energy resources have several impacts such as depletion of fossil fuels and emission of greenhouse gases into the environment.

Green technology innovation has become a critical solution to these environmental concerns. Green technology is the branch of science that aims at reducing human's impact on the environment [1]. In other words, green technology innovation employs ecological-economic principles to conserve natural resources and energy, eliminate or minimize environmental pollution and degradation while developing long-lasting sustainable development solutions that produce economic, environmental, and social benefits [2]. These green technology innovations are being increasingly utilized in various sectors, including healthcare services, energy industry, and transportation. These solutions not only contribute to a healthier and more sustainable environment but also improve the standard of living for individuals.

Recently, there has been increased interest in green technology for clean energy. The latest trends in green technology for clean energy include the development of renewable energy sources such as solar and wind to generate electricity. According to the International Energy Agency (IEA), solar systems accounted for 14.7% of the total electricity generation in 2023 which shows a significant increase from 2022

DOI: 10.1201/9781032656830-17

while wind energy accounted for 11.4% of the total electricity generation [3]. The use of solar photovoltaic system for power generation is expected to surpass coal (which is currently the most utilized resource for power generation) by 2027. Another trend in green technology is the utilization of efficient energy storage system, the use of renewable energy has called for the need for storing energy [4].

According to [4], there are various techniques for storing energy depending on the type of renewable energy used. There is a need to study various techniques to achieve an efficient storage system. Such systems help address the intermittent and fluctuating nature of renewable energy sources resulting in a more reliable and consistent supply of clean energy. Another trend for green technology is the use of smart grid systems. The advent of renewable energy brought about the need to quantify the benefits of renewable energy technologies. Smart grid system is a key driver for employing intelligent devices (such as the Internet of Things (IoT) and smart communication systems) and technologies (the use of artificial intelligence (AI) and big data analytics) for effective energy management and monitoring of clean energy systems [5]. One key driver behind these developments is the urgent need to limit greenhouse gases emissions and mitigate the harmful effects of climate change. Thus, it is important to harness these renewable energy sources to achieve a sustainable world.

The major reasons for these recent developments in green technology for clean energy are the depletion of non-renewable energy resources and the need to limit the emissions of carbon dioxide. Also, the growing demand for a sustainable environment and the heightened awareness of the effects of climate change have played a critical role in the adoption of green technology for clean energy systems. Generally, the major applications of green technology for clean energy include the harnessing of renewable energy sources, the development of energy-efficient devices, efficient energy storage system, etc. Although these recent trends for renewable energy systems show tremendous benefits, they have numerous drawbacks.

According to Ref. [6], one of the major limitations of renewable technology is the cost of deployment. Due to the high cost of utilizing renewable energy systems, the transition into clean energy particularly for developing countries is subject to limitations. Furthermore, the fluctuating and variable patterns of clean energy due to variable weather conditions for solar and wind technologies make it difficult to have consistent supply of clean energy [7]. In addition, solar and wind energy sources require a large amount of land to generate a large amount of energy [7]. Thus, this is a challenge for areas or locations with limited land. Additionally, there have been growing concerns regarding the environmental impacts of solar panels despite their limited emission of greenhouse gases [7]. Hence, the drawbacks of renewable energy technologies will be discussed extensively in this chapter.

This chapter presents an overview of the current trends of green technology for clean energy system. Also, we will focus on the potential benefits and challenges of these new technologies. Also, we will discuss the integration of smart grid systems as well as smart devices IoT devices for the development of renewable energy systems, addressing the limitations of green energy systems and future work on green energy systems. In this chapter, we use the terms "clean energy," "sustainable energy," "smart energy," and "renewable energy" interchangeably.

17.1.1 Key Contributions of the Chapter

The following are the significant contributions of this chapter:

i. The chapter explores green technology innovation for renewable energy systems.
ii. The chapter presents the challenges of green technology for sustainable energy.
iii. This chapter reviews the current research trends of green technology for clean energy systems.

17.1.2 Chapter Organization

Section 17.2 shows the related works on the application of green technology for clean energy systems. Section 17.3 presents the applications of green technology, and Section 17.4 presents green technology for sustainable energy. Section 17.5 discusses the limitations of green technology for smart energy sources. Section 17.6 presents the discussion. Finally, Section 17.7 concludes the chapter.

17.2 RELATED WORK

The advent of sustainable development goals (SDGs) has necessitated the need to tackle the current economic, environmental, and political challenges faced globally. With the aid of the development of green technologies, there has been a wide increase in utilizing renewable energy technologies, particularly for power generation to achieve an eco-sustainable world. The most common sources of renewable energy sources include sun, hydro, and wind. Renewable energy sources are often used to support traditional energy sources. For solar systems, more individuals are adopting off-grid systems where the primary support systems are battery storage systems.

Halkos et al. [7] evaluated the different types of renewable energy sources, their possible implications, challenges as well as their effect on the economy, human health, and climate change. The paper highlighted the current state of renewable energy and how to mitigate climate change and achieve sustainable future. For the successful integration of renewable energy sources, there is a need to understand the consumers' perspective on adopting this system. The paper sheds light on customer's willingness to pay for these renewables across various countries as well as customer's attitude toward renewable energy technology. In addition, the study also proposed frameworks for policies and strategies for a successful transition to renewable energy.

For the swift adoption of renewable, the government needs to implement different policies to support the utilization of renewable energy. Kabel et al. [8] analyzed the policies used by governments to promote the development of renewable energy and their environmental and socioeconomic effects. These policies include tax incentives, loans, feed-in tariffs, and renewable portfolio standards. Furthermore, Akadir et al. [9] suggested that the use of renewable energy in 28 European countries has led to not only environmental sustainability but also economic growth and other growth factors. The study also highlighted the relationship between renewable energy, economic growth, and environmental sustainability in 28 European countries [9]. Hence, the result of these studies could initiate the implementation of policies for renewable energy technologies to foster the growth of these technologies globally. Thus, researchers have found that renewable energy promotes socio-economic growth while reducing the emission of greenhouse gases.

Renewable energy plays a significant role in achieving sustainable development. The authors of [9] explored various factors including social, economic, and environmental factors that affect the sustainable development index. Moreover, this study discusses various mathematical models for examining factors that contribute to sustainable development. One of the key factors that affect the sustainable development index is the gross national savings. Hence, the result of the study shows that sustainable development can be achieved through the use of non-renewable energy particularly in developing countries; however, this comes with increased negative impacts on the environment unlike renewable energy which has lesser negative effects on the environment.

Overall, the major barriers to adopting these renewable energy sources, particularly in developing countries include the cost of implementation as well as ineffective policies [9]. Hence, the growth of adopting renewable energy systems is at a slower pace in developing countries compared to the rest of the world. Therefore, it is essential that governments of these developing countries design robust policies to encourage the utilization of clean energy systems.

Another factor that can hinder the growth of adoption of clean energy technologies is fossil fuel subsidies and reserves. According to IMF, $7 trillion total was spent on subsidies in 2022 only, which is 2$ trillion more than what was spent in 2020 [10]. The authors of [11] suggested that these fossil fuel subsidies be directed toward renewable energy—to reduce the cost of renewables as well as to reduce carbon emissions—as this will have a remarkable effect on the transition to renewable energy. In addition, the authors of [12] discussed the significance of fossil fuel subsidies swap to fund renewable energy while providing socioeconomic benefits, improving public access to healthcare facilities, job opportunities, and gender equality and reducing greenhouse gases emissions. According to Ref. [12], four countries—India, Indonesia, Zambia, and Morocco—are leading in the fossil fuel to clean energy swaps. Since fossil fuel subsidies are not properly directed, it is important for governments to create effective reforms and strategies for the transition to clean energy that would benefit the target market/audience.

Although the transition to clean energy is noteworthy, Ref. [13] emphasized the distinction between "energy transition" and "energy addition". In this study, the author of [13] argued that the implementation of renewable energy sources paved the way for higher energy consumption rather than replacing fossil fuel; thus, the use of new energy sources such as wind farms and solar farms is an indication of energy addition, not energy transition. The authors of [13] also suggest the possible implications of energy additions. For example, the energy transition to renewables may not show the true picture of how carbon emission is reducing rather misleading the general public on the proportion rather than its true impacts.

This is evidenced by the carbon emission data (Figure 17.1) from the International Energy Agency (IEA) between 2019 and 2022 where global energy carbon emission

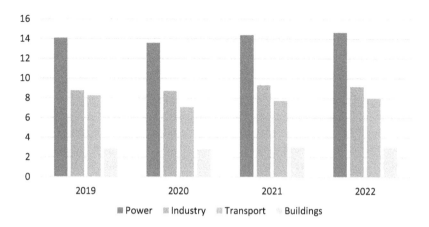

FIGURE 17.1   Carbon emission by sector (2019–2022).

grew despite the efforts to reduce fossil fuel usage [14]. The years 2021–2022 have witnessed a significant increase in carbon emission due to the COVID-19 pandemic [14]. In addition, oil also accounted for the carbon emission in 2023 with a slight increase in emission in 2023.Overall, it is indicative that the implementation of renewable energy is yet to have a major effect in reducing carbon emissions unless there is robust plan backed by government policies for reducing fossil fuel use as well as limiting the entire electricity production.

There are various techniques and/or policies that can be established to reduce greenhouse gases and encourage the over-reliance on fossil fuels. Such policies/techniques that support de-carbonization include: remove barriers that hinder energy-efficient systems, deploy renewable energy technologies and utilizing home solar-powered appliances, and create the right environment for investing in renewables, as well as supporting carbon, capture, utilization, and storage (CCUS) systems [14]. These initiatives require keen attention and have the potential to reduce carbon emissions.

In recent years, there has been a considerable increase in the number of CCUS projects particularly in the United States. CCUS technologies can help cut CO_2 and greenhouse gases emissions. CCUS systems capture, utilize, as well as store carbon for various applications. These CCUS systems go through several processes ranging from industrial separation, pre-combustion, post-combustion, oxy-fuel combustion, chemical looping combustion, direct air capture [15–18]. CCUS technologies can be used for various applications such as for obtaining aviation fuels, chemicals, cement, and energy production [19].

Energy storage system addresses the issue of power fluctuations—specifically for areas with frequent blackouts—and variable weather patterns. The incorporation and deployment of energy storage system such as battery system with renewable energy sources enable constant and reliable power for users. Thus, the integration of efficient energy storage system with clean energy systems is significant for decarbonization [20]. Furthermore, Jiang et al. [2] highlighted the key benefits of the rapid development of renewable energy technology. This study has outlined the following major benefits of utilizing renewable energy: reduction of environmental pollution, mitigation of

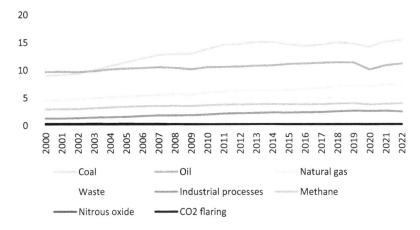

FIGURE 17.2   Global energy carbon emission (2000–2022).

climate change effects, conservation of resources such as fossil fuel, and promoting sustainable development.

17.3   GREEN TECHNOLOGY APPLICATIONS

Green technology includes diverse technology that is used to lessen the effects of humans on the environment. Green technology encompasses various sectors which ranges from energy industry to transportation sector to agriculture and financial sector, among others. Various industrial sectors employ green technologies which are shown in Figure 17.3:

17.3.1   RENEWABLE ENERGY

According to Ref. [34], the main criterion for green technology includes social equitability, economic viability, and sustainability. The most significant goals of green technology are to limit greenhouse gases and to achieve SDGs. Studies have found that the use of green technology—such as the utilization of renewable energy sources for power generation—is critical for limiting carbon emissions [23]. The commonly used renewable energy technologies include solar, wind, and hydro energies. In recent years, there has been growing interest in utilizing renewable energy for electricity generation due to the negative impacts of traditional energy sources—fossil fuels—on the environment. In addition, the unstable nature of renewable energy systems calls for efficient storage systems. Although renewable energy sources have minimal carbon emissions, they require more land use than the traditional energy sources which poses danger to wildlife conservation and the ecological systems [35]. Most studies tend to focus on carbon emission reduction for renewable energy technologies but fail to consider the effects of these technologies on biodiversity. Hence, more research needs to be carried out on the impacts of renewables on the ecosystem.

TABLE 17.1
Summary of Green Technologies for Clean Energy

References	Clean Energy Technologies/Application	Focus	Gaps
[21]	• Clean-tech startup • Net zero buildings • Internet-enabled clean-tech startups • Cities lead climate charge focusing on reducing greenhouse gases • Utilities in the United States investing in solar and storage system	The article predicted how utilization of clean energy systems will increase as well as the revenues from these systems. Discussed various technologies that contribute to the usage of clean energy systems	Finance and policies were the major challenges for adopting clean energy system
[22]	• Green Internet of Things (IoT) • Sensor • Cloud computing • Smart metering	Review of various devices and technologies for green innovation	
[23] [24] [25] [26] [27] [28]	• Unit root test • ARDL-bound testing technique • Granger causality test • Stationary unit root test • Co-integration test • Cross-section-augmented autoregressive distributed lag (CS-ARDL) • Smart grid • Increased use of clean technology and energy storage system • Development of green roof • Smart lighting system • Utilization of digital technologies for energy and transport • Increased use of low-energy devices • ARDL model • The panel quantile model with additive fixed • Novel quantile regression method	This paper explores the correlation between green technology innovation and renewable energy and carbon dioxide emissions based on the STIRPAT model in Turkey during the period of 1990–2018 The next eleven (N-11) economies are in line with the global phenomena of environmental degradation; the present study addresses the gap and investigates green technology and renewable energy with CO_2 emission from 1980 to 2018. Presents a comprehensive discussion on smart city development particularly for buildings, transport, and energy sectors Emphasized the importance of effective policies for adopting clean energy solutions Investigated the significance of green innovation and education on environmental sustainability in highly polluted Asian cities from the period of 1991 to 2019 Discovered that an increase in green technology reduces CO_2 emission in China, India, and Japan in the long run while an increase in clean energy investment and education tends to reduce CO_2 emission in Russia and Japan The study concluded that green technology, clean energy investment, and education improve environmental conditions in the long run Analyzes the effect of clean energy policies on green technology using panel data of 102 countries This study highlights the effects of financial inclusion, green innovation, and energy efficiency on environmental challenges. The study presented recent discoveries and implications for structural change of economy, green innovation, and industry-related factors in terms of sustainable development	Energy consumption, population, and per capita enhance carbon emissions

Ref.	Methods	Description	Results
[29]	• The second-generation econometric techniques • The Dumitrescu and Hurlin (D–H) causality test	The study investigates the impacts of green technology, financial development, and utilization of renewable energy for reducing carbon emission for the 12 notable emitters according to statistics from 1991 to 2018 The study recommended policies implementation, use of green technology, and renewable energy for limiting greenhouse gases	The empirical results in the study indicated that financial development increases carbon emission while green technology innovations and renewable energy use reduce carbon emission
[30] [31]	• The fully modified OLS • Dynamic OLS • Classical co-integration regression • Bayer–Hanck co-integration • ARDL bounds test	This article designed a three-level classification system of green technology (CSGT)—which includes primary, secondary, and tertiary categories—through a hybrid method that incorporates top-down and bottom-up techniques All methods used in the study indicated that renewable energy consumption and financial innovation enhance environmental quality while economic growth and inflation worsen environmental quality in the both short and long run Also, green innovation had no a significant effect on the environment in the short run but it had significant effect on the environmental quality of economic development in E7 countries The article recommended policies for green innovation and renewable energy particularly for developing countries	
[32] [33]	• Westerlund and Edgerton's panel cointegration test	The study analyzes the impact of green technology innovation on energy intensity in 14 industrial sector in 17 OECD countries between the years 1975 and 2005. The results of the study showed that green innovation caused a significant reduction in energy intensity across most sectors The findings of this study are intended to educate policymakers about the significance of green innovation on energy intensity as well as developing policies that aid green research and development (R&D) The article investigates the impacts of green technology and investment in the energy industry in China's provincial and regional data between 1995 and 2017 period The study also found that income, environmental innovation, investment in energy industry, and renewable energy consumption are all determining factors to account for CO_2 emissions subsequently The study recommended that investments in clean energy and green innovation will aid in reducing carbon emissions	

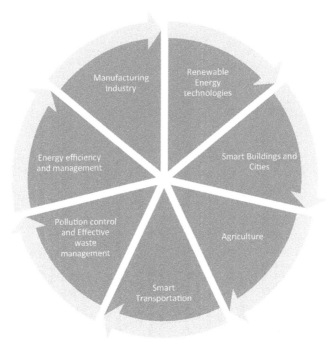

FIGURE 17.3   Green technology applications.

17.3.2   SUSTAINABLE CITIES—NET ZERO BUILDINGS

Another way to address the impacts of clean energy technologies on the ecosystem is to incorporate these technologies into smart buildings and smart cities' designs. For example, the authors of [36] highlighted the importance of integrating renewable energy systems into buildings. These buildings are aimed at achieving zero carbon emissions. In addition, the authors of [36] described the combination of various renewable energy sources—such as solar, wind, biomass, and hydro-power—in high-rise buildings. Another contributing factor to consider for net zero buildings includes the use of energy-efficient devices in proposed buildings. According to US Energy Information Administration (EIA) [37], the cooling and heating systems, refrigeration and lighting systems, accounted for 52% of total energy consumption for residential buildings in 2020. This is shown in Figure 17.4.

The integration of smart and intelligent appliances, energy-efficient cooling and heating system, natural lighting can help reduce energy consumption in buildings and cities. For instance, the use of natural lighting for buildings can help reduce the overdependence on artificial lighting system [38]. Due to limited spaces in densely populated cities, the installation of solar panels on roof-tops in high-rise buildings and installation of carports can minimize the impacts of renewable energy systems on the ecosystem. In addition, with the aid of new technologies, researchers employ AI techniques—artificial neural networks (ANNs) for predicting renewable energy production for proposed projects. While these technologies seem promising, several factors need to be considered for achieving net zero energy buildings, including the

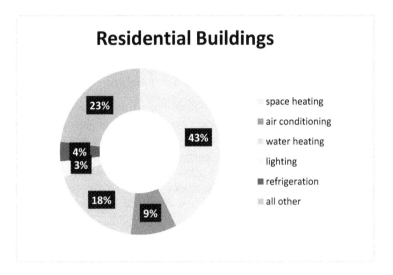

FIGURE 17.4   2020 US residential energy consumption chart.

(Source: U.S. Energy Information Administration, 2020 Residential Energy Consumption Survey (RECS).)

economic, social, and environmental factors. Thus, researchers are calling on architects and designers to develop and incorporate net zero initiatives into buildings and other infrastructural designs.

17.3.3 GREEN TRANSPORTATION

The authors of [39] highlighted the need for adopting alternative fuels for vehicles. Another way to achieve sustainable cities is to employ alternative fuels for vehicles. This has been seen in the recent development of electric vehicles. The use of electric vehicles could help reduce CO_2 emissions. However, this new technology has economic implications. Thus, a holistic approach needs to be developed to ensure sustainable transportation. This includes the limiting of private vehicles, particularly in urban areas and the use of technology to develop cars that are environmentally friendly. Since the transportation sector constitutes pollution to the environment particularly in cities, there is an urgent need to address the transportation sector in order to achieve sustainable development. To achieve sustainable transportation, the following factors are considered:

- Economic
- Environment
- Social

The authors of Ref. [39] discussed the possible implications of implementing sustainable transportation which are as follows:

- Limits risk, accident pollution, and traffic congestion.
- Enhanced security and safety.
- Improved energy and resource management.

In addition, the authors of [40] discuss the application of IoT and machine learning (ML) techniques for efficient and reliable traffic congestion in smart cities.

17.3.4 WASTE MANAGEMENT AND GREEN TECHNOLOGY

Waste management systems have a significant impact on the environment and ecosystem as a whole. Poor waste management system can pose dangers to humans, biodiversity, animals, and the environment at large. According to Ref. [41], poor waste management can worsen greenhouse gases emission, environmental degradation, poor crop yield, and epidemic diseases, particularly in developing countries. Green technology techniques through the 5Rs, reduce, reuse, recycle, recover, and repair, has numerous advantages such as reduction of carbon emissions, job and wealth creation, and conservation of resources [41].

For example, sustainable products include reuse of plastics bottles for agriculture and other useful products, among others. Similarly, the use of recyclable and sustainable products aids to reduce humans' impacts on the environment as well as increase the products' life cycle [34]. The authors of [41] call for initiatives to extract renewable energy from waste. This is as evidenced in Ref. [41] where the Renewable Resource Centre (RRC) at the University of Ibadan extracts biogas from poultry, piggery, and cattle waste for the purpose of lighting as well other "waste to wealth" and "waste to energy" applications.

17.4 ROLE OF GREEN TECHNOLOGY FOR SUSTAINABLE ENERGY SOLUTIONS

In this section, we discuss the role of green technology in renewable technology and sustainable development. The effects of green technology for waste management include reducing loss of biodiversity, limiting epidemic diseases, reducing environmental pollution, and reducing carbon emission. As mentioned in Ref [41], waste can be transformed into energy and wealth—for generating electricity and job creations in developing countries. According to Ref. [42], renewable energy power generation is predicted to be 40% of global power generation by 2040. The following factors are to be considered for renewable energy technologies: technological innovation, energy storage, and consumer demands [42]. Similarly, the most common effects of green technology for renewable energy system are described by [41–44] which include the reduction of carbon emission as well as creation of job and wealth as well as conservation of resources. As mentioned in Ref. [42], the following cutting-edge technologies—IoT, cloud computing, and ML—are employed for the incorporation of renewable energy systems into smart grid system for achieving smart cities. Due to the variable nature of renewable technologies, smart grid is the combination of renewable energy, distribution, traditional energy sources, and consumer demand using smart communication system for effective grid management [42]. The role of green technologies in clean energy systems is discussed in Table 17.2.

TABLE 17.2

The Role of Green Technologies for Clean Energy Systems

References	Role of Green Technology for Sustainable Energy Solutions	Enabling Technologies	Applications
[40]	• Effective and reliable congestion management	• IoT • Machine learning	Design and implementation of an ML and IoT-based adaptive traffic-management system for smart cities
[41]	• Transforming waste to wealth: job creation and health in developing countries • Transforming waste to energy: generating energy from biogas, etc. • Reduction of carbon emissions • No loss of biodiversity • Conservation of resources		Green technology techniques for solid waste management in developing countries
[42]	• Address carbon emission and resource efficiency • Support the development of new sustainable products. • Encourage economic growth and job creation	• Digital technologies • Electronics • Communications technology • Manufacturing systems	Enabling technologies for future trends in renewable energy
[43]	• Management and control of smart energy systems • Smart usage of renewable energy storage	• Cloud computing • IoT • Blockchain technology • Artificial intelligence	Smart grid mechanism for green energy management
[44]	• Encompasses smart integration of traditional energy source, power distribution, renewable energy systems, and energy storage • The use of communication for achieving and monitoring of energy system.	• Communication technologies • Sensors	Recent green energy technologies and resilient future grid

17.5 LIMITATIONS OF GREEN TECHNOLOGY FOR SMART ENERGY SOURCES

The drawbacks of employing green technology for clean energy sources include the following:

- Economic impacts: the cost implications of utilizing renewable energy and green innovation specifically for developing countries.
- Policies enforcement: the lack of effective policies and strategies for supporting renewable energy technologies particularly for developing countries makes it difficult to adopt these renewables.
- Limited research on the main impacts of renewable energy technologies on biodiversity and environment.

17.6 DISCUSSION

The chapter suggests that the applications of green technology innovations in various industries not only can lead to reduction in carbon emission, but also reduce the loss of biodiversity as well as conservation of resources. As mentioned previously, energy-efficient appliances play a significant role in reducing energy consumption. In addition, the utilization of effective storage system and off-grid systems can help reduce the overreliance traditional energy sources. Although there has been a rapid increase in the adoption of renewable energy recently, carbon emission has increased between 2021 and 2022. This chapter provides an overview of the possible implications of green technology for clean energy systems. This chapter also presents the relationship between economic growth, green innovation, and inflation on environmental quality. The limitations of green technology for green energy are also discussed.

17.7 CONCLUSION

The major factors to be considered for employing green technology include environmental, social, and economic factors. From the review of various literatures, it is evident that effective policies are required to ensure the successful implementation of renewable energy technologies. Thus, it is essential that policies are implemented for effective use of these technologies. Similarly, more research and development are needed for measuring the major impacts of green technology on the environment and biodiversity. Furthermore, to address the economic implications of renewable energy particularly for developing economies, effective initiatives as well as subsidies from the government need to be carried out to encourage stakeholders to employ these technologies. Thus, it is imperative that proper education is done for citizens to understand the impacts of these technologies. Also, to encourage the adoption of green energy, governments can use incentives to support stakeholders in the adoption of these technologies. For a sustainable future, more research needs to be carried out for the implementation of effective and efficient energy storage system as well as renewable energy technologies.

REFERENCES

[1] Bradu P, Biswas A, Nair C, Sreevalsakumar S, Patil M, Kannampuzha S, Mukherjee AG, Wanjari UR, Renu K, Vellingiri B, Gopalakrishnan AV. Recent advances in green technology and Industrial Revolution 4.0 for a sustainable future. *Environmental Science and Pollution Research*. 2022 Apr 9:1–32.

[2] Jiang T, Ji P, Shi Y, Ye Z, Jin Q. Efficiency assessment of green technology innovation of renewable energy enterprises in China: A dynamic data envelopment analysis considering undesirable output. *Clean Technologies and Environmental Policy*. 2021 Jul;23:1509–19.

[3] Aydin M, Bozatli O. The impacts of the refugee population, renewable energy consumption, carbon emissions, and economic growth on health expenditure in Turkey: new evidence from Fourier-based analyses. *Environmental Science and Pollution Research*. 2023 Mar;30(14):41286–98.

[4] Olabi AG. Renewable energy and energy storage systems. *Energy*. 2017 Oct 1;136:1–6.

[5] Bui N, Castellani AP, Casari P, Zorzi M. The internet of energy: a web-enabled smart grid system. *IEEE Network*. 2012 Jul 23;26(4):39–45.

[6] Hansen JP, Narbel PA, Aksnes DL. Limits to growth in the renewable energy sector. *Renewable and Sustainable Energy Reviews*. 2017 Apr 1;70:769–74.

[7] Halkos GE, Gkampoura EC. Reviewing usage, potentials, and limitations of renewable energy sources. *Energies*. 2020 Jun 5;13(11):2906.

[8] Safwat Kabel T, Bassim M. Literature review of renewable energy policies and impacts. *KABEL, Tarek Safwat*. 2019 May 31:28–41.

[9] Ponkratov VV, Kuznetsov AS, Muda I, Nasution MJ, Al-Bahrani M, Aybar HŞ. Investigating the Index of Sustainable Development and Reduction in Greenhouse Gases of Renewable Energies. *Sustainability*. 2022 Nov 10;14(22):14829.

[10] https://www.imf.org/en/Topics/climate-change/energy-subsidies

[11] Qadir SA, Tahir F, Al-Fagih L. Impact of fossil fuel subsidies on renewable energy sector. In12th Int. Exergy, Energy Environ. Symp.(IEEES-12), Doha, Qatar 2020 Dec.

[12] Bridle R, Sharma S, Mostafa M, Geddes A. *Fossil Fuel to Clean Energy Subsidy Swaps*. International Institute for Sustainable Development: Winnipeg, MB, Canada. 2019 Jun.

[13] York R, Bell SE. Energy transitions or additions?: Why a transition from fossil fuels requires more than the growth of renewable energy. *Energy Research & Social Science*. 2019 May 1;51:40–3.

[14] Daszkiewicz K. Policy and regulation of energy transition. *The geopolitics of the global energy transition*. 2020:203–26.

[15] Hong WY. A techno-economic review on carbon capture, utilisation and storage systems for achieving a net-zero CO_2 emissions future. *Carbon Capture Science & Technology*. 2022 Jun 1;3:100044.

[16] Metz B, Davidson O, De Coninck HC, Loos M, Meyer L. *IPCC special report on carbon dioxide capture and storage*. Cambridge: Cambridge University Press; 2005.

[17] Wang X, Song C. Carbon capture from flue gas and the atmosphere: A perspective. *Frontiers in Energy Research*. 2020 Dec 15;8:560849.

[18] Leung DY, Caramanna G, Maroto-Valer MM. An overview of current status of carbon dioxide capture and storage technologies. *Renewable and Sustainable Energy Reviews*. 2014 Nov 1;39:426–43.

[19] Khouibiri, N., et al., "How can cloud BI contribute to the development of the economy of SMEs? Morocco as model", *Lecture Notes in Networks and Systems*, 2024;837:149–159, DOI: 10.1007/978-3-031-48465-0_20

[20] Kittner N, Lill F, Kammen DM. Energy storage deployment and innovation for the clean energy transition. *Nature Energy*. 2017 Jul 31;2(9):1–6.

[21] Folorunso, S.O. et al., "Prediction of Student's Academic Performance Using Learning Analytics", *Lecture Notes in Networks and Systems*, 2024;837:314–325. DOI:10.1007/978-3-031-48465-0_41

[22] Kaur N, Aulakh IK. Clean technology: an eagle-eye review on the emerging development trends by application of IOT devices. In *2018 IEEE International Conference on Smart Energy Grid Engineering (SEGE)* 2018 Aug 12 (pp. 313–320). IEEE.

[23] Shan S, Genç SY, Kamran HW, Dinca G. Role of green technology innovation and renewable energy in carbon neutrality: A sustainable investigation from Turkey. *Journal of Environmental Management*. 2021 Sep 15;294:113004.

[24] Shao X, Zhong Y, Liu W, Li RY. Modeling the effect of green technology innovation and renewable energy on carbon neutrality in N-11 countries? Evidence from advance panel estimations. *Journal of Environmental Management*. 2021 Oct 15;296:113189.

[25] Razmjoo A, Gandomi AH, Pazhoohesh M, Mirjalili S, Rezaei M. The key role of clean energy and technology in smart cities development. *Energy Strategy Reviews*. 2022 Nov 1;44:100943.

[26] Li L, Li G, Ozturk I, Ullah S. Green innovation and environmental sustainability: Do clean energy investment and education matter? *Energy & Environment*. 2023 Nov;34(7):2705–20.

[27] Yang QC, Zheng M, Chang CP. Energy policy and green innovation: A quantile investigation into renewable energy. *Renewable Energy*. 2022 Apr 1;189:1166–75.

[28] Singh AK, Raza SA, Nakonieczny J, Shahzad U. Role of financial inclusion, green innovation, and energy efficiency for environmental performance? Evidence from developed and emerging economies in the lens of sustainable development. *Structural Change and Economic Dynamics*. 2023 Mar 1;64:213–24.

[29] Habiba UM, Xinbang C, Anwar A. Do green technology innovations, financial development, and renewable energy use help to curb carbon emissions? *Renewable Energy*. 2022 Jun 1;193:1082–93.

[30] Guo R, Lv S, Liao T, Xi F, Zhang J, Zuo X, Cao X, Feng Z, Zhang Y. Classifying green technologies for sustainable innovation and investment. *Resources, Conservation and Recycling*. 2020 Feb 1;153:104580.

[31] Hao Y, Chen P. Do renewable energy consumption and green innovation help to curb CO_2 emissions? Evidence from E7 countries. *Environmental Science and Pollution Research*. 2023 Feb;30(8):21115–31.

[32] Wurlod JD, Noailly J. The impact of green innovation on energy intensity: An empirical analysis for 14 industrial sectors in OECD countries. *Energy Economics*. 2018 Mar 1;71:47–61.

[33] Guo J, Zhou Y, Ali S, Shahzad U, Cui L. Exploring the role of green innovation and investment in energy for environmental quality: An empirical appraisal from provincial data of China. *Journal of Environmental Management*. 2021 Aug 15;292:112779.

[34] Soni GD. Advantages of green technology. *Social Issues and Environmental Problems*. 2015 Sep;3(9):1–5.

[35] Gibson L, Wilman EN, Laurance WF. How green is 'green' energy? *Trends in Ecology & Evolution*. 2017 Dec 1;32(12):922–35.

[36] de Oliveira RS, de Oliveira MJ, Nascimento EG, Sampaio R, Nascimento Filho AS, Saba H. Renewable energy generation technologies for decarbonizing urban vertical buildings: A path towards net zero. *Sustainability*. 2023 Aug 29;15(17):13030.

[37] https://www.eia.gov/energyexplained/use-of-energy/homes.php Accessed 25 February, 2024.

[38] Gago EJ, Muneer T, Knez M, Köster H. Natural light controls and guides in buildings. Energy saving for electrical lighting, reduction of cooling load. *Renewable and Sustainable Energy Reviews*. 2015 Jan 1;41:1–3.

[39] Shah KJ, Pan SY, Lee I, Kim H, You Z, Zheng JM, Chiang PC. Green transportation for sustainability: Review of current barriers, strategies, and innovative technologies. *Journal of Cleaner Production*. 2021 Dec 1;326:129392.

[40] Lilhore UK, Imoize AL, Li CT, Simaiya S, Pani SK, Goyal N, Kumar A, Lee CC. Design and implementation of an ML and IoT based adaptive traffic-management system for smart cities. *Sensors*. 2022 Apr 10;22(8):2908.

[41] Nandy S, Fortunato E, Martins R. Green economy and waste management: An inevitable plan for materials science. *Progress in Natural Science: Materials International*. 2022 Feb 1;32(1):1–9.

[42] Salvarli MS, Salvarli H. For sustainable development: Future trends in renewable energy and enabling technologies. In *Renewable energy-resources, challenges and applications* 2020 Sep 9. IntechOpen.

[43] Fakhar A, Haidar AM, Abdullah MO, Das N. Smart grid mechanism for green energy management: a comprehensive review. *International Journal of Green Energy*. 2023 Feb 19;20(3):284–308.

[44] Salkuti SR. Emerging and advanced green energy technologies for sustainable and resilient future grid. *Energies*. 2022 Sep 13;15(18):6667.

18 Enhancing Wireless Network Performance through High-Quality Fiber Optic Deployment in Highway Environments

Augustus Ehiremen Ibhaze, Ayodele A. Afolabi,
and Agbotiname Lucky Imoize
University of Lagos, Lagos, Nigeria

18.1 INTRODUCTION

A fiber optic backbone channel is a high-capacity fiber optic network that connects various sub-networks, data centers, and other network components. The backbone channel acts as the main pipeline for data transmission, and any degradation in the backbone channel can significantly impact the entire network's performance. Fiber optic has become the most preferred mode of communication for many industries due to its unique bandwidth, reliability, scalability, and security. Furthermore, fiber provides a unique opportunity to connect smart grids reliably with sufficient capacity for both current and future utilization. To guarantee a stable performance of all aspects of the network, a fiber backbone system is the best solution.

As high-speed data requirement continues to increase on a daily basis, the deployment of high-performance broadband technologies will optimize the overall performance of the system [1]. Deployment of new fiber solutions could be capital-intensive and requires comprehensive planning and expectations of future use; however, its benefits outweigh the starting capital as it is the longest-durable network asset [2]. Fiber is the backbone of modern communications, facilitating advancement beyond conventional use and opening great opportunities for revenue generation and ideal business models. The rise in fiber optic in Nigeria was driven by the extent of Internet demand and penetration across the world [3].

Galaxy Backbone – the National Information and Communication Technology Infrastructure Backbone (NICTIB) had foresight about the long-term benefits of fiber optic and invested heavily in deploying fiber optic channels to provide high-quality services to government parastatal in various states and the Federal Capital

DOI: 10.1201/9781032656830-18

Territory. The results of the study indicate that the implemented solution has significantly improved the network's reliability, uptime, and performance, reducing the risk of fiber degradation and related issues. A fiber optic communication channel is designed to transport a fiber optic signal from the transmitter to the receiver with no traces of distortion [4].

Lightwave system uses fiber as the communication channel because fiber has the capacity to transmit light with a loss as low as 0.21 dB/km. Fiber optic is a strand of thin pure glass about the size of human hair which transmits signals over a long distance. When light hits the cladding at an angle greater than the critical angle (angle of incidence at which maximum refraction occurs), it will be reflected, a process known as total internal reflection. Light reflected from glass to air boundary is called Fresnel reflection [5]:

Recall from Snell's law given in (18.1):

$$n_1 \sin \theta_1 = n_2 \sin \theta_2 \tag{18.1}$$

$$\text{where } \theta_1 = \sin^{-1}\left(\frac{n_2 \sin \theta_2}{n_1}\right)$$

Given the refractive index of glass as $n_1 = 1.5$ and air as $n_2 = 1.0$ with critical angle of $\theta_2 = 90°$, $\theta_1 = 41.8°$ which implies that higher refractive index materials have small critical angle. The core is the central conducting part that transmits light while the cladding is a mirror-like material that traps light in the core. Buffer/coating is the material used to protect optical fiber from moisture and physical damage while the strength member protects the optical fiber from strain. It should be noted that the outer jacket provides environmental and mechanical protection.

Fiber optic cables come in various sizes depending on the mode. For the single-mode 9/125 μm, the 9 μm is the diameter of the center layer of the fiber cable while the 125 μm refers to the diameter of the cladding and the center layer inclusive. The same description applies to the multimode structure of $\frac{50}{125}$ μm and $\frac{62.5}{125}$ μm, respectively. The color codes for fiber optics are defined by the International Industry Association TIA 598 [6]. Color codes are used in fiber optics to identify fiber, premises cable, OSP cable, buffer tubes, and connectors.

Among the numerous advantages of fiber optics are its exceptional bandwidth, low loss, ability to carry multiple signals over long distances, immunity against electromagnetic interference, effective cost over distances, lightweight, and ease of handling. Fiber optics find wide application in military/data security, surgical/dentistry, sensors/scientific, mechanical inspection, telephony, and Internet/computer networks. Tables 18.1 and 18.2 depict the classification of fiber optics based on the mode and index, respectively.

The goal of this study is to present a practical technique for fiber loss minimization through a stepwise implementation technique. A step-by-step approach for fiber deployment with fiber loss tracking was articulated. A 2.18 dB performance gain was achieved relative to the theoretical estimate of the fiber loss budget.

TABLE 18.1

Fiber Optic Classification Based on Mode

Types of Optical Fiber	Definition	Core/ Cladding Diameter	Wavelength (nm)	Attenuation Coefficient (dB/km)	Where Applicable
Single mode	Allows propagation of light in one direction	9/125 microns	1550 1310 1625	0.2–0.23 0.3–0.35 0.18–0.20	Metro & Backbone (FTTH, Telco, CATV, long data links, and high-speed LAN Backbones)
Multimode	Allows propagation of light in more than one direction	50/125 microns and 62.5/125 microns	1300 850	1 3	Data center, premises cabling (LAN, CCTV, security, and fiber to the desk/PC)

The higher the wavelength(nm), the lower the attenuation coefficient(km/dB), and the better the bandwidth.

The attenuation A measured in decibels (dB) at wavelength λ of a fiber between two cross-sections x and y at a distance L apart is modeled as equation (18.2) [7]:

$$A(dB) = 10\log\frac{P_x}{P_y} \qquad (18.2)$$

where P_x is the optical power traversing the cross-section x at the wavelength λ.

P_y is the optical power traversing the cross-section y at the wavelength λ.

Attenuation coefficient a is expressed as attenuation in dB per unit length as shown in (18.3):

$$a(dB/km) = \frac{A}{L} \qquad (18.3)$$

The unit of optical power is expressed in dBm, the letter "m" in dBm refers to the reference power relative to 1 milliwatt given by (18.4)

$$Power(dBm) = 10\log_{10}\frac{measured\ power}{1mW} \qquad (18.4)$$

When the measured power is 1 mW, then the optical power becomes zero dB.

TABLE 18.2
Fiber Optic Index Classification

Multimode Fiber	Class	Distance (m)	Outer Sheath Color		Connector Color	Core/Cladding Diameter (μm)	Where Applicable
Step Index	OM1 (plastic optical fiber)	33	Orange		Beige	62.5/125	100 Megabit Ethernet applications
Graded Index	OM2	82	Orange		Black	50/125	LAN/building applications and supports Ethernet transmission with a maximum of 10 GB
	OM3	300	Aqua		Aqua		Data centers and supports 10G or even a 40/100G high-speed Ethernet transmission
	OM4	450	Aqua	Violet	Aqua		40/100G high-speed data center
	OM5	550	Lime		Lime		100G/200G/400G ultra-large data centers

18.2 METHODOLOGY

18.2.1 OPTICAL CHANNEL IMPLEMENTATION

A fiber optic communication channel converts an electrical signal to an optical signal as shown. This signal is transmitted through fiber optic and the fiber channel acts as a waveguide and transmits the optical signals toward the receiver by the principle of total internal reflection. The optical detector receives the fiber optic signals/pulses and converts them into electrical pulses which are amplified by the amplifier. A fiber project is technically implemented based on a pre-defined series of steps shown in Figure 18.1. This is particularly pivotal to fiber projects which involve a mixture of civil and telecoms activities, where any slightest error in one phase can lead to an extension of errors in the subsequent phases.

In order to achieve quality fiber optic deployment at each phase of implementation, line marking/setting-out had to be carried out as the initial step with the use of rope, pegs, and sawdust as shown in Figure 18.2 to ensure a straight excavation

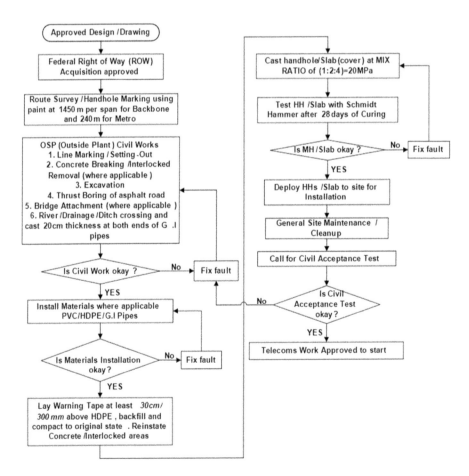

FIGURE 18.1   Fiber optic channel implementation procedure.

FIGURE 18.2 Marking, setting-out, and excavation.

layout as bends (macro/micro) may lead to an increased attenuation which automatically degrades system performance.

Excavation was done at 30 cm width and 1.2 m, 1 m, 0.8 m, and 0.6 m based on the soil type and importantly based on NICTIB/Galaxy Backbone technical specification depicted in Table 18.3.

Having certified excavation depth, width, and straightness, approval was granted to lay 32 mm diameter HDPE using NICTIB/Galaxy Backbone color configuration (red, black, yellow, and green), with red color HDPE laid closest to the road as shown in Figure 18.3.

Thrust boring was done under asphalts roads, railways, and structures without having to disturb the normal activities on the surface. A rectangular pilot pit of at least 10.9 meters long was first dug to suit the combined length of steel pipes (6 m length each) and thrust boring machine. A drilling bit was attached to the shaft of the machine and

TABLE 18.3
Nigeria NICTIB E-Government OSP Soil Type Technical Specification

Soil Type	Excavation Depth (m)	Excavation Width (m)	Remarks
Normal soil	≥1.2	0.3	Ok
Hard soil	≥1.0	0.3	Ok
Gravel	≥0.8	0.3	20 cm thickness casting required
Soft/hard Rock	≥0.6	0.3	20 cm thickness casting required

FIGURE 18.3 Laid HDPE.

FIGURE 18.4   Installation of galvanized industrial pipes, warning tape, and backfilling.

TABLE 18.4
NICTIB Outside Plant Handhole Technical Specification

Nigeria NICTIB E-Government OSP Technical Requirements

Handhole Concrete Strength (MPa)	C20
Handhole outer dimension (length × width × depth) – Urban	900 mm × 900 mm × 1100 mm
Handhole outer dimension (length × width × depth) – Suburban	900 mm × 700 mm × 820 mm
Thickness of manhole side	100 mm
Thickness of manhole bottom	100 mm
Type of cover for urban manhole	Foundry/iron
Type of cover for suburban manhole	Concrete

manually driven; this process was repeated until the shaft was driven through the entire thrust boring length. Afterward, finding was conducted and galvanized industrial pipes and HDPE were loaded as shown in Figure 18.4 and warning tape was laid, backfilled, and compacted.

Handhole is an underground vault that houses closures and fiber optic cable slack. It aids closure/cable access during operational/maintenance work. Handholes were constructed using reinforced concrete structures precast with a mix ratio of 1 cement:2 sharp sands:4 granites. Table 18.4 gives the NICTIB outside plant handhole technical specification. Figure 18.5 shows handhole casting and testing phase.

Figure 18.5 shows that the Schmidt hammer was impacted horizontally on the handhole in order to take four different readings on each surface of the four sides of

FIGURE 18.5   Handhole casting and testing phase.

FIGURE 18.6   Handhole installation.

the handhole. Handholes were deployed as shown in Figure 18.6 after compressive strength was certified. Figure 18.7 is a stepwise approach to fiber optics installation.

18.2.2   FIBER OPTICS INSTALLATION STEPWISE APPROACH

An optical fiber cable planner and route survey were carried out to identify OFC drum placement and blowing. Fiber optic cable drum test was conducted by fusion-splicing fiber strands to pigtail and tested with OTDR to ascertain attenuation and continuity of the optical fiber and also clear doubt of any unforeseen factory error as detailed in Figure 18.7. Satisfied by the OTDR test result (0.18 dB/km to ≤ 0.23 dB/km), HDPE/duct calibration and optical fiber cable blowing commenced simultaneously. The meeting point of two drums of fiber cable was a Splice-Joint, 20 m optical fiber cable was left in joint handhole while 10 m slack was left in other handholes for future maintenance work. Figure 18.8 shows optical fiber cable blowing and termination.

At different meeting points of each of the two drums of fiber optic cable, the two ends of optical fiber cable were prepared into closure and spliced color to color at low splice loss ≤ 0.01 dB using fusion-splicing machine to enhance better link performance and fiber optic continuity from Onitsha to Owerri switch. Termination was done in such a way that the bare fiber optic strands (pigtails) were spliced to strands of optical fiber cable blown/installed and the connector head of pigtails was patched accordingly inside the patch panel to prevent core/fiber misalignment as shown in Figure 18.8.

OTDR is a testing instrument that characterizes/verifies the integrity of an optical fiber. The OTDR can operate just like a RADAR by injecting/sending a series of high-power laser/optical pulses into the fiber under test and some of the optical pulses transmitted through the optical fiber cable will scatter while some will be reflected and returned to the OTDR [8]. The OTDR uses the returned Rayleigh back-scatter and reflections to determine fiber break, splice loss, distance, connector loss, termination/pigtail (spike), micro bend, macro bend, reflectance, or faults. The strength of the returned pulses is measured, integrated as a function of time, and then plotted against fiber optic length [9]. Bi-directional OTDR trace was conducted on

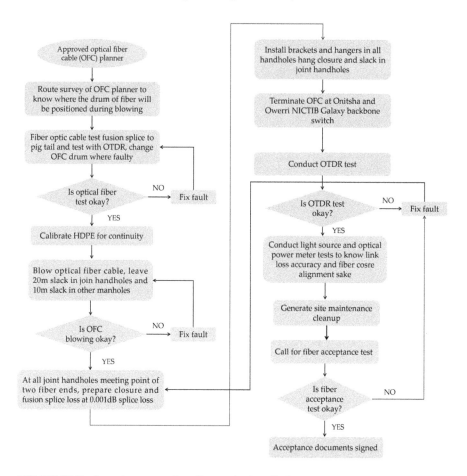

FIGURE 18.7   Stepwise approach to fiber optics installation.

Onitsha to Owerri fiber plant using 1550 nm (wavelength) to check the quality of fiber cable blown, splicing, and termination. The overall attenuation was verified by injecting a known level of light at Onitsha and then measured the power level at Owerri. The difference in the two levels measured gave dB value (often called "insertion loss").

18.2.3   Loss Budget

A loss budget is a workable threshold that guides against loss of signal. Before measuring the overall optical system loss with LSPM, it is crucial to first calculate the loss budget to have an idea of what the optical system loss should be if the components are truly in good condition. Loss budget is the sum of cable loss, splice loss, and connector loss. Based on the investigated terrain, the given parameters are:

Onitsha-Owerri fiber cable length = 90.5 km, attenuation coefficient = 0.20 dB/km,

FIGURE 18.8 Fiber optic cable blowing splicing and termination.

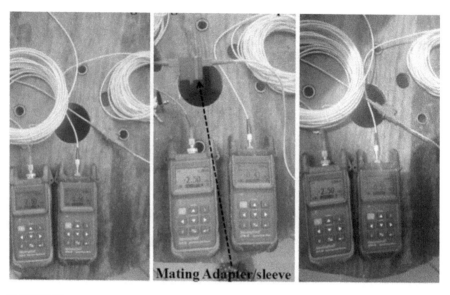

FIGURE 18.9   Light source and power meter referencing.

Splice loss = 0.02 dB, wavelength = 1550 nm, and connector loss = 0.5 dB (one connector at Onitsha and one at Owerri switch). Note that there were 16 splice joints and 2 additional splice points (termination to a patch panel at Onitsha and Owerri) making 18 splices. See Equation (18.5).

$$\text{Loss budget} = \text{cable loss} + \text{splice loss} + \text{connector loss} \qquad (18.5)$$

$$= (90.5\,\text{km} \times 0.2\,\text{dB}/\text{km}) + (18 \times 0.02\,\text{dB}) + (0.5\,\text{dB} \times 2)$$

$$= 19.46\,\text{dB}(\text{maximum allowable optical fiber system loss})$$

Referencing of LSPM was done to cancel out any losses caused by connectors or patch cords by connecting two patch cords, one to light source port and another to optical power meter port while a mating adapter/sleeve was used to couple light source directly to optical power meter as shown in Figure 18.9.

18.3   RESULTS AND DISCUSSION

From the compressive strength test conducted on 25 handholes, 16 readings were taken for each of the handholes, and the 3 lowest (A, B, and C) and 3 highest (N, P, and O) readings shown in Table 18.5 were discarded [10] and the average of the remaining 10 readings were taken with the corresponding compressive strength checked from the Schmidt hammer graph [11–12].

From Table 18.5, the rebound value is estimated as an average value given by:

$$\text{Average of 10 rebound values} = \frac{30 + 27 + 31 + 31 + 28 + 29 + 28 + 29 + 32 + 29}{10} = 29.4$$

TABLE 18.5
Results of Compressive Strength Test Conducted on Twenty-Seven Handholes

S/N	HH	A	B	C	D	E	F	G	H	I	J	K	L	M	N	O	P	Average Rebound Value	Corresponding MPa Value	Remark
										Sub-Urban HH										
1	HH01	27	26	27	30	27	31	31	28	29	28	29	32	29	38	34	37	29.4	C23	OK
2	HH02	26	25	26	28	28	33	32	26	27	26	31	29	30	33	34	33	29	C23	OK
3	HH03	26	27	27	30	29	29	27	29	29	28	28	29	28	33	33	38	28.6	C22	OK
4	HH04	27	26	27	29	29	31	29	30	36	34	31	28	36	37	36	38	31.1	C27	OK
5	HH05	26	25	26	29	29	33	28	32	27	27	26	31	28	32	33	32	28.8	C22	OK
6	HH06	26	28	27	29	29	33	30	28	29	30	29	30	33	44	33	35	30	C25	OK
7	HH07	27	26	27	27	31	28	29	28	32	29	30	29	29	35	33	33	29.2	C23	OK
8	HH08	27	27	27	29	30	30	30	29	31	29	30	31	29	32	33	34	29.8	C23	OK
9	HH09	24	24	27	33	27	32	29	33	34	32	33	27	28	34	35	43	30.8	C25	OK
10	HH10	26	26	27	27	30	27	27	27	31	29	30	28	28	34	33	34	28.4	C22	OK
11	HH11	24	26	24	30	26	26	26	27	27	27	31	29	32	34	33	33	28.2	C22	OK
12	HH12	26	26	26	28	27	28	33	27	28	27	28	31	28	34	33	34	28.5	C22	OK
13	HH13	27	24	27	29	28	27	32	27	27	28	27	29	29	33	33	33	28.3	C22	OK
14	HH14	27	27	28	30	30	30	28	30	30	28	29	28	29	37	33	45	29.2	C23	OK
15	HH15	25	27	27	27	27	29	29	27	28	28	29	29	28	35	33	33	28.2	C22	OK
16	HH16	27	26	25	29	30	27	30	27	33	28	29	28	27	34	33	35	28.7	C22	OK
17	HH17	24	26	26	27	28	27	29	26	27	27	26	27	27	30	33	30	27	C20	OK
18	HH18	25	24	26	29	27	26	27	26	32	31	27	26	30	33	33	33	28.1	C22	OK
19	HH19	26	26	27	26	27	28	28	27	27	28	27	26	27	30	33	30	27.2	C20	OK
20	HH20	25	26	24	28	27	27	28	27	28	27	27	26	27	28	33	29	27.2	C20	OK
21	HH21	27	26	27	27	27	27	28	27	27	30	28	29	28	30	33	32	27.8	C20	OK
22	HH22	26	27	27	32	30	29	33	32	29	27	27	27	32	34	33	34	29.3	C23	OK
23	HH23	24	27	27	27	27	27	27	32	27	28	30	27	27	36	33	32	28.4	C22	OK
24	HH24	25	26	26	27	27	28	27	27	28	27	27	27	26	30	33	30	27.1	C20	OK
S/N											Urban/Metro HH										
25	HH25	26	27	25	27	34	32	30	35	29	30	30	27	29	36	38	36	30.3	C25	OK

TABLE 18.6
OTDR Event Table When Total Loss Is 16.902 dB

OTDR Events Table

S. No.	Event Type	Distance (km)	Loss (dB)	Reflectance (dB)	Attenuation (dB/km)
1	Begin	0	0.001	−43.416	18.677
2	Splice loss	8.034	0.17		0.184
3	Splice loss	19.339	0.136		0.187
4	Splice loss	25.367	0.167		0.183
5	Splice loss	55.712	0.219		0.183
6	Splice gain	61.773	−0.104		0.184
7	Reflect	73.6	0.122	−63.922	0.187
8	End	90.582	28.481	−14.721	0.154

The 29.4 average of 10 rebound values from Table 18.5 corresponds to C23 which is 23 MPa (Mega Pascal) in the Schmidt hammer graph [11–12]. This result of the compressive strength test shows handholes met Galaxy Backbone standard specification of which the supremum of C20 [20 MPa] is required. Table 18.6 gives the OTDR event table when the total loss is 16.902 dB. Figure 18.10 presents OTDR trace acquisition from Onitsha to Owerri with an attenuation of 0.187 dB/km.

The OTDR report is shown in Figure 18.10. The vertical axis of the plot measures the attenuation and reflection in the unit of decibels (dB) while the horizontal axis measures the length/distance of the fiber optic cable in kilometers. Advanced setting

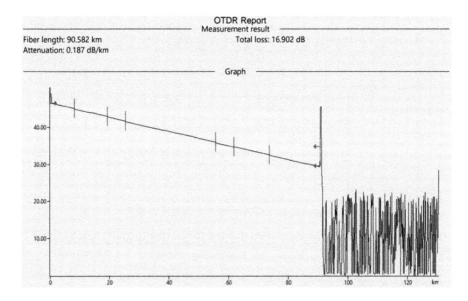

FIGURE 18.10 OTDR trace acquisition from Onitsha to Owerri with an attenuation of 0.187 dB/km.

was done first by selecting 1550 nm (nanometer) wavelength for single mode fiber, 5 μs pulse width, and 1 minute acquisition time. It was observed that as soon as the OTDR laser was shot, Rayleigh scattering occurred and the reflected light was able to characterize eight events as shown in Table 18.6. The first event is the initial pulse/ dead zone or the connector point having a reference distance of 0.00 km while the reflectance was negative 43.416 dB which is optimal when considering the reflectance value of a quality connector as shown as depicted in Table 18.6. The second event is a splice point (where two ends of fiber were spliced with a splicing machine) at a distance of 8.034 km and resulted in a loss of 0.17 dB and an attenuation of 0.184 dB/km way below industry standard of 0.25 dB (the lower the attenuation, the higher the performance of any optical fiber channel).

The third event to the fifth event are splice points with good losses (0.167 dB, 0.167 dB, and 0.219 dB) and attenuation (0.187 dB/km, 0.183 dB/km, and 0.183 dB/km). The sixth event is also a splice point but has a splice gain (two fibers with different backscatter coefficients spliced together) of negative 0.104 dB. It can be seen from the OTDR trace of Figure 18.11 that while events two to five experienced negligible drop as the graph went linear, event six experienced an insignificant shoot-up (−0.104 dB) before it maintained balance at event seven. Event seven is a reflective event because it has a reflectance of negative 63.922 dB and a loss of 0.122 dB. Event eight is the end of fiber at Owerri Telecoms shelter, event eight has a high reflective peak (pigtail termination with connector) and dropped down to noise. The reason for the reflective peak is the discontinuity that exists between a connector pair (there is a minute gap between the connectors). A splice does not have discontinuity and that is why it only has attenuation associated with it. Table 18.7 provides useful information on the OTDR event table when the total loss is 16.902 dB.

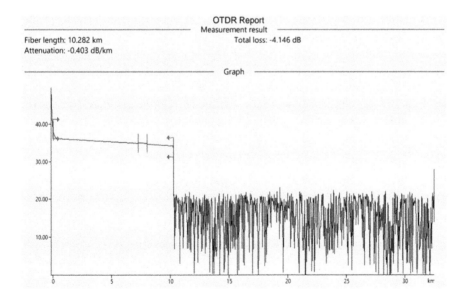

FIGURE 18.11 OTDR trace showing a fiber cut at 10.282 km from Onitsha toward Owerri.

TABLE 18.7
OTDR Event Table When Total Loss Is 16.902 dB

OTDR Events Table

S. No.	Event Type	Distance (km)	Loss (dB)	Reflectance (dB)	Attenuation (dB/km)
1	Begin	0.000 Km			
2	Splice gain	0.041 Km	−6.134 dB		
3	Splice gain	7.280 Km	−0.148 dB		0.189 dB/Km
4	Splice loss	8.029 Km	0.258 dB		0.157 dB/Km
5	End	10.282 Km	13.021 dB		0.174 dB/Km

It can be deduced from the OTDR trace of Figure 18.11 and Table 18.7 that there was continuity when the OTDR laser was first shot until it got to 10.282 km and dropped down abruptly. In an effort to correct the one core failure, a bidirectional OTDR field test was conducted at the closest joint to the suspected point of failure, the test result gave 10.1 km toward Onitsha telecoms shelter and 182 m complete cut toward Owerri. Once the 178 m measurement was carried out using the measuring wheel, the failure was observed along the fiber length. Consequently, error correction evaluation index was conducted by the following relation: let N be the total number of cores while n be the number core failure. Since $N = 48$ and $n = 1$, therefore, Error Correction Evaluation Index $= \dfrac{n}{N} \times 100$. Error Correction Evaluation index = 2% (negligible). Table 18.8 shows the light source power meter test result on Onitsha-Owerri route.

The implemented fiber optic channel was tested for degradation and overall attenuation with the use of an optical time domain reflectometer and light source power meter with findings detailed in Table 18.8. Having compared the link loss budget of 19.46 dB with the maximum power meter value of 17.28 dB of the implemented optical fiber channel shown in Table 18.8, it was inferred that there was an improvement level of 2.18 dB (see equation (18.2)). Additional information related to the key results presented is available in the following reference materials [13–22]:

Percentage of optical fiber channel improvement level

$$= \frac{\text{Level of improvement}}{\text{Loss Budget}} \times 100 \tag{18.6}$$

Percentage of optical fiber channel improvement level $= \dfrac{2.18\,(\text{dB})}{19.46\,(\text{dB})} \times 100$

Percentage of optical fiber channel improvement level $= 11.20\%$

TABLE 18.8
Light Source Power Meter Test Result on Onitsha-Owerri Route

Onitsha-Owerri Fiber Optic Light Source and Power Meter Result

Fiber Route: ONITSHA-OWERRI   **Test Date: 27/05/2023**

No. of joints in a link: Equipment: **Fujikura Fusion Splicer/JDSU LSPM**

Spec. cable 0.25 dB/km @1550 nm   Cable type: G652

Splice loss 0.10 dB/km   Wavelength (1310 nm/1550 nm):

Total no. of joints: Cable length: 90.582 km

   Max. attenuation (cable +joints):

Fiber No.	Tx. Power (A) in dB	Rx. Power (B) in dB	Attenuation (A–B) in dB	Fiber No.	Tx. Power (A) in dB	Rx. Power (B) in dB	Attenuation (A–B) in dB
1	−10.54	−27.42	16.88	25	−10.54	−27.03	16.49
2	−10.54	−27.28	16.74	26	−10.54	−26.9	16.36
3	−10.54	−27.33	16.79	27	−10.54	−27.34	16.80
4	−10.54	−27.4	16.86	28	−10.54	−26.76	16.22
5	−10.54	−27.46	16.92	29	−10.54	−27.4	16.86
6	−10.54	−27.72	17.18	30	−10.54	−27.06	16.52
7	−10.54	−27.43	16.89	31	−10.54	−27.64	17.10
8	−10.54	−27.29	16.75	32	−10.54	−27.41	16.87
9	−10.54	−27.38	16.84	33	−10.54	−27.54	17.00
10	−10.54	−27.24	16.70	34	−10.54	−27.17	16.63
11	−10.54	−26.66	16.12	35	−10.54	−27.81	17.27
12	−10.54	−27.8	16.26	36	−10.54	−27.02	16.48
13	−10.54	−27.11	16.57	37	−10.54	−27.51	16.97
14		**FIBER CUT**		38	−10.54	−27.53	16.99
15	−10.54	−27.28	16.74	39	−10.54	−27.67	17.13
16	−10.54	−27.46	16.92	40	−10.54	−27.62	17.08
17	−10.54	−27.14	16.60	41	−10.54	−27.23	16.69
18	−10.54	−27.05	16.51	42	−10.54	−27.55	17.01
19	−10.54	−27.52	16.98	43	−10.54	−27.79	17.25
20	−10.54	−27.01	16.47	44	−10.54	−27.56	17.02
21	−10.54	−27.18	16.64	45	−10.54	−27.61	17.07
22	−10.54	−27.82	17.28	46	−10.54	−27.39	16.85
23	−10.54	−27.73	17.19	47	−10.54	−27.52	16.98
24	−10.54	−27.21	16.67	48	−10.54	−27.43	16.89

18.4   CONCLUSION

High-quality implementation of the NICTIB fiber optic channel resulted in a 2.18 dB improvement; hence, the fiber optic channel would perform optimally. It is recommended that each phase of fiber optic channel implementation be done with strict adherence to telecommunication standard specifications to avert prevailing and future fiber optic channel degradation. High-quality implementation of a fiber optic channel is instrumental to ensuring low loss, reliable and high-speed data transfer, downtime reduction, and user experience improvement. The case study presented in this research paper shows that the implementation of fiber optic channel for Galaxy Backbone (NICTIB) Nigeria's Onitsha-Owerri network has significantly improved the performance and reliability of the network, reduced downtime, and improved user experience. The stepwise approach adopted for high-quality implementation provides a model for other telecommunication providers looking for the best technique to minimize fiber degradation for their proposed fiber channel expansion.

REFERENCES

[1] Ibhaze, A.E., Orukpe, P.E., and Edeko, F.O., "High capacity data rate system: Review of visible light communications technology," *Journal of Electronic Science and Technology*, Vol. 18, No. 3, p. 84, 2020.

[2] NRTC, NRECA and ERICSSON, "The value of a broadband backbone for America's electric cooperatives – A benefit assessment study," *Strategy & Corporate Development*, pp. 1–25, p. 5, 2017.

[3] Okorodudu, F., and Okorodudu, P., "Fiber optics communication in Nigeria and a wider internet penetration: the Nexus," *International Journal of Innovative sciences, Engineering and Technology*, Vol. 3 No. 12, pp. 238–245, 2016.

[4] Agrawal, G.P., *Fiber-Optic Communication Systems*, John Wiley and Sons, New York, 1997.

[5] Brientin, A., Leduc, D., Gaillard, V., Girard, M., and Lupi, C., "Numerical and experimental study of a multimode optical fiber sensor based on Fresnel reflection at the fiber tip for refractive index measurement," *Optics & Laser Technology*, Vol. 143, p. 32, 2021.

[6] ANSI/TIA/EIA 598-A "Optical Fiber Cable Color Coding," NDS Information telecom system, pp. 1–8, 1994.

[7] Han, Z., Jeong, S., Jang, J., Woo, J.H., and Oh, D., "Ultrasonic attenuation characteristics of glass-fiber-reinforced polymer hull structure," *MDPI Applied Sciences*, Vol. 11, p. 3, 2021.

[8] Mai, T.V., Molnar, J.A., and Tran, L.H., "Fiber optic test equipment - evaluation of OTDR dead zones and ORLM return loss," *Proceedings AUTOTESTCON*, 2004.

[9] Sante, R. D., and Donati L., "Strain monitoring with embedded Fiber Bragg Gratings in advanced composite structures for nautical applications," *Measurement*, Vol. 46, No. 7, pp. 2118–2126, 2013.

[10] ASTM International, ASTM C805 / C805M-18, Standard Test Method for Rebound Number of Hardened Concrete, ASTM International, West Conshohocken, PA, 2018.

[11] Shamim, Rejuwana, and All, "Enhancing Cloud-Based Machine Learning Models with Federated Learning Techniques", Lecture Notes in Networks and Systems, Volume 838 LNNS, Pages 594 – 6062024, DOI: 10.1007/978-3-031-48573-2_85.

[12] Dey A., Miyani G., Debroy S., and Sil A., "In-situ NDT investigation to estimate degraded quality of concrete on existing structure considering time-variant uncertainties," *Journal of Building Engineering*, Vol. 27, p. 2, 2020.

[13] Ren, Z., Cui, K., Li, J., Zhu, R., He, Q., Wang, H., Deng, S., and Peng, W., "High-quality hybrid TDM/DWDM-based fiber optic sensor array with extremely low crosstalk based on wavelength-cross-combination method," *Optics Express*, Vol. 25, No. 23, pp. 28870–28885, 2017.

[14] Ughegbe, G.U., Adelabu, M.A., and Imoize, A.L., "Experimental data on radio frequency interference in microwave links using frequency scan measurements at 6 GHz, 7 GHz, and 8 GHz," *Data in Brief*, Vol. 35, p. 106916, 2021.

[15] Adelabu, M.A., Imoize, A.L., and Ughegbe, G.U., "Statistical analysis of radio frequency interference in microwave links using frequency scan measurements from multi-transmitter environments," *Alexandria Engineering Journal*, Vol. 61, No. 12, pp. 11445–11484, 2022.

[16] Marhic, M.E., Andrekson, P.A., Petropoulos, P., Radic, S., Peucheret, C., and Jazayerifar, M., "Fiber optical parametric amplifiers in optical communication systems," *Laser & Photonics Reviews*, Vol. 9, No. 1, pp. 50–74.

[17] Fan, J.C., Lu, C.L., and Kazovsky, L.G., "Dynamic range requirements for microcellular personal communication systems using analog fiber-optic links," *IEEE Transactions on Microwave Theory and Techniques*, Vol. 45, No. 8, pp.1390–1397, 1997.

[18] Morais, R.M., "Implementation of fiber optic technology in naval combatants," *Marine Technology and SNAME News*, Vol. 24, No. 01, pp. 59–71, 1987.

[19] Dubey, S., Kumar, S., and Mishra, R., "Simulation and performance evaluation of free space optic transmission system," In *2014 International Conference on Computing for Sustainable Global Development (INDIACom)* (pp. 850–855). IEEE, 2014, March.

[20] Prisco, J., "Fiber optic regional area networks in New York and Dallas," *IEEE Journal on Selected Areas in Communications*, Vol. 4, No. 5, pp. 750–757, 1986.

[21] Roslyakov, A., "Fiber development index to implement the FTTE concept in F5G networks." In *Optical Technologies for Telecommunications 2021* (Vol. 12295, pp. 7–12). SPIE, 2022, July.

[22] Kareem, F.Q., Zeebaree, S.R., Dino, H.I., Sadeeq, M.A., Rashid, Z.N., Hasan, D.A., and Sharif, K.H., "A survey of optical fiber communications: challenges and processing time influences," *Asian Journal of Research in Computer Science*, Vol. 7, No. 4, pp. 48–58, 2021.

Index

Pages in *italics* refer to figures and pages in **bold** refer to tables.